Lecture Notes in Artificial Intelligence 934

Subseries of Lecture Notes in Computer Science
Edited by J. G. Carbonell and J. Siekmann

Lecture Notes in Computer Science

Edited by G. Goos, J. Hartmanis and J. van Leeuwen

Springer

Berlin
Heidelberg
New York
Barcelona
Budapest
Hong Kong
London
Milan
Paris
Tokyo

Pedro Barahona Mario Stefanelli
Jeremy Wyatt (Eds.)

Artificial Intelligence in Medicine

5th Conference on Artificial Intelligence
in Medicine Europe, AIME '95
Pavia, Italy, June 25-28, 1995
Proceedings

 Springer

Series Editors

Jaime G. Carbonell
School of Computer Science
Carnegie Mellon University
Pittsburgh, PA 15213-3891, USA

Jörg Siekmann
University of Saarland
German Research Center for Artificial Intelligence (DFKI)
Stuhlsatzenhausweg 3, D-66123 Saarbrücken, Germany

Volume Editors

Pedro Barahona
Dep. de Informática, Universidade Nova de Lisboa
Quinta da Torre, P-2825 Monte da Caparica, Portugal

Mario Stefanelli
Dip. di Informatica e Sistemistica, Università degli Studi di Pavia
Via Abbiategrasso 209, I-27100 Pavia, Italy

Jeremy Wyatt
Biomedical Informatics Unit, Imperial Cancer Research Fund
Lincoln's Inn Fields, London WC2A 3PX, United Kingdom

CR Subject Classification (1991): I.2, I.4, J.3, H.4

ISBN 3-540-60025-6 Springer-Verlag Berlin Heidelberg New York

CIP data applied for

© Springer-Verlag Berlin Heidelberg 1995
Printed in Germany

Typesetting: Camera ready by author
SPIN: 10486282   06/3142 – 543210 – Printed on acid-free paper

Preface

The European Society for Artificial Intelligence in Medicine (AIME) was established in 1986 after a highly successful workshop held in Pavia the year before. The aims of AIME are to foster fundamental and applied research in the application of Artificial Intelligence techniques to medical care and medical research, and to provide a forum for reporting significant results at biennial conferences. In accordance with this latter goal, this volume contains the proceedings of AIME'95, the 5th Conference on Artificial Intelligence in Europe, which follows previous conferences held in Marseille (1987), London (1989), Maastricht (1991) and Munich (1993).

In the announcement of the conference, authors were encouraged to submit original contributions to the development of theory, techniques, and applications of AI in medicine. Contributions to theory should include a presentation or an analysis of the properties of novel AI methodologies potentially useful to solve relevant medical problems. Papers on techniques should describe the development or extension of AI methods and their implementation, and discuss the assumptions and limitations which characterize the proposed methods. Application papers should describe the implementation of AI systems to solve significant medical problems, and present sufficient information to allow evaluation of practical benefits of using the system.

This call resulted in 64 papers and 10 posters being submitted to the conference. Each of them was evaluated by at least two members of the Programme Committee and were classified according to their originality, relevance, quality and clarity, as well as an overall impression. Based on these marks the papers were sorted, and the best 32 were eventually selected for presentation in oral presentations, and published in the proceedings. All of them were rated as Good or Very Good by the evaluators from the Programme Committee, and some of them even received an Excellent mark in some of the above items. Some papers also rated as Good could not be accepted for presentation due to time limitations. Given their quality, all these and others considered as Fair by the Programme Committee will be presented in special poster sessions during the conference. An extended abstract is included in this volume.

As a whole, it is our impression that the effort of organising the AIME Conferences is paying off well, and that the quality of the papers has been steadily increasing. We are thus confident that this steady increase will continue in the future and that this Conference will continue to help in this process.

We would like to finish by thanking all that contributed to the success of this Conference: the authors, the members of the Programme Committee and the Organizing Committee; the invited speakers, Mark Musen and Alexandre Herold; the organisers of the tutorials, Robert Macura and Katarzyna Macura (Case-Based Reasoning), Colin Gordon and Ian Herbert (Guidelines and Protocols), Oivind Braaten (Genetic Algorithms) and Mark Musen (Knowledge Sharing and Reuse) for their relevant contribution to the conference; and the institutions that sponsored the conference, namely Consiglio Nazionale delle Ricerche, University of Pavia, Istituto di Analisi Numerica, C.N.R., Pavia, I.R.C.C.S. Policlinico San Matteo, Pavia and Consorzio di Bioingegneria e Informatica Medica, Pavia.

May 1995

Pedro Barahona
Mario Stefanelli
Jeremy Wyatt

Programme Committee Chairman
Pedro Barahona

Organizing Committee Chairman
Mario Stefanelli

Tutorials Chairman
Jeremy Wyatt

Programme Committe

Pedro Barahona (Portugal), Chairman

Steen Andreassen (Denmark)	Rory O'Moore (Ireland)
Jan van Bemmel (The Netherlands)	Alan Rector (United Kingdom)
Jytte Brender (Lingby, Denmark)	Jean-Louis Renaud-Salis (France)
Enrico Coiera (United Kingdom)	Niilo Saranummi (Finland)
Luca Console (Italy)	Mario Stefanelli (Italy)
Rolf Engelbrecht (Germany)	Jan Talmon (The Netherlands)
John Fox (United Kingdom)	Mario Veloso (Portugal)
Catherine Garbay (France)	Ove Wigertz (Sweden)
Werner Horn (Austria)	Jeremy Wyatt (United Kingdom)
Elpida Keravnou (Cyprus)	Peter Zanstra (The Netherlands)

Organizing Committee

Mario Stefanelli (chairman)

Giovanni Barosi	Cristiana Larizza
Riccardo Bellazzi	Liliana Ironi
Silvana Quaglini	Angelo Rossi Mori
Giordano Lanzola	Franco Sicurello

Table of Contents

Keynote Address

A Component-Based Architecture for Automation of Protocol-Directed Therapy   3
Mark A. Musen, Samson W. Tu, Amar K. Das and Yuval Shahar

Medical Records

Coordinating Taxonomies: Key to Re-Usable Concept Representations   17
A.L. Rector

Generating Personalised Patient Information Using the Medical Record   29
K. Binsted, A. Cawsey and R. Jones

Analysis of Medical Jargon: The RECIT System   42
*A.-M. Rassinoux, C. Juge, P.-A. Michel, R.H. Baud, D. Lemaitre, F.-C. Jean,
P. Degoulet and J.-R. Scherrer*

Medical Knowledge Representation for Medical Report Analysis   53
J.F. Smart and M. Roux

Temporal Reasoning and Simulation

Modelling Medical Concepts as Time-Objects   67
E.T. Keravnou

Modeling Medical Reasoning with the Event Calculus: An Application to the   79
Management of Mechanical Ventilation
L. Chittaro, M. Del Rosso and M. Dojat

A General Framework for Building Patient Monitoring Systems   91
C. Larizza, G. Bernuzzi and M. Stefanelli

Semi-Qualitative Models and Simulation for Biomedical Applications   103
P. Barahona

Generating Explanations of Pathophysiological Systems Behaviors from   115
Qualitative Simulation of Compartmental Models
L. Ironi and M. Stefanelli

Probabilistic Models

An Information-Based Bayesian Approach to History Taking   129
G. Carenini, S. Monti and G. Banks

Medical Decision Making Using Ignorant Influence Diagrams   139
M. Ramoni, A. Riva, M. Stefanelli and V. Patel

Dynamic Propagation in Causal Probabilistic Networks with Instantiated   151
Variables
O.K. Hejlesen, S. Andreassen and S.K. Andersen

Patient Management and Therapy Planning

Alerts as Starting Point for Hospital Infection Surveillance and Control 165
E. Safran, D. Pittet, F. Borst, G. Thurler, M. Berthoud, P. Schulthess, P. Copin, V. Sauvan, A. Alexiou, L. Rebouillat, M. Lagana, J.-P. Berney, P. Rohner, R. Auckenthaler and J.-R. Scherrer

Cooperative Software Agents for Patient Management 173
G. Lanzola, S. Falasconi and M. Stefanelli

High Level Control Strategies for Diabetes Therapy 185
A. Riva and R. Bellazzi

Therapy Planning Using Qualitative Trend Descriptions 197
S. Miksch, W. Horn, C. Popow and F. Paky

Adaptation and Abstraction in a Case-Based Antibiotics Therapy Adviser 209
R. Schmidt, L. Boscher, B. Heindl, G. Schmid, B. Pollwein and L. Gierl

Evaluation of Knowledge Based Systems

Field Evaluations of a Knowledge-Based System for Peripheral Blood 221
Interpretation
L.W. Diamond, D.T. Nguyen, P. Ralph, B. Sheridan, A. Bak, C. Kessler and D. Muncer

Functional Evaluation of SETH: An Expert System In Clinical Toxicology 231
S.J. Darmoni, P. Massari, J.-M. Droy, T. Blanc and J. Leroy

Evaluating a Neural Network Decision-Support Tool for the Diagnosis of Breast 239
Cancer
J. Downs, R.F. Harrison and S.S. Cross

Knowledge-Based Systems for Lymph Node Pathology: A Comparison of Two 251
Approaches
D.T. Nguyen, I.A. Park, P. Cherubino, P.B. Tamino and L.W. Diamond

Diagnostic Support Systems

Mapping Laboratory Medicine onto the Select and Test Model to Facilitate 265
Knowledge-Based Report Generation in Laboratory Medicine
H. Kindler, D. Densow, B. Fischer and T.M. Fliedner

Machine Learnig Techniques Applied to the Diagnosis of Acute Abdominal Pain 276
C. Ohmann, Q. Yang, V. Moustakis, K. Lang and P.J. van Elk

Reflections on Building Medical Decision Support Systems and Corresponding 282
Implementation in Diagnostics Shell D3
B. Puppe

Models for Clinical Information Systems

Decision Models for Cost-Effectiveness Analysis: A Means for Knowledge   295
Sharing and Quality Control in Health Care Multidisciplinary Tasks
S. Quaglini, M. Stefanelli and F. Locatelli

Model-Based Application: The Galen Structured Clinical User Interface   307
*L. Alpay, A. Nowlan, D. Solomon, C. Lovis, R.H. Baud, T. Rush and J.-R.
Scherrer*

A Knowledge-Based Moddeling of Hospital Information Systems Components   319
H. Kanoui, M. Joubert and R. Favard

Use of a Conceptual Semi-Automatic ICD-9 Encoding System in an Hospital   331
Environment
C. Lovis, P.-A. Michel, R.H. Baud and J.-R. Scherrer

Neural Networks and Image Interpretation

Quality Assurance and Increased Efficiency in Medical Projects with Neural   343
Networks by Using a Structured Development Method for Feedforward Neural
Networks (SENN)
*T. Waschulzik, W. Brauer, M. Förster, K. Kirchner, R. Engelbrecht, T. Shütz,
T. Koschinsky and G. Entenmann*

A Prototype Neural Network Decision-Support Tool for the Early Diagnosis of   355
Acute Myocardial Infarction
J. Downs, R.F. Harrison and R.L. Kennedy

Integration of Neural Networks and Rule Based Systems in the Interpretation of   367
Liver Biopsy Images
N. Bianchi and C. Diamantini

A Cooperative and Adaptive Approach to Medical Image Segmentation   379
C. Spinu, C. Garbay and J.M. Chassery

Posters

COBRA: Integration of Knowledge-Bases with Case-Databases in the Domain   393
of Congenital Malformation
S. Tsumoto, H. Tanaka, H. Amano, K. Ohyama and T. Kuroda

Case-Based Medical Multi-Expertise: An Example in Psychiatry   395
I. Bichindaritz

TIME-NESIS: A Data Model in Managing Time Granularity of   397
Natural-Language Clinical Information
C. Combi, F. Pinciroli and G. Pozzi

Induction of Expert System Rules from Clinical Databases Based on Rough Set   399
Theory and Resampling Methods
S. Tsumoto and H. Tanaka

Sequential Knowledge Acquisition: Combining Models and Cases
401
B. Brigl, A. Grau, P. Ringleb, Th. Steiner, W. Hacke and R. Haux

Medical Fuzzy Expert Systems and Reasoning about Beliefs
403
P. Hájek and D. Harmancová

Diagnosis of Human Acid-Base Balance States via Combined Pattern
405
Recognition of Markov Chains
M. Kurzynski, M. Wozniak and A. Blinowska

Intelligence Formation Problems in Children at an Early Age Applying New
407
Computer Technologies under Conditions of Rehabilitation Center
I.F. Olkhovsky and S.I. Blokhina

Telecardiology
409
I. McClelland, K. Adamson and N. Black

Modelling a Sharable Medical Concept System: Ontological Foundation in
411
GALEN
G. Steve, A. Gangemi and A. Rossi Mori

A Graph-Based Approach to the Structural Analysis of Proliferative Breast
413
Lesions
V. Della Mea, N. Finato and C.A. Beltrami

A Workstation for Clinical Decision Support in a Local Area Network for
415
Cardiology
A. Taddei, M. Niccolai, M. Raciti, C. Michelassi, M. Emdin, P. Marzullo and
C. Marchesi

Knowledge-Based Education Tool to Improve Quality in Diabetes Care
417
E. Salzsieder, U. Fischer, A. Hierle, U. Oppel, A. Rutscher and C. Sell

NEPHARM: A Pharmacokinetic Database for Adjusting Drug Dosage to
419
Impaired Renal Function
F. Keller, R. Arnold, T. Frankewitsch, D. Zellner and M. Giehl

A Hybrid Architecture for Knowledge-Based Systems
421
E. Christodoulou

Representing Medical Context Using Rule-Based Object-Oriented
423
Programming Techniques
M. Dojat and F. Pachet

Integration of Neural Networks and Knowledge-Based Systems in Medicine
425
A. Ultsch, D. Korus and T.O. Kleine

Generated Critic in the Knowledge Based Neurology Trainer
427
F. Puppe, B. Reinhardt and K. Poeck

An Approach to Analysis of Qualitative Data with Insufficient Number of
429
Quantization Levels
N. Polikarpova

Inductively Learned Rule for Breast Cancer Domain with Improved   431
Interobserver Reproducibility
D. Gamberger

Development and Evaluation of a Knowledge-Based System to Support   433
Ventilator Therapy Management
N. Shahsavar and O. Wigertz

A Neural Support to the Prognostic Evaluation of Cardiac Surgery   435
F. Fiocchi, A. Gamba, R. Pizzi and F. Sicurello

DECISion-Support System for Radiological Diagnostic   437
G. Zeilinger, J. De Mey, G. Gell and G. Vrisk

A Preliminary Investigation into the Analysis of Electromyographic Activity   439
Using a System of Multiple Neural Networks
P. Caleb, P.K. Sharpe and R. Jones

Knowledge-Based System to Predict the Effect of Pregnacy on Progression of   441
Diabetic Retinopathy
C. Sell, S. Herfurth, A. Rutscher, E. Salzsieder, A. Hierle, U. Oppel, M. Förster,
G. Müller and DIADOQ-Group

A Software to Evaluate Multislices Radiotherapic Treatment Planning   443
R. Anselmi, G. Paoli, G. Ghiso, F. Foppiano, R. Martinelli and L. Andreucci

TKR-tool: An Expert System for Total Knee Replacement Management   444
J. Heras and R.P. Otero

Author Index   447

Keynote Address

A Component-Based Architecture for Automation of Protocol-Directed Therapy

Mark A. Musen, Samson W. Tu, Amar K. Das, and Yuval Shahar

Section on Medical Informatics
Stanford University School of Medicine
Stanford, California U.S.A. 94305-5479

The automation of protocol-based care requires reasoning about a patient's situation over time and about how the standard protocol plan can be adapted to address the patient's current clinical situation. The EON architecture brings together (1) a skeletal-planning reasoning method, ESPR, that can determine appropriate clinical interventions by instantiating an abstract protocol specification, (2) a temporal-reasoning system, RÉSUMÉ, that can infer from time-stamped patient data higher-level, interval-based concepts, and (3) a historical database system, Chronus, that can perform temporal queries on a database of interval-based patient descriptions. The modular problem-solving elements of EON operate on knowledge bases of clinical protocols that clinicians enter into domain-specific knowledge-acquisition tools generated by the PROTÉGÉ-II system. The EON architecture provides an integrated framework for development, execution, and maintenance of clinical-protocol knowledge bases.

1 Automated Support of Protocol-Based Care

Once limited to therapy for controlled clinical trials, protocol-based care is becoming increasingly common in medical practice. All participants in the health-care system now recognize the importance of appropriate use of protocols to provide optimum patient management and to ensure a high quality of care. The American Medical Association's *Directory of Practice Parameters* now lists over 1500 entries, and new guidelines continue to be promulgated by government agencies, professional-specialty organizations, payers, hospitals and other health-care institutions, physician group practices, and patient-advocacy groups. There is no question that written descriptions of guidelines and protocols are filling up loose-leaf binders in the libraries of many practitioners at an accelerating rate. The text of many protocols and guidelines also is becoming available via the World Wide Web and other electronic information sources. Despite the increasing accessibility of these guideline descriptions, there are few data to suggest that guidelines are having an appreciable affect on most aspects of medical practice (Lomas et al., 1989). Whereas in many cases it may be appropriate for a provider not to follow guidelines—particularly if those guidelines are not based on empirical evidence—often guidelines are ignored merely because they never reach the practitioner's consciousness.

Since the 1970s, workers in medical informatics have recognized the importance of communicating guidelines to practitioners directly at the point of patient care. For example, experience with the Regenstrief Medical Information System and the HELP system has documented significant changes in clinician behavior when computer systems can offer situation-specific advice. Recently, a group of investigators, working through the American Society for Testing and Materials (ASTM), has defined a standard procedural language, known as the Arden syntax (Hripcsak et al., 1994), for encoding situation–action rules like those in the HELP system. Developers of the

Arden syntax have promoted this Pascal-like language because of pressing needs to facilitate exchange of guideline logic among health-care institutions using existing software technology. The developers recognize, however, that this new standard has significant limitations: The language currently supports only atomic data types, lacks a defined semantics for making temporal comparisons or for performing data abstraction, and provides no principled way to represent clinical guidelines that are more complex than individual situation–action rules (Shwe et al., 1992; Musen, 1992).

Despite the sustained popularity of rule-based approaches to knowledge representation, researchers in artificial intelligence (AI) have long recognized that interdependencies among individual rules can lead to unpredictable system behavior (Heckerman and Horvitz, 1988) and that the procedural semantics of rule-based representations can severely complicate the maintenance of large knowledge bases (Clancey, 1983). Developers of knowledge-based systems increasingly have sought to define software architectures in which the purpose of each entry in the knowledge base is unambiguous and in which discrete, well-understood problem-solving procedures operate on explicit, declarative models of the relevant application domain (Musen and Schreiber, in press). In our laboratory, we have developed a framework for building knowledge-based systems known as PROTÉGÉ-II (Musen, 1992; Musen et al., 1995). PROTÉGÉ-II provides a methodology and an extensive set of tools for building knowledge-based systems from reusable components in a principled manner. As part of our evaluation of the PROTÉGÉ-II approach, we have constructed a number of problem-solving components that we have put to use in a computer system known as THERAPY-HELPER (or "T-HELPER"), which facilitates protocol-based care for patients who have AIDS (Musen et al., 1992a).

T-HELPER includes a knowledge-based system, constructed using PROTÉGÉ-II, that (1) recommends therapy consistent with predefined protocols and guidelines and (2) informs its physician-users when a given patient is potentially eligible for enrollment in additional protocols. We refer to this knowledge-based system for protocol-directed therapy as EON (Musen et al., 1992b). The current EON architecture comprises (1) a generic problem solver for determining appropriate protocol-directed therapy, (2) a subsystem that takes as input primary time-stamped patient data and that generates as output relevant time-dependent abstractions of those data, and (3) a database system that stores time-dependent patient information in an extended relational format and that supports the queries required for protocol-related reasoning (Figure 1). EON can be viewed as a task-specific architecture for protocol-based care that operates on the protocol knowledge bases developers create using PROTÉGÉ-II. We currently embed this architecture within the T-HELPER system to provide protocol-directed decision support for patients with AIDS. We anticipate, however, that the three components of the existing EON architecture will serve as building blocks for a general system that can automate protocol-directed problem solving for a variety of clinical domains.

2 Domain Ontologies and Problem-Solving Methods

Contemporary methodologies for the development of knowledge-based systems emphasize the use of conceptual abstractions that define problem-solving behaviors inpendently from the programming constructs (e.g., production rules) that might be

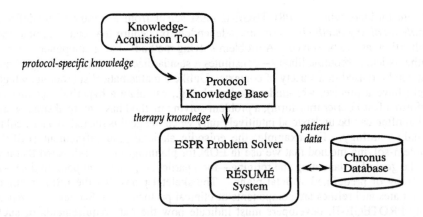

Figure 1: The EON architecture includes the ESPR problem-solving method, which incorporates the RÉSUMÉ temporal-abstraction system. The Chronus database system supports time-related queries on historical patient data. The ESPR method operates on a knowledge base of clinical protocols that developers create using a knowledge-acquisition tool generated by the PROTÉGÉ-II system.

used to encode those behaviors. For medical applications, approaches such as PROTÉGÉ-II and GAMES (van Heijst et al., 1994) allow developers to structure knowledge as explicit models of the relevant application area, which they then can relate to domain-independent problem-solving procedures.

In PROTÉGÉ-II, system builders first specify a domain model that defines the terms and relationships of the general application area—for example, the domain of protocol-based care for AIDS patients. Because the domain model indicates the general constructs of the relevant application area (e.g., the relationships between classes of protocols and classes of drugs) but does not define specific instances of those classes (e.g., how the drug zidovudine might be used in a particular protocol), we refer to the model as an *ontology* of the application area. Our ontology of AIDS protocols, for example, defines concepts such as clinical trials, drug regimens, prescriptions, laboratory tests, and so on. In the PROTÉGÉ-II approach, we use the domain ontology[1] to generate automatically a domain-specific knowledge-acquisition tool that nonprogrammers can use to enter the details of specific protocols (Eriksson et al., 1994). Thus, the generated knowledge-acquisition tool is used to define the particular sequences of interventions that should be administered to patients who will be treated according to particular protocols. If the developers should revise their domain ontology for a given area of medicine, it is straightforward for them to use PROTÉGÉ-II to generate an updated knowledge-acquisition tool that reflects their current view of the clinical domain.

---

[1]Technically speaking, we make a distinction between the *domain ontology*, which is designed to be somewhat general and thus maximally reusable in new application tasks, and the *application ontology*, which enhances the domain ontology with additional distinctions that are required to model the task at hand (Gennari et al., 1994). For simplicity, we use the term *domain ontology* in this paper to refer to both kinds of models.

System builders who use PROTÉGÉ-II must select from a library of predefined *problem-solving methods* a domain-independent procedure that can automate the application task to be solved. A problem-solving method—like a component from a mathematical subroutine library—constitutes a standardized computational algorithm that can be reused in a variety of contexts. Unlike a mathematical subroutine, which might have a precise, well-understood goal (e.g., calculate a hyperbolic cosine or perform a fast Fourier transform), a problem-solving method has a more abstract goal, which often can be understood intuitively only when the goal is related to a particular domain ontology. For example, the episodic skeletal-plan refinement (ESPR) problem-solving method that we use in EON for planning protocol-directed therapy has the goal of instantiating and refining—for a particular patient and point in time—a hierarchy of predefined skeletal plans. The skeletal plans that the ESPR method instantiates and refines are components of clinical protocols (see Section 3). When using PROTÉGÉ-II, developers must indicate how the data requirements of such domain-independent problem-solving methods relate to the various concepts defined in the relevant domain ontologies. Construction of a knowledge-based system is thus a matter of defining explicit mappings between the concepts represented in the domain ontology and the abstract data requirements of the particular problem-solving method that has been chosen to solve the application task at hand (Gennari et al., 1994).

In PROTÉGÉ-II, many problem-solving methods are themselves made up of more primitive building blocks. A problem-solving method such as ESPR may define a number of *subtasks* that require additional methods to solve. These smaller-grained problem-solving methods may themselves pose subtasks that must be solved by other problem-solving methods. Eventually, subtasks can be solved by atomic problem-solving methods that do not pose subtasks, called problem-solving *mechanisms* (following a naming convention adopted by other authors; Marques et al., 1992). A developer who uses PROTÉGÉ-II must configure a general problem-solving method for the task to be automated by assembling appropriate problem-solving methods and mechanisms (Eriksson et al., in press). For each subtask in the initial problem-solving method, the developer selects from our library of problem-solving methods a method or mechanism that addresses the computational requirements of that subtask.[2] The method-selection process is repeated recursively until all the subtasks are satisfied. The process of configuring the problem-solving method takes place during the knowledge-acquisition phase. Because the data requirements of the configured methods and mechanisms must be mapped to concepts in the domain ontology, the final method configuration is static and cannot be changed—unless the developer later should choose to define new mappings.

Thus, in the PROTÉGÉ-II approach, system builders define a domain ontology and configure a problem-solving method, and then map the domain ontology onto the data requirements of the method. The domain ontology also provides the basis for generation of a domain-specific knowledge-acquisition tool that application specialists can use to enter the content knowledge required for individual knowledge bases. The result is that intelligent systems can be constructed from reusable building blocks in a flexible, yet principled, manner. The ensuing componential architecture clarifies the

---

[2]The developer alternatively may adapt an existing method for the new situation or may program the necessary method from scratch.

role that each knowledge element plays in problem solving, thus aiding validation of the knowledge base and ongoing maintenance of the system.

3 Episodic Skeletal-Plan Refinement

The EON architecture for automating protocol-based therapy is constructed from reusable problem-solving methods that are elements of the PROTÉGÉ-II method library. At the core of EON is a problem-solving method known as episodic skeletal-plan refinement (ESPR; Figure 2). The ESPR method, which was inspired by the behavior of the ONCOCIN system for protocol-based care in oncology (Tu et al., 1989), has been the subject of ongoing research in our laboratory for several years. Using the ESPR method, administration of therapy according to predefined protocols can be construed as an abstract (skeletal) plan that can be decomposed into one or more constituent plans (*planning entities*) that are each more detailed than is the abstract plan. These planning entities, however, may themselves be skeletal in nature, requiring further decomposition. The output of the planning process is a fully specified (instantiated) plan—which, in the case of the protocol-based–care domain, represents a treatment recommendation for the practitioner to follow (Tu et al., in press).

The ESPR method has three classes of input data: (1) a skeletal plan to be instantiated, (2) the data that define the current case, and (3) the current time. The skeletal plan may entail algorithms that take place over time (e.g., a clinical protocol that specifies a sequence of interventions), making the appropriate instantiation of the skeletal plan time-dependent. Because the ESPR method can be invoked repeatedly at different times for a given case (e.g., to plan therapy for multiple clinic visits for a given patient), the method is said to be *episodic*. As will become evident, both the need to represent the situation for a given case over time and the time-dependent nature of the skeletal-planning process itself make temporal reasoning an important element of the ESPR method.

The ESPR method entails three subtasks: (1) propose plan, (2) identify problem, and (3) revise plan (see Figure 2). The *propose plan* subtask involves determining the standard plan given the results of the previous planning episodes and the current point in time. The *identify problem* subtask identifies characteristics of the current case that might require the problem solver to modify the standard plan. The *revise plan* subtask alters the standard plan in response to any problems that have been identified. The three subtasks of the ESPR method are in turn solved by additional problem-solving methods, as indicated in Figure 2. (The knowledge-based temporal-abstraction method, which solves the *identify problem* subtask, is described in Section 4.)

When applied to the problem of determining therapy for patients being treated according to clinical protocols, the ESPR method deduces a treatment plan following a strategy analogous to the propose–critique–modify approach that has been applied to a number of design tasks (Chandrasekaran, 1990). First, the method examines the basic protocol algorithm to determine the set of clinical interventions that normally should be administered to a patient, given the patient's history of previous treatment; this "standard plan" may be appropriate for an uncomplicated patient, but frequently there are special situations that mandate some sort of plan modification (e.g., a reduction in the usual dose of zidovudine in an AIDS patient who develops anemia). Thus, the *identify problem* subtask determines whether any predefined patterns in the data that

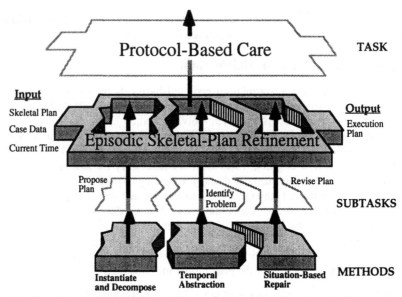

Figure 2: The Episodic Skeletal-Plan Refinement method solves the task of generating a protocol-based treatment plan by calling three substasks, each of which are addressed by appropriate problem-solving methods.

are to be avoided are present. If so, then the *revise plan* subtask makes an appropriate adjustment to the basic treatment plan that initially had been proposed.

4 Knowledge-Based Temporal Abstraction

Identification of the special situations that may dictate modification of the standard treatment plan generally must be inferred by analyzing time-stamped data stored in a clinical database. Whereas the database stores raw data values, the conditions that predicate alterations to a patient's treatment typically are described at a higher level of abstraction. The *identify problem* subtask of the ESPR method (see Figure 2) is such a temporal-abstraction task, and it can be solved by a method known as *knowledge-based temporal abstraction* (Shahar and Musen, 1993). This method allows the EON system to evaluate and summarize the state of the patient over a time interval (possibly, the entire record of the patient). The method takes as its input time-stamped data (e.g., individual hemoglobin values and therapy events) and returns as its output abstractions of those data that are interpreted over specific time points or intervals (e.g., periods of anemia or of normal hemoglobin levels). The knowledge-based temporal-abstraction method, like other methods in the PROTÉGÉ-II library, is a modular building block. The method itself identifies a number of subtasks, each of which is solved by other methods or mechanisms from the library.

The knowledge-based temporal-abstraction method entails five subtasks: (1) temporal context restriction (creation of relevant interpretation contexts crucial for focusing and limiting the scope of inference); (2) vertical temporal inference (inference from contemporaneous propositions regarding raw or abstract data into higher-level concepts); (3) horizontal temporal inference (inference from similar-type propositions associated with time intervals that cover different, but meeting or overlapping, time periods); (4) temporal interpolation (union of nonmeeting points or intervals, associated with propositions of similar type); and (5) temporal pattern matching (creation of intervals by matching of patterns over disjoint intervals, associated with propositions of various types). The subtasks of the temporal-abstraction method have been described in detail elsewhere (Shahar and Musen, 1993; Shahar, 1994).

Each of the five subtasks is solved by a different temporal-abstraction problem-solving mechanism, also stored in the PROTÉGÉ-II library. Collectively, these temporal-abstraction mechanisms comprise a subsystem known as RÉSUMÉ (Shahar and Musen, 1993). Workers in our laboratory have applied the RÉSUMÉ system not only to the temporal-abstraction problems of protocol-based care, but also to the monitoring of children's growth (Kuilboer et al., 1993) and to the interpretation of data from patients with diabetes (Shahar et al., 1994).

Thus, the RÉSUMÉ system provides a method that solves temporal abstraction tasks—such as the *identify problem* subtask of the ESPR method, which in turn solves the more general task of recommending therapy according to predefined protocols. Each time the ESPR method executes, it first calls on the *propose plan* subtask to establish a putative treatment plan for the current patient visit based on the standard protocol. The ESPR method then turns to the *identify problem* subtask, and invokes the knowledge-based temporal-abstraction method in RÉSUMÉ. This method activates the temporal-abstraction mechanisms, which create a set of generalizations that provide a detailed model of the patient's condition over time.

Once the relevant temporal abstractions have been generated, they activate the *situation-based repair* method, which has been configured for the *revise plan* subtask; the situation-based repair method takes as input protocol-specific knowledge of various contingencies that might affect the ultimate treatment recommendation, and generates as output modifications to the standard plan that adjust for any patient-related problems that RÉSUMÉ might have detected.

5 Storing Interval-Based Data

The EON architecture uses a relational database to store the time-dependent patient data that drive decision support for protocol-based therapy. These data include past and present patient symptoms and problems, and a record of all treatment decisions. This account of prior patient data is important for a variety of reasons, including planning current therapy and determining whether there are new protocols or guidelines for which the patient might be eligible (Tu et al., 1993). Time-dependent data are stored in the database as *historical relations* that have a *start time* and an *end time* associated with each entry (Figure 3).

Standard relational databases are not well suited for storing such historical information. The Structured Query Language (SQL) is awkward for processing queries based on temporal relationships that must be derived from the time-interval interdependencies among the individual tuples. More important, standard relational algebra, which defines the semantics of SQL, is not *closed* for data stored as historical relations. For example, if the user should PROJECT away the time stamps associated with data such as those in Figure 3, the result is not a valid historical relation; if the user should JOIN two relations such as those in Figure 3, the result is a relation that has *four* time stamps, not two. To address these problems, we have created a new

Start Time	End Time	Patient	Drug	Dose
Jan 3, 1995	March 6, 1995	1111	Zidovudine	500
March 7, 1995	April 5, 1995	1111	Zidovudine	300
Sept 4, 1994	May 14, 1995	2222	Zidovudine	500
May 14, 1995	now	2222	ddI	400
May 14, 1995	now	2222	Septra DS tablet	1

Figure 3: Chronus operates on relationl tables in which each n-tuple has been augmented with a start time and an end time. Data that occur at a single time point are denoted by entries that have an identical start time and end time. Events that are ongoing have a stop time of *now*.

temporal query system, called Chronus (Das and Musen, 1994), that is supported by a temporal relational algebra that confers special status to the time stamps associated with each tuple. The algebra thus ensures that any operation on a historical relation returns a new relation in which each tuple similarly is associated with exactly one start time and exactly one end time. The Chronus algebra thus is an alternative to standard relational algebra. The Chronus algebra retains the SELECTION operator from the standard algebra, but substitutes a special TEMPORAL PROJECTION operator that, unlike the standard PROJECTION operator, disallows the removal of time stamps from any historical relation. CARTESIAN PRODUCT is not allowed in the Chronus algebra; instead, the algebra introduces three new TEMPORAL JOINs that allow the contents of two relational tables to be combined in different ways, depending on temporal relationships between the tuples of the two tables (Das and Musen, 1994).

The Chronus algebra introduces an additional operator, CATENATION, that has no correlate in standard relational algebra. This new operator merges the data in two temporally adjacent tuples of a relation when (1) the nontemporal elements of the tuples are identical, (2) the temporal intervals associated with the tuples meet or overlap, and (3) the semantic properties of the nontemporal elements of the tuples make it sensible to merge them. Thus if, after a series of data manipulations, a relation is created in which there are two tuples indicating that a particular drug was administered to a patient at a specific dose, and the time stamps of the two tuples denote contiguous periods of drug administration, then the CATENATION operator will replace the two tuples with a single tuple, with new time stamps that denote the complete period during which the drug was administered.

As in traditional relational database systems, users do not query data using the algebra directly, but rather interact with a more abstract query language that translates the users' requests into a sequence of appropriate algebraic operations. Chronus supports a query language known as TimeLine SQL (TLSQL), which has much the same syntax as SQL, but which incorporates a new WHEN clause that allows the user to select data based on a variety of temporal comparisons (Das and Musen, 1994).

6 Discussion

The EON architecture comprises a number of components: the ESPR problem-solving method and the other problem-solving methods that satisfy the subtasks of ESPR—including the RÉSUMÉ system for performing knowledge-based temporal abstraction; the Chonus historical database system, supporting the TLSQL temporal query language; and the particular knowledge bases of protocols and guidelines that developers create using the knowledge-acquisition tools generated by PROTÉGÉ-II.

This approach to automating protocol-based care certainly involves more computational machinery than would be the case if we were to attempt to encode protocols using approaches such as the Arden syntax (Hripcsak et al., 1994). At first glance, the EON architecture may seem unduly intricate. Nevertheless, all systems that support protocol-based care must be able to reason about patient therapy, must be able to abstract generalizations about the patient from point data, must support archival storage and retrieval of time-related patient data, and must provide some means by which developers can review the knowledge of existing protocols and can encode the knowledge of new treatment specifications. Computational approaches to guideline-directed therapy that do not support these functions have significant limitations. Venturing to represent substantial protocols in the Arden syntax, for example, currently requires developers to write complex programs for large numbers of medical logic modules that may interact with one another in rather arcane ways.

A significant advantage of the EON architecture stems from its relationship with the PROTÉGÉ-II system. PROTÉGÉ-II provides both a methodology and a set of tools for configuring problem-solving method such as ESPR, and for constructing declarative ontologies of clinical application areas. In the case of the T-HELPER system, we have used PROTÉGÉ-II to construct a knowledge-acquisition tool tailored to the requirements of HIV-related protocols, and have used this tool to enter specifications for a dozen clinical-trial protocols. Because the knowledge-acquisition tool presents the contents of each protocol knowledge base in domain-specific terms, nonprogrammers can enter new information into the tool and browse through existing protocol specifications. At the same time, the modular configuration of problem-solving building blocks used to construct the ESPR method makes the computational elements of the architecture easier for system developers to maintain and enhance.

We have tested the EON architecture within the context of the T-HELPER system. Our experience demonstrates that EON's ability to reason about protocol-based care is sufficient to provide therapy recommendations for a wide range of AIDS-related clinical trials, including antiretroviral studies and trials of antibiotics for primary and secondary prophylaxis of opportunistic infections. We have not yet used EON to provide advice for clinical trials in other areas of medicine, although our approach is sufficiently general that we do not anticipate major obstacles. Ongoing work in our laboratory centers around extension and further integration of the components in EON, and application of the approach to a wider range of clinical protocols.

To date, most of our experience has been in using EON to reason about clinical-trial protocols. Although most clinical-trial protocols are far more complex than are most practice guidelines, clinical trials represent an unusual situation in medicine: Patients with significant comorbidity generally are excluded from clinical trials, and therapy specifications in clinical trials are as precise as possible. Clinical trials represent controlled experiments, and the corresponding protocols must be as reproducible as possible. Practice guidelines, on the other hand, tend to be far less "prescriptive," and may allow for a fair degree of lattitude—both in defining the clinical situations that may trigger application of a guideline and in administering the interventions that are mandated by the particular clinical situation. Such leeway is appropriate in situations when there is no normatively preferred treatment and when it is impossible to enumerate the myriad contingencies that may influence the provider's decision making. The ambiguity of many guidelines, however, precludes our ability to model them as fully specified therapy plans. Although the knowledge-based temporal-abstraction method in RÉSUMÉ does allow identification of clinical situations that may not be fully specified, the planning subtasks of the ESPR method currently assume that treatment recommendations should be fully delineated; the method is not designed to generate as output a class of plan recommendations. Nevertheless, in situations when a practice guideline is ambiguous regarding appropriate therapy, it may be more helpful to offer a *critique* of a provider's intended treatment (van der Lei and Musen, 1991) than to suggest a specific therapy recommendation. Our group consequently plans to develop new problem-solving methods that can compare current therapy to that suggested by applicable guidelines and that can identify significant discrepencies.

Acknowledgments

This work has been supported in part by grant HS06330 from the United States Agency for Health Care Policy and Research, and by grants LM05157, LM05304, and LM05708 from the National Library of Medicine. Dr. Musen is the recipient of National Science Foundation Young Investigator Award IRI-9257578.

References

Chandrasekaran, B. (1990). Design problem solving: A task analysis. *AI Magazine* **11**(4):59-71.

Clancey, W.J. (1983). The epistemology of a rule-based expert system: A framework for explanation. *Artificial Intelligence* **20**:215–251.

Das, A.K. and Musen, M. A. (1994). A temporal query system for protocol-directed decision support. *Methods of Information in Medicine*, **33**:358–370.

Eriksson, H., Puerta, A.R., and Musen, M.A. (1994). Generation of knowledge-acquisition tools from domain ontologies. *International Journal of Human–Computer Studies*, **41**:425–453.

Eriksson, H., Shahar, Y., Puerta, A.R., Tu, S.W., and Musen, M.A. (in press). Task modeling with reusable problem-solving methods. *Artificial Intelligence*.

Gennari, J.H., Tu, S.W., Rothenfluh, T.E., and Musen, M.A. (1994). Mapping domains to methods in support of reuse. *Internaitonal Journal of Human-Computer Studies,* **41**:399–424.

Heckerman, D. and Horvitz, E. (1988). The myth of modularity in rule-based systems for reasoning with uncertainty. In Lemmer, J.F. and Kanal, L.N., eds. *Uncertainty in Artificial Intelligence 2*, pp. 23–34. Amsterdam: North-Holland, 1988.

Hripcsak, G., Ludemann, P., Pryor, T.A., Wigertz, O.B., and Clayton, P.D. (1994). Rationale for the Arden syntax. *Computers and Biomedical Research* **27**:291–324.

Kuilboer, M.M., Shahar, Y., Wilson, D.M., and Musen, M.A. (1993). Knowledge reuse: Temporal-abstraction mechanisms for the assessment of children's growth. *Proceedings of the Seventeenth Annual Symposium on Computer Applications in Medicine*, pp. 449–453. Washington, D.C.

Lomas, J., Anderson, G.M., Domnick-Pierre, K., et al. (1989). Do practice guidelines guide practice? The effect of a consensus statement on the practice of physicians. *New England Journal of Medicine* **321**:1306–1311.

Marques, D., Dallemagne, G., Klinker, G., McDermott, J., and Tung, D. (1992). Easy programming: empowering people to build their own applications. *IEEE Expert* 7:16–29.

Musen, M.A. (1992). Dimensions of knowledge sharing and reuse. *Computers and Biomedical Research* 25:435–467.

Musen, M.A., Carlson, C.W., Fagan, L.M., Deresinski, S.C., and Shortliffe, E.H. (1992a). T-HELPER: Automated support for community-based clinical research. *Proceedings of the Sixteenth Annual Symposium on Computer Applications in Medical Care*, pp. 719–723. Baltimore, MD.

Musen, M.A., Tu, S.W., and Shahar, Y. (1992b). A problem-solving model for protocol-based care: From e-ONCOCIN to EON. In *Proceedings of MEDINFO '92, Seventh World Congress on Medical Informatics*, pp. 519–525. Amsterdam: North-Holland.

Musen, M.A., Gennari, J.H., Eriksson, H., Tu, S.W., Puerta, A. R. (1995). PROTÉGÉ-II: Computer Support For Development Of Intelligent Systems From Libraries of Components. *Proceedings of MEDINFO '95,* Eighth World Congress on Medical Informatics, Vancouver, BC.

Musen, M.A. and Schreiber, A.T. (in press). Architectures for intelligent systems based on reusable components. *Artifical Intelligence in Medicine*.

Shahar, Y., Das, A.K., Tu, S.W., Kraemer, F.B, and Musen, M.A. (1994). Knowledge-based temporal abstraction for diabetic monitoring. *Proceedings of the Eighteenth Annual Symposium on Computer Applications in Medical Care,* pp. 697-701. Washington, D.C.

Shahar, Y. and Musen, M.A. (1993). RÉSUMÉ: A temporal-abstraction system for patient monitoring. *Computers and Biomedical Research* 26:255–273.

Shahar, Y. (1994). A knowledge-based method for temporal abstraction of clinical data. Ph.D. Dissertation, Program in Medical Information Sciences, Stanford University, Computer Science Dpartment Technical Report STAN-CS-TR-94-1529.

Shwe, M., Sujansky, W., and Middleton, B. (1992). Reuse of knowledge represented in the Arden syntax. *Proceedings of the Sixteenth Annual Symposium on Computer Applications in Medical Care*, pp. 47–51. Baltimore, MD.

Tu, S.W., Kahn, M.G., Musen, M.A., Ferguson, J.C., Shortliffe, E.H., and Fagan, L.M. (1989). Episodic skeletal-plan refinement on temporal data. *Communications of the ACM* 32:1439–1455.

Tu, S.W., Kemper, C.A., Lane, N.M., Carlson, R.W., and Musen, M.A. (1993). A methodology for determining patients' eligibility for clinical trials. *Methods of Information in Medicine* 32:317–325.

Tu, S.W., Eriksson, H., Gennari, J., Shahar, Y., Musen, M.A. (in press). Ontology-based configuration of problem-solving methods and generation of knowledge-acquisition tools: Application of PROTÉGÉ-II to protocol-based decision support. *Artificial Intelligence in Medicine.*

van der Lei, J. and Musen, M.A. (1991). A model for critiquing based on automated medical records. *Computers and Biomedical Research* 24:344–378.

van Heijst, G., Schreiber, G., Lanzola, G. and Stefanelli, M. (1994). Foundations for a methodology for medical KBS development. *Knowledge Acquisition* 6:395–434.

Medical Records

Coordinating Taxonomies:
Key To Re-usable Concept Representations

AL Rector

Medical Informatics Group, Department of Computer Science
University of Manchester, Manchester M13 9PL, England
Tel +44-161-275-6133/6188, FAX: +44-161-275-6932, Internet: rector@cs.man.ac.uk

Abstract: A unified controlled medical vocabulary has been cited as one of the grand challenges facing Medical Informatics. We would restate this challenge as 'achieving a re-usable and application-independent representation of medical concepts.' Achieving a re-usable representation of medical concepts is a pre-requisite for meeting two key strategic goals of the next decade of the development in medical informatics: interoperability and cumulative development. A key strategy for achieving re-usability is to separate concepts into their component parts, organise those parts in nearly pure hierarchies, and then recombine into composite representations which can be classified flexibly and automatically. This paper explores the means and consequences of this strategy as implemented in the GALEN project. It discusses both the strengths — providing greater detail, greater computer support, and avoiding many arguments which are endemic in discussions of classification systems — and the limitations intrinsic in such a formal approach.

1. Introduction

1.1. A Re-usable representation of medical concepts

Achieving a unified controlled medical vocabulary has been cited as one of the grand challenges facing Medical Informatics. We would restate this challenge as 'achieving a re-usable and application-independent representation of medical concepts.' Achieving such a re-usable representation of medical concepts is a pre-requisite to meeting two key strategic goals of the next decade of the development in medical informatics: interoperability and cumulative development.

Complete application independence is, of course, a chimera. Any representation will only be re-usable within a limited area and for limited purposes. In this paper we will concern ourselves primarily with concepts being used for 'patient centred information systems' — systems used directly by clinicians as a routine part of clinical care to capture information which may then be re-used in other secondary application. We will discuss a specific strategy for achieving important aspects of re-use — separating taxonomies cleanly and then re-coordinating them by defining composite concepts and classifying those composite concepts automatically — and we will explore a number of issues raised by this strategy.

1.2. The Barriers to Re-use and application independence

Why is it so difficult to achieve a re-usable representation of medical concepts — or of the concepts used in any other field of endeavour? The aspiration to re-use and application independence is not new. Attempts to produce standard classifications and to formalise their logical foundations go back at least to Aristotle. The first attempts at standardised classification and coding systems for medicine date back to the middle of the nineteenth century.

In this paper we argue that one key problem limiting re-use and application-independence is that the formalisms used force application-specific choices on developers of vocabularies and classifications. We argue that one of the key application-specific decisions which formalisms force on users is the organisation of a fixed taxonomy based on indivisible representations of concepts. Inevitably, indivisible representations of complex concepts conflate several ideas, and since any fixed organisation of such complex ideas is fundamentally arbitrary, choices can only be made on the basis of the current application. We contend that avoiding such decisions is an important technique for achieving re-use. The general technique is:

a) To separate the concepts into their apparently elementary parts for this representation;

b) To form clean taxonomies of the parts

c) To recombine the parts by defining the original concepts in terms of the elementary parts

d) To classify the composite concepts according to formal rules.

Our contention is that it should only very rarely be necessary for an elementary entity to have more than one parent within a taxonomy, though composite entities may have indefinitely many parents formally identified on the basis of their definitions. Our experience is that arguments about where something should be classified almost invariably indicate that the concept has not been adequately decomposed. When decomposed and defined in terms of more elementary concepts, the arguments usually disappear.

The problem of conflating several ideas into a unitary representation is particularly serious in standard mono-axial classification systems such as ICD-10, ACRNEMA or the Read Clinical Classification, but is common in other representations. Consider for example the nomenclatures for Internist/QMR and HELP [1].

A second related mechanism for avoiding application-specific decisions — bridging the levels of detail needed in different applications — is mentioned briefly but will be the subject of a separate paper.

These ideas are not necessarily new. Other authors have suggested using compositional terminological systems as an interlingua for reconciling clinical terminologies, *e.g.* [1-4] or for mediating between databases, *e.g.* [5-7]. The purpose of this paper is to emphasise the specific issue implicit but often not emphasised in these writings — that the use of compositional systems allows clean separation of taxonomies and principled construction of multiple classifications — and to explore some of the specific consequences of this policy raised by GALEN's implementation.

1.3. Background

This paper presents techniques and a point of view developed during the GALEN project[1] which is developing 'Terminology servers' based on compositional models of

[1]   General Architecture for Languages Encyclopædias and Nomenclatures in Medicine. The members of the GALEN consortium are: University of Manchester (UK, Coordinator), Hewlett-Packard UK Ltd, Hôpital Cantonal Universitaire de Genève (Switzerland), Consiglio Nazionale delle Ricerche (Italy), University of Liverpool (UK), Katholieke Universiteit Nijmegen (Netherlands), University of Linköping (Sweden), The Association of Finnish Local Authorities (Finland), The Finnish Technical Research Centre (Finland), GSF-Medis Institut, (Germany), Conser Sistemi Avanzati (Italy)

medical concepts to act as a means for integration and interoperability of clinical applications. The examples are drawn from the GALEN CORE Model, version 5, in its form as of autumn 1994 [8]. The work has been used as the basis for clinical user interfaces and for conversion amongst existing coding systems. Work to test the wider re-usability of the representation is in progress.

GALEN is based on the assumption that there is a distinctive 'concept level' or 'terminology level' which can usefully be represented separate from both the 'linguistic level' and more 'general inference' or 'assertional level'. GALEN is an attempt to realise a strong notion of such a concept level — a single concept model which can be re-used by different linguistic systems and different general inference systems. This is a strong sense of re-use — not merely that one representation can be transformed into another, but that a common model can be established which can serve as a single basis for many different applications.

GALEN uses the GALEN Representation and Integration Language (GRAIL) Kernel [9, 10], a 'description logic' or 'knowledge representation language' related to KL-ONE [11], CLASSIC [12], or BACK [13] and to Conceptual Graphs [14] but with special features to support the requirements of coordinating taxonomies, particularly for coordinating the 'kind-of' (subsumption) relation with other transitive relations. Most of the discussion is equally applicable to other related systems. A formulation of the relationship between part-whole relations and subsumption analogous, though not identical, to that presented here has been developed for conceptual graphs by Bernauer [15]. Related formulations for representing SNOMED and other medical records have also been developed using Conceptual Graphs by Campbell [3]and a series of models with some features in common have been developed by the CANON group and presented as a series of papers [2, 4, 16]. GRAIL is strictly terminological, it does not, itself, provide any 'assertional' or 'general inferential' component — in the language of the KL-ONE community [17, 18] it provides only a 'T-Box', albeit an extended one, but no 'A-Box'. The goal is that the same 'T-Box' should be re-usable by various inferential mechanisms.

2. Fundamental Ideas

2.1. Taxonomies: Kinds, parts, and other relations

A 'taxonomy' is a hierarchical structure. Most coding and classification systems are taxonomies, but taxonomies are also widely found in biology and elsewhere in engineering and science. Thesauri, which are one kind of taxonomy, are almost universally used as means of accessing bibliographic literature.

A taxonomy may be organised according to a single relationship or a combination of relationships. Thesauri typically avoid defining the precise relationship between different levels of the taxonomy by describing them simply as 'broader than' and 'narrower than'. Other systems are more explicit and specify the specific relationship used in a taxonomy.

The two most important types of relationship used to organise taxonomies are 'kind of' and 'part of' — sometimes called 'generic' and 'partitive' relations respectively. For example, 'automobile' is a kind of 'vehicle' but 'wheel' is a part of 'automobile'. Other taxonomies may be based on other relationships, e.g. causation. The 'kind of' relation is often referred to as 'subsumption' and we may say that 'vehicle subsumes automobile'. In most knowledge representation systems, subsumption plays a special role — anything which is true of a concept is also true of any other concept which it subsumes.

2.2. Description and Classification

Most traditional classification or coding systems consist of a set of indivisible entities explicitly classified in 'taxonomies'. In many cases a single entity can only be placed in only one place in the taxonomy, *i.e.* any one code can have only one 'parent'. The arrangement of the taxonomy is typically related to the original use of the classification — hence aetiology and infections play a major role in the International Classification of Diseases because of their importance in traditional epidemiology. Similarly, anatomical structure plays a more important role in ACRNEMA because of the importance of anatomical structures to radiographic findings. Even if more than one parent is allowed, few systems record the reason for different parents, hence it is difficult or impossible to follow a particular alternative view through consistently. Cross references are provided in SNOMED and to a lesser extent in ICD, but they must be maintained laboriously by hand, and there is no way to check that they are complete.

The fundamental idea of the GALEN representation is that most concepts are represented by descriptions as 'composite entities' rather than by elementary indivisible entities. For example, "pulmonary tuberculosis" might be represented as "an inflammation of the lungs caused by an infection by tubercle bacillus".

Composite entities can be classified according to their compositional structure in whatever way is appropriate for a given application. For example, pulmonary tuberculosis can be classified either as a "disease of the lung" or a "inflammation caused by an infection by tubercle bacillus" or both as required.

The classifications need not be specified in advance. Classification can be performed as required — *e.g.* according to anatomy, morphology or aetiology. For any system representing a wide range of concepts, the total number of possible classifications is indefinitely large.

2.3. Natural Kinds

While the idea of GALEN is that most concepts be represented by defining descriptions, not everything can be defined. The system must start with elementary entities; Furthermore, many concepts are best treated as 'natural kinds' — sometimes referred to as 'polythetic categories'. Roughly speaking, natural kinds are recognised rather than inferred. For example, "Rheumatoid Arthritis" is not well enough understood to be defined, but is a recognised syndrome. Within a model, natural kinds are represented as elementary entities rather than composite entities. However, the decision to represent a given concept by an elementary entity does not imply any philosophical commitment to view it as a natural kind, merely the pragmatic decision not to define it within this model — *i.e.* not to identify a set of *sufficient* conditions for its recognition.

Note that even though there is no set of sufficient conditions in the model, there may still be necessary conditions, *e.g.* that "Rheumatoid Arthritis is caused by an autoimmune process." From this it would follow that "rheumatoid arthritis" would be classified as an "autoimmune disease" but being an auto-immune arthritis would not be sufficient for some other form of arthritis to be classified as "rheumatoid arthritis". The ability to distinguish between representations of necessary and sufficient criteria for concepts is one of the distinctive features of the GRAIL Kernel. Necessary conditions affect how an entity is classified but do not affect what further entities are classified under it.

The taxonomy of natural kinds forms the skeleton of the model — the elementary entities from which composite descriptions are built up.

2.4. Transitive relationships

Part-whole and causal relationships play a key role in medical concepts. Any system hoping to represent medical concepts adequately must deal with such transitive relations in a consistent way. Transitive relations present special problems to a programme of representing most concepts by composite entities in a model because they interact in specific ways with the kind-of or 'subsumption' taxonomy.

For example, the "shaft of the femur" is a part of the "femur" rather than a kind of the femur. However, a "fracture of the shaft of the femur" is a kind of "fracture of the femur". Conversely, if we focus on the femur rather than the fracture, a "Femur which has a fractured shaft" is a kind of "fractured femur". Similar patterns hold for direct causation, though not necessarily for loose causal associations. An important and novel features of the GRAIL Kernel is that it implements coordination of transitive and kind-of taxonomies in this way.

2.5. Managing granularity and bridging level of abstraction

Any one application abstracts away many details as irrelevant and emphasises others which are particularly relevant. For example for purposes of most applications, "Gastric Ulcer" is sufficient detail. However, for a pathology system, one might wish to organise a classification of lesions according to the tissues affected, in which it is critical to represent the fact that ulcers actually affect the "mucosa of the wall of the Stomach". To be re-usable, a representation must allow applications to ignore details which are irrelevant to them and allow other applications to extend the representation where they require more detail without disturbing existing applications.

3. Special Issues: Roles, Signs & Symptoms, Systems, Dualities, and Proper Names

3.1. Roles

What is a "vitamin"? a "hormone"? a "steroid"? a "protein"? Clearly, "vitamin" and "hormone" are functionally defined whereas "steroid" and "protein" are structurally defined. There are many analogous situations in which the detail of the functional processes are beyond the scope of the model but where a clear separation of taxonomies is necessary. Such distinctions are represented in the GALEN CORE model by 'roles'. For example:

(ChemicalSubstance which playsPhysiologicalRole-VitaminRole) name Vitamin

ChemicalSubstance which playsPhysiologicalRole-HormoneRole) name Hormone

This use of 'roles' has become one of the fundamental mechanisms of the model. It has virtually replaced explicit statements that an entity has more than one parent. It is almost always preferable to represent the subsumption formally using a role as above, because the use of the role explains *why* the one entity is subsumed by the other whereas the use of an explicit subsumption link to two parents does not.

3.2. Signs, Symptoms and Diagnoses — method of observation

The distinction between "sign", "symptom" and "diagnosis" plagues many medical classification schemes. Frequently the same phenomenon can play more than one role — "wheeze" may be reported by the patient and hence be a symptom or be seen by the doctor and hence be a sign. Some types of "seizure" may be observed and treated as signs or reported as symptoms; in other situations it is only after long investigation that a particular phenomenon is diagnosed as being a seizure. The GALEN CORE Model surmounts this problem by separating the taxonomy of methods of observation from the taxonomy of pathophysiological conditions. For example:

(Disorder which isShownBy PatientReport) name Symptom.

(Disorder which isShownBy MedicalObservation) name Sign

etc.

This approach has the advantage that it extends naturally to other related distinctions, such as whether a wheeze was heard with a stethoscope, heard without a stethoscope by the doctor, or reported by the patient.

3.3. Functional parts and Functional systems

There are at least two uses of the word system in common usage: those based on function such as "the digestive system", "the circulatory system", or the "endocrine system" and those based on anatomical structure such as "the skeletal system" or the "gastrointestinal tract". The CORE Model distinguishes these two uses clearly but forms both compositionally. Functional systems are composed of those entities with the requisite function; anatomical systems are treated simply as collections.

3.4. Dualities

Many medical phenomena come in pairs. The most common pairing is process and lesion, *e.g.* 'ulceration' and 'ulcer' . Often the same word is used for both members of the pair, *e.g.* 'erosion' (a process) and 'erosion' (a lesion). Often common usage blurs the distinction, for example we speak both of 'gastritis' as if it were a process — 'worsening', 'progressive', 'chronic', etc. — and of 'gastritis' as something which can be seen through a gastroscope — clearly a physical lesion. At least two cases exist, those where the physical object is the outcome of a process and those where one physical object or process is caused by another.

The phenomenon is quite general and not limited to pathological lesions, *e.g.* consider the process of 'secretion' and the 'secretion' which is its outcome. Language blurs these distinctions in some cases and emphasises them in others — *e.g.* there is a clear distinction in most languages between 'viral hepatitis' and 'hepatitis virus'. A formal model must either mirror this pattern blurring and distinguishing or provide formal transformations between the different forms. If it is to be independent of surface language, it must do so in a way which does not commit it to the usage in any one linguistic community.

3.5. Proper Names and the Definite Article

There are many anatomical structures named by the combination of a general type and a specific organ or location — *e.g.* "the hepatic artery", "the common ileac vein", or "the islets of Langerhans". These concepts present special problems. The first two examples are fundamentally different from the third. For example, "the hepatic artery" is not the only artery supplying the liver, hence it is a different concept from the category "artery which supplies the liver" which subsumes "the hepatic artery" along with a number of other less important arteries. By contrast, the "islets of Langerhans" is a simple eponym — "Langerhans" has no other meaning in the system. The phrase "islets of Langerhans" can be treated as special string or a pseudo-attribute can be created to link "islet" and "Langerhans" without danger of misunderstanding.

In an early version of the CORE Model, to emphasise this distinction, all general entities were labelled with plurals — "arteries which supply the liver", "bones", "hands" etc. This convention avoided the confusion with specific named entities such as "hepatic artery" but was felt to be awkward and at variance with standard practice in the English speaking community.

There is a fundamental conflict in these cases between the needs of linguistic systems and the needs of concept systems. To the concept system proper, both *"Hepatic Artery"* and *"Islets of Langerhans"* are specific named entities best treated as natural kinds and represented as elementary entities. However, to ignore the common compositional structure of the phrases in most languages reduces the value of a compositional model to parsing systems. The current model compromises. A special set of subattributes, *servesSpecifically, hasSpecificCause, hasSpecificOutcome,* etc. are provided which are convey the strong partly linguistic linkage while still retaining the compositional structure.

4. An Extended Example

4.1. Anatomy, morphology and process

The most obvious area for use of compositional models coordinating several independent clean taxonomies is in the basic description of lesions, their location, form and cause.

For example, classifications of burns can be made by whether they are thermal or chemical, their location, their penetration, their extent, and their circumstances. There is little *a priori* reason to choose one ordering of these features over another — different applications require different information. An occupational health system is likely to be concerned with the circumstances whereas an intensive care unit is much more likely to be concerned with the extent and depth of the burn. Furthermore, the occupational health unit may be more concerned with 'circumstances in which burns occur' than the 'burns occurring in particular circumstances'.

The GALEN CORE allows statements such as such as:

> *BurnLesion which <*
> *   hasLocation Arm*
> *   hasDepth halfThickness*
> *   hasExtent 4cm2*
> *   hasCircumstances KitchenAccident*
> *   hasCause Heat>*

There are separate classifications for anatomical entities such as *Arm,* for energy such as *Heat, and* for the depth of burns, areas, and types of circumstances in which accidents may occur. Any of the dimensions may be refined further, *e.g.*:

> *BurnLesion which <*
> *   hasLocation (AnteriorSurface which isDivisionOf LowerArm)*
> *   hasCircumstances (KitchenAccident which involves FatOrOil)*
> *   hasCause Heat>*

BurnLesion <u>which</u> *hasCause Chemical*
    BurnLesion <u>which</u> <*hasCause Chemical hasLocation UpperExtremity*>
      BurnLesion <u>which</u> <*hasCause Chemical hasLocation Hand*>
        BurnLesion <u>which</u> <*hasCause Chemical hasLocation PalmarSurfaceOfHand*>
        ...

  ...

  BurnLesion <u>which</u> *hasCause Acid*
    BurnLesion <u>which</u> <*hasCause Acid hasLocation UpperExtremity*>
      BurnLesion <u>which</u> <*hasCause Acid hasLocation Hand*>
        BurnLesion <u>which</u> <*hasCause Acid hasLocation PalmarSurfaceOfHand*>
       ...

  BurnLesion <u>which</u> *hasCause Alkali*
    ...   ...

...

BurnLesion which *hasCause Heat*
  BurnLesion <u>which</u> *hasCause Heat*
    BurnLesion <u>which</u> <*hasCause Heat hasLocation UpperExtremity*>
    BurnLesion <u>which</u> <*hasCause Heat hasLocation Hand*>
      BurnLesion <u>which</u> <*hasCause Heat hasLocation PalmarSurfaceOfHand*>

...

Figure 1: Part of taxonomy derived by composition and classification

Given these compositional structures we can classify them as required in more abstract categories such as:

    BurnLesion <u>which</u> *hasLocation UpperExtremity*

i.e. "burns of the upper extremity", or

    BurnLesion <u>which</u> <
        hasCircumstances HouseholdAccident
        hasCause Heat>

i.e. "thermal burns due to household accidents". Since the classification of burns does not have to be established in advance, data entry need be in no specific order. Rather than look up and down a fixed hierarchy, the user may be presented with an overall 'predictive data entry' form using and the ability to expand any item as discussed in detail in [19-21]

We may further generate fixed classifications along any of the potential axes, for example by circumstances then depth then anatomy as shown in Figure 1. .

Note that even in this simple example, depending on interest, there are at least $2^4=16$ possible lesions and $4!=24$ possible orderings. If the internal structure of the axes such as 'involving OilOrFat' are taken into account there are many more. Any one of these orderings may be particular useful for a specific application, but there is no fundamental reason for choosing one over the other in general. Any choice amongst them biases a representation in favour of one group of applications and makes it awkward for use by others.

Note that in this example we have chosen to treat *BurnLesion* as elementary, as a natural kind. *BurnProcess* is defined in the GALEN models as the composite entity *Process* <u>which</u> *hasOutcome BurnLesion*. We have chosen to make lesions rather than

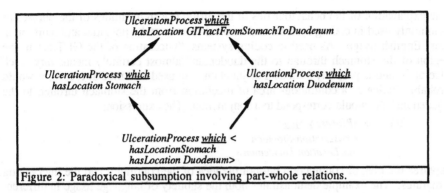

UlcerationProcess which
hasLocation GITractFromStomachToDuodenum

UlcerationProcess which
hasLocation Stomach

UlcerationProcess which
hasLocation Duodenum

UlcerationProcess which <
hasLocationStomach
hasLocation Duodenum>

Figure 2: Paradoxical subsumption involving part-whole relations.

processes elementary consistently throughout the model as it seems to produce a slightly simpler structure. However, this choice is fundamentally arbitrary.

5. Paradoxes and Limitations

5.1. Choices concerning dualities

The GRAIL Kernel is strongly typed and lacks disjunction. Given a duality such as *BurnLesion* and *BurnProcess* it is always necessary to choose one or the other. Yet many statements in natural language do not make this distinction clear. In many situations the distinction is unnecessary because it makes no difference. For example, "The burn was caused by acid" makes no distinction between process and lesion — acid is the cause of both the process and the lesion. By contrast, "The erosion lasted for three months" is ambiguous — was it the erosive process which lasted for three months, possibly resulting in an erosive lesion at the end of that time or has there been an erosive lesion for three months? Without context, and possibly even with it, there is no way of being confident of the answer.

On the one hand the ability to make such distinctions can be valuable. On the other it leads to problems. The first problem is that there are many situations in which we want to retrieve all cases or categories pertaining to either half of a duality — *e.g.* all cases of or all kinds of burns, erosions, etc. The second, more serious problem, is that it may lead to recording 'over interpretations' of doctors' intended meaning. If there is no means of disambiguating the natural language phrase, then forcing a disambiguation risks distorting meaning. No complete answer is available at the moment within GRAIL for this problem, although several 'work arounds' are in use and extensions are being considered.

5.2. Paradoxes with conjunctions and regions

The rules for the interaction of part-whole relations and subsumption produce the taxonomy shown in Figure 2.:

This occurs because the stomach and duodenum are parts of the "GI tract from stomach to duodenum". Ulceration of either organ is a kind of ulceration of that section of the GI tract. However, ulceration of both individually is a kind of each, *i.e.* "ulceration of both the duodenum and stomach" is a kind of "ulceration of the stomach". The authors were surprised when one of the members of the consortium requested almost precisely this behaviour. They were gratified to see the formalism correspond to users' intuitions, but remained concerned that the construction seemed 'odd'.

The explanation of the behaviour lies in the difference in meanings of the phrases as commonly used in coding systems and medical records and a potential ambiguity in at least English usage. As used in coding systems, "Ulceration of the GI Tract in the region of the stomach through to the duodenum" almost certainly means *any* such ulceration and is properly seen as a disjunction. As used in a medial record, it would probably indicate "a continuous area of ulceration from the stomach through to the duodenum" and would correspond to a conjunction. The expression:

> *UlcerationProcess* <u>*which*</u>
>   *<hasLoation-Stomach*
>   *hasLocation-Duodenum>*

captures the idea of conjunction but not the idea of the continuity and the intervening structures. The example demonstrates both the subtlety of language usage in different contexts and a previously unrecognised difference between the behaviour of discrete and diffuse lesions not catered for in GRAIL or, as far as we are aware, any related formalism.

6. Discussion

The mechanism of dividing the representation into separate taxonomies of elementary entities and then recombining these entities through composition has proven fruitful. The extent to which it has proved possible to separate taxonomies into simple hierarchies has been one of the major surprises of the GALEN project. Originally, the requirement that any entity, whether elementary or composite, should be able to have multiple parents in the subsumption hierarchy was fundamental to the design of the representation language. In early models twenty-five to thirty per cent of all elementary entities had more than one parent in the subsumption hierarchy. This was consistent with our earlier experience in the PEN&PAD project and with results from examining independently compiled hierarchies in the Oxford System of Medicine project.

In the current representation only one or two percent of all elementary entities have more than one parent. Multiple parents occur almost solely amongst the very abstract concepts at the top of the taxonomy where they serve specific purposes such as allowing a single abstraction notion of "disease" to cover pathological "lesions", "processes" and "states". All other multiple classification is now done formally through classification of composite concepts.

However, the power of the approach is not without cost. Issues are raised which do not arise in traditional representations, such as forced distinction between process and structure in natural dualities and problems in dealing with proper nouns which appear descriptive such as "the hepatic artery". Amongst the most powerful ideas are the use of 'roles' and the flexible handling of "sign", "symptom" and "diagnosis" which follows from separating the observation method from the phenomena observed.

Of the broader claims at the beginning of this paper, the usefulness of a separate concept layer seems to us increasingly well established. The examples given here of its use to clarify the classification of concepts from independent taxonomies is only one important area where this separation, empirically, provides a manageable level of complexity. Subjectively, it has provided important for focusing for attention of the knowledge engineers compiling the model and for providing a general test for resolving arguments. The GALEN model strives to capture the expected behaviour of the concepts in applications for data entry and retrieval. It attempts to capture neither the detailed diagnostic criteria nor the idiosyncrasies of linguistic usage.

It must be emphasised that the strategy of 'coordinating taxonomies' is not, in itself, sufficient to achieve re-use and sharing of terminology. We have touched briefly, for example, on the issue of bridging the levels of detail needed in different applications which must also be addressed to achieve effective terminological re-use. Nor is terminological re-use sufficient to achieve complete sharing of knowledge and interoperability — for example terminological methods are based on the structural composition and definition of terms and do not deal with issues requiring numerical calculations such as transformations amongst different systems of units. Furthermore, while a sharing of knowledge and interoperability at this level are necessary to achieve comparability of data, they are not in themselves sufficient, since clinical diagnostic criteria are explicitly excluded from this terminological model. If this seems paradoxical, consider the fact that two groups of clinicians can talk to each other about a condition without prior agreement on precise diagnostic criteria.

The contention of this paper is, however, that 'coordinating taxonomies' provides one key strategy for achieving re-use of a central core of terminological knowledge and a foundation on which to build for further layers of inference and information management for decision support and medical records. In practice, this approach has allowed a representation which has served a range of applications in data entry, medical records, drug interaction [22], and the authoring of decision support systems. So far these applications have been relatively isolated from each other, so that the fundamental claim to re-usability has only been tested for the general abstract layers of the model rather than for specific subdomains. Such tests in specific domains are just now beginning.

References

1. Masarie Jr F, Miller R, Bouhaddou O, Giuse N, Warner H. An interlingua for electronic interchange of medical information: using frames to map between clinical vocabularies. Computers in Biomedical Research 1991;24(4):379-400.
2. Evans DA, Cimino J, Hersh WR, Huff SM, Bell DS, The Canon Group. Position Statement: Towards a Medical Concept Representation Language. Journal of the American Medical Informatics Association 1994;1(3):207-217.
3. Campbell KE, Das AK, Musen MA. A logical foundation for representation of clinical data. JAMIA 1994;1(3):218-232.
4. Cimino J. Controlled Medical Vocabulary Construction: Methods from the Canon Group. Journal of the American Medical Informatics Association 1994;1(3):296-197.
5. Bergamachi S, Sartori C. On taxonomic reasoning in conceptual design. ACM Transactions on Database Systems 1992;17(3):385-421.
6. Maida AS. Knowledge representation requirements for description-based communication. in: Nebel B, Rich C, Swartout W, (ed). Third International Conference on Principles of Knowledge Representation and Reasoning, KR '92. Cambridge MA: Morgan Kaufmann, 1992: 232-243`.
7. Borgida A. A new look at the foundations and utility of Description Logics (or Terminology Logics are not just for the Flightless Birds). Internal report. Rutgers University Department of Computer Science, 1992
8. Rector A, Gangemi A, Galeazzi E, Glowinski A, Rossi-Mori A. The GALEN CORE Model Schemata for Anatomy: Towards a re-usable application-independent model of medical concepts. in: Barahona P, Veloso M, Bryant J, (ed). Twelfth International Congress of the European Federation for Medical Informatics, MIE-94. Lisbon, Portugal, 1994: 229-233.

9. Rector A, Nowlan W, Glowinski A. Goals for Concept Representation in the GALEN project. 17th Annual Symposium on Computer Applications in Medical Care (SCAMC-93). McGraw Hill, 1993: 414-418.

10. Rector A, Nowlan W. The GALEN Representation and Integration Language (GRAIL) Kernel, Version 1. The GALEN Consortium for the EC AIM Programme. (Available from Medical Informatics Group, University of Manchester), 1993

11. Brachman R, JB S. An overview of the KL-ONE knowledge representation system. Cognitive Science 1985;9(2):171-216.

12. Borgida A, Brachman RJ, McGuiness DL, Resnick LA. CLASSIC: A Structural Data Model for Objects. ACM SIGMOD International Conference on Management of Data. ACM, 1989: 58-67.

13. Nebel B. Computational Complexity of Terminological Reasoning in Back. Artificial Intelligence 1988;34:371-383.

14. Sowa J. Conceptual Structures: Knowledge Representation in Mind and Machine. New York: John Wiley & Sons, 1985

15. Bernauer J, Goldberg H. Compositional classification based on conceptual graphs. Artificial Intelligence in Medicine Europe (AIME-93). Munich: , 1993:

16. Tuttle MS. The position of the canon group: A reality check. Journal of the American Medical Informatics Association 1994;1(3):298-299.

17. MacGregor R. The evolving technology of classification-based knowledge representation systems. in: Sowa J, ed. Principles of Semantic Networks: Explorations in the representation of knowledge. San Mateo, CA: Morgan Kaufmann, 1991: 385-400.

18. Brachman R, Fikes R, Levesque H. An essential hybrid reasoning system; knowledge and symbol level accounts of KRYPTON. International Joint Conference on Artificial Intelligence (IJCAI-85). Morgan Kaufman, 1985: 532-539.

19. Nowlan W, Rector A. Medical Knowledge Representation and Predictive Data Entry. in: Stefanelli M, Hasman A, Fiesch M, Talmon J, (ed). Artificial Intelligence in Medicine Europe (AIME-91). Maastricht: Springer-Verlag, 1991: 105-116.

20. Nowlan W, Rector A, Kay S, Horan B, Wilson A. A Patient Care Workstation Based on a User Centred Design and a Formal Theory of Medical Terminology: PEN&PAD and the SMK Formalism. in: Clayton P, (ed). Fifteenth Annual Symposium on Computer Applications in Medical Care. SCAMC-91. Washington DC: McGraw- Hill, Inc, 1991: 855-857.

21. Horan B, Rector A, Sneath E, et al. Supporting a Humanly Impossible Task: The Clinical Human-Computer Environment,. in: Diaper D, (ed). Interact 90. Elsevier Science Publishers, B.V.North-Holland, 1990: 247-252.

22. Solomon W, Heathfield H. Conceptual modelling used to represent drug interactions. in: Barahona P, Veloso M, Bryant J, (ed). Twelfth International Congress of the European Federation for Medical Informatics, MIE-94. Lisbon, Portugal: , 1994: 186-190.

Generating Personalised Patient Information Using the Medical Record

Kim Binsted[1]*, Alison Cawsey[1]** and Ray Jones

[1] Department of Computing Science, University of Glasgow
[2] Department of Public Health, University of Glasgow

Abstract. This paper presents an approach for providing patients with personalised explanations of their medical record. Simple text planning techniques are used to construct relevant explanations based on information in the record and information in a general medical knowledge base. We discuss the results of the evaluation of our system with diabetes patients at three diabetes clinics in Scotland.

1 Introduction

Patients in Britain now have the legal right to access to an explained version of their medical record. This right, however, is rarely exercised. Heavy demands on doctors' time mean that they seldom have time to explain even the most important aspects of a patient's record in detail; and because medical records are complex documents, often in a paper file containing a semi-organised jumble of letters, hand-written notes, and uninterpreted results, they are useless and even dangerous without such explanation.

This situation is especially unfortunate for patients with chronic problems. There is considerable evidence that providing personalised information about a patient's condition can significantly improve the treatment of some chronic problems. A study by Osman et al (1994) involving 801 asthma patients in Scotland showed that, by providing asthma patients with personalised booklets about their condition and treatment, hospital admissions were reduced by 54%. Moreover, patients with chronic problems tend to have complex records, with long lists of complications and treatments, making explanation all the more necessary (see Berry et al (1994) for an analysis of patient information needs).

Medical records are increasingly stored in electronic form. The problem of providing explained access to such records seems ideally suited to the techniques of automatic explanation generation, a subfield of artificial intelligence. By 'automatic explanation generation' we mean the production, by a computer, of a piece of natural language text on a subject which allows the reader to understand that subject more clearly.

Our goal in this project was therefore to build an interactive system which generates natural language explanations of items in patient records, personalised

---

* E-mail address is kimb@aisb.ed.ac.uk. More information on the work presented here is available from the WWW page http://www.dcs.gla.ac.uk/~www/Piglet.html.
** E-mail address is alison@dcs.gla.ac.uk.

to the patient in question. The explanations produced by the system are brief and simple, so that the patient is not put off by too much complex information; yet all the medical terms in the explanations are 'click-able' (i.e. the explanations are in hypertext), so that the patient can seek more information if desired. Both the content and the form of the explanation are tailored to suit the needs of the patient.

Such a system has a role both as general medical information tailored to the needs of a particular patient, and as a way of providing explained access to the patient's medical record. We would anticipate that such a system would be of most use in a waiting room, to be used before a session with the doctor. Here it might serve as a "warm up" for the consultation, refamiliarising patients with their records, giving them basic relevant information, and reminding them of any questions they might have.

We believe that using techniques to generate an explanation *automatically*, from a simple medical knowledge base and the medical record, has the following possible advantages:

- Because the explanations are generated from information in both the patient record and in a general medical knowledge base, they change automatically as the information in the record and in the knowledge base is updated.
- The style of explanation can be adjusted, according to the preferences of the doctor and of the patient, by changing the text plans used to generate the text – individual explanations need not be rewritten. Moreover, the style of explanation will be consistent over the whole session.
- Unlike with more general patient education materials, information particularly relevant to this patient can be emphasised (encouraging patient compliance), and irrelevant information minimised.
- There is the possibility of multilingual generation, for patients who do not share a first language with the medical staff. The knowledge base, patient record, and content-specifying rules can be made language independent, with a single module for translating the output into a specific language (see Rosner (1994) for a discussion of multilingual explanation generation).

We realised, however, that certain potential disadvantages of automatic explanation generation had to be avoided. In this kind of application, there is no room for unclear or misleading information. It is essential that a medical expert be able to check through the generated explanations – that is, the explanations must not vary in unpredictable ways. The system must not 'put words in the doctor's mouth' by adding information to the record, either explicitly or implicitly. Finally, the knowledge base, the record, and the text plans used to generate the explanations must all be easily modified by medical personnel.

2 Related Work

There is a significant amount of past research demonstrating the acceptability of computer-based patient education and the efficacy in particular of personalised patient education. Jones et al (1992), for example, have demonstrated the feasibility of giving patients access to general health information and to their medical record (and associated hand-crafted explanations). By allowing patients direct access to their records, through a suitably simple interface, patients can obtain information that is (partially) selected by them and not by the system, thus hopefully obtaining more relevant and appropriate advice. Recent studies involving 70 people, allowed to access information in their medical record (and associated simple fixed explanations), showed that 84% found it helpful and would use such a system again (Jones 1992). However, there is no attempt to personalise the details of the information presented to the user.

Osman et al (1994) gave patients in their study personalised computer-generated booklets, based on information in the patient record and on responses to patient questionnaires. They showed that this information could lead to a dramatic reduction in hospital admission rates. However, the techniques for selecting and organising the information were primitive, based on the use of spreadsheets and mail-merge software. The patient did not interact directly with the computer system to obtain the information. There are clearly potential advantages with interactive systems, and more sophisticated generation techniques.

There has been only limited work on using explanation generation techniques in this area. De Rosis et al (1994) generate explanations of drug specifications, to be printed out and given to the patient. The explanations are modified according to the prescriber's preferences, rather than the patient's, but text generation techniques are used to ensure that the explanation is not stilted. Because the explanation is printed out, the patient cannot interact with the system directly.

The most directly relevant related research is the MIGRAINE project (Buchanan et al 1992, Carenini et al 1993, and Buchanan and Moore 1994). The MIGRAINE system generates explanations of migraine, using the patient's medical history (also taken by the computer, as MIGRAINE does not have access to the patient record) to personalise the text to the patient's needs. The system uses a slightly more sophisticaed text generation strategy than in our work. For example, extensive use is made of the discourse history (the record of the 'conversation' between the system and the user) to modify the text plans chosen for generation. A more general theory of discourse structure is used as the basis for an explanation planner, and the system has a capacity for re-explaining things that weren't understood first time.

We have taken a more pragmatic approach in our work. We felt, and our evaluation indicated, that patients would be uncomfortable with medical information that is in a significantly different form each time it is presented, so we don't attempt to make (much) use of the discourse history. Our text planner uses domain specific text plans for different broad classes of medical issue (e.g., problem vs. treatment). We don't provide for issues to be re-explained another way by the system, but have a facility for patients to mark issues to be discussed further with their doctor. Our system is somewhat broader in scope than MIGRAINE, more closely linked with the medical record, and with its develoment driven by repeated evaluation and feedback from doctors and patients.

3 System Architecture

PIGLET (Personalised Intelligent Generator of Little Explanatory Texts) has four main parts: a set of electronic patient records, a general medical knowledge base, a text generator, and a display module. The text generator uses the information in the knowledge base to provide the specifications for a hypertext explanation of the patient record, and the display module uses this specification to display the hypertext on the screen.

The initial form and content of the electronic records was taken from a set of Nottingham Diabetes Clinic records (Jones 1983). Each record was made up of:

- A set of personal information (name, age, address, occupation etc.)
- A list of problems, each with date of diagnosis and current activity
- A list of treatments, each with start and end date and dosage (for medication)
- A list of tests and measurements, each with date and result

Although medical records generally contain much more information than listed above, this is a reasonable minimum amount of information to expect from an electronic summary. Because no such computer records were available from local diabetes clinics, PIGLET's medical records were hand-built from real paper records from diabetes clinics in the Glasgow area; however, the system should work equally well with any computerised record system which can provide the above information about the patient.

PIGLET's medical knowledge base is an 'isa' hierarchy of objects, with multiple inheritance of slot values from ancestors. The structure of this hierarchy is loosely based on that of the Read codes (the standard method of encoding medical concepts in Britain), and each object is named according to the Read system; however, since the Read hierarchy is not a strict 'isa' hierarchy (e.g. complications of diabetes are given as subtypes of diabetes) and does not allow for multiple ancestors, we have modified the hierarchy as necessary to suit our purposes. Again, although this knowledge base was handcrafted, we believe that most of the information PIGLET requires is of a general enough nature that it should be available from other general-purpose medical knowledge bases. See figure 1 for a small part of the hierarchy.

The text generator uses domain-dependent text plans, as used in Cawsey (1992). A text plan is associated with the class(es) of objects, and represents the way the objects are explained. Each text plan is used to generate a list of text specifications (the explanation of an issue), which are then translated by the display module into click-able hypertext on the screen. Text plans are defined in terms of the class of object they apply to, preconditions concerning when they apply, and subgoals. These subgoals can refer to lower level subplans, and thus dictate the structure and content of the text, or be actual text templates, specifying the actual text or hypertext to be used to realise a specific goal. This is an example text plan which illustrates the use of templates to construct a sentence within an explanation.

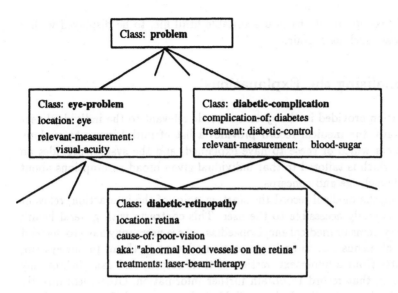

Fig. 1. A small subsection of the knowledge base.

```
(defplan patient-measurement (i)
  :class measurement
  :preconds (in-record i)
  :goals ((template 'plain "Your")
          (template 'hypertext (name i) (do-window i))
          (template 'plain "was last measured at")
          (measurement-value (in-record i))
          (measurement-date (in-record i)))))
```

This plan will apply only to a measurement which is in the patient record, and will generate (as part of a larger explanation) the specifications of a sentence giving the measurement value and date. For example, it could generate:

Your cholesterol level was last measured at 3.8 mmol/L (15 July 1994).

The phrase 'cholesterol level' can be clicked on for further explanation.

There can be several text plans of the same name. Which is applied first is determined by the class of the object (PIGLET attempts to apply the most specific plan first) and the number of preconditions (PIGLET attempts to apply the plan with the most preconditions first).

The display module takes a list of such text specifications and turns it into readable text on the screen, in accordance with both the generated specifications and a set of machine specific parameters. These text specifications can, of course, be used by other display modules to display the text in different ways. For example, we have also implemented a piglet2html translator, which turns a set of PIGLET text specifications into a suitable html file, to be displayed with a WWW browser such as Mosaic.

4 Personalising the Explanations

The information provided in the system is made relevant to the individual user in several ways: the medical record provides a list of initial issues to explore, the user selects which of these to have explained, and the system provides an explanation which is tailored to that individual given broad assumptions about their likely knowledge and concerns.

By making the medical record the starting point of the interaction, relevant information is easily accessible to the user. This contrasts with general health information systems or medical encyclopaedias, where users may have to descend many levels of menus etc. to find information relevant to them. In our system, the user starts from a hypertext version of their record, and can click on any of the issues in that record to obtain further information. Other, less directly relevant medical issues will only be available to be explained if they are related in some way to the user's own medical problems and treatments.

Within this context, the user can select for themselves the issues they want to have explained, by clicking on the relevant hypertext words. This contrasts with written information or more rigid computer-based patient education materials where the user may be taken through a fixed series of screens, which may or may not be relevant.

More interestingly from an AI point of view, the way a selected issue is explained will also depend on the user. The detailed content of the explanation is determined by the user's likely concerns (as inferred from their record), while the way the information is conveyed (in terms of relations to related objects etc.) is influenced by a simple model of what different categories of user are likely to know about.

The information content is influenced by the user's concerns as follows. First, we assume that the user is concerned about all the issues (e.g., problems, treatments etc.) mentioned in their record. We then use that information to decide when to add additional personal reminders or elaboration concerning these key issues whenever we are explaining a related issue. For example, when explaining what humulin is (a kind of insulin) we might remind the user about the type of insulin they are taking, and remind them what their blood sugar level is. When describing a particular drug, if the patient is taking this drug, we remind them of exactly how the drug should be taken (e.g., one tablet, three times a day, with food), and mention any possible side-effects or warning signs.

The way something is explained is influenced by a simple user model indicating, for different categories of user, which medical issues are likely to be familiar. User categories are currently based on the user's major chronic problem, rather than, say, education level, as we felt that knowledge of medical issues depended primarily on whether users had direct experience of the disorders etc. in question. So, we represent which issues a typical diabetes patient might know about (e.g., insulin), and which issues all patients should know about (tablet, pain etc.). As we expand the system to deal with other problems we will extend this model.

When selecting between different possible ways of describing an object, we aim to find a description that is given in terms of related known issues (according to the simple user model). For example, to describe a issue, one method first identifies it as an instance of a general class. The general class should be a known issue for the particular class of user (diabetes patient). This method might generate:

Humulin is a kind of insulin.

since "insulin" is a term familiar to all diabetes patients, but it would not generate:

Amlodipine is a coronary vasodilator.

since "coronary vasodilator" is not considered to be an understood term. This restriction prevents the 'dictionary problem' — the uselessness of explaining an unfamiliar issue only in terms of other unfamiliar issues — by ensuring that at least some of the terms in an explanation are commonly understood.

Although PIGLET makes minor use of the discourse history to make explanations flow more smoothly, a term is not assumed to be understood just because it has been explained before. We felt that patients would be disturbed by explanations that changed radically on reviewing, and wanted explanations of any given subject to remain fairly static over the course of a session (although not necessarily between sessions or between patients).

In summary, information in PIGLET is made relevant by: making the user's own patient record the starting point of the interaction; allowing the user to select what they want to see, through a hypertext style interface; adapting the detailed content of the explanations given information in the record; and adapting the way something is explained given a model of issues the user is likely to know about. No existing system combines all these features.

5 Example

A patient using our system would be given, following introductory screens, the opportunity to browse through their personal information, their problem list, their treatment list, or their recent tests and measurements. If they chose problems, they might get a screen containing the following text:

PROBLEMS ON RECORD
Type one diabetes mellitus Diagnosed: 1965.
Angina pectoris Diagnosed: 1975.
Background diabetic retinopathy Diagnosed: 1988.
Nephrotic syndrome Diagnosed: 1988.
Hyperlipidaemia Diagnosed: 1992.

Each of the underlined bold face words are buttons that can be touched or clicked on to obtain further information. Suppose hyperlipidaemia was selected. The following screen of information would be displayed:

HYPERLIPIDAEMIA
"Hyperlipidaemia" is the term for having a high level of lipids (fats), particularly cholesterol, in the blood. According to your **record**, you have this problem. Possible causes include **diabetes mellitus** and **alcoholism**. Some common treatments include **a special diet** and **a lipid-lowering drug**. It increases the risk of **atherosclerosis** and **coronary artery disease**. **Your cholesterol level** was last measured at 3.8 mmol/L (15 July 1994). You have a lipid-lowering drug in your record: **bezafibrate**.

Note how the user is given both general information, and reminders of specific information in their record relevant to this problem. Now, if the user next touched the word "bezafibrate", they would obtain the following explanation.

BEZAFIBRATE
Bezafibrate is a cardiovascular drug which reduces the amount of some kinds of fat in the bloodstream. According to your **record**, you are currently undergoing this treatment. It is often used to treat **hyperlipidaemia**. This could have some side-effects; in particular, **nausea**. Your prescription of bezafibrate comes in 200 mg tablets. It is to be taken three times each day (morning, afternoon, and tea time).

Again, note how the user is both given general information about the drug, and personalised reminders of how they should take the drug.

If, following the explanation of hyperlipidaemia, the patient had clicked on cholesterol level, the following explanation would be given, again combining general explanations with reminders of patient specific data.

CHOLESTEROL LEVEL
A patient's cholesterol level (measured with a blood test) is an indication of the amount of fatty food in their diet. It can be influenced by heredity and diseases such as diabetes. A normal level is 2-4 mmol/L. Your cholesterol level was last measured at 3.8 mmol/L (15 July 1994). Important: A high cholesterol level (greater than 6.5 mmol/L) increases the risk of **atherosclerosis** and **heart disease**.

In all these explanations, buttons are available at the bottom of the screen allowing the patient to obtain further help; go back to previous explanations; obtain a list of written materials they may consult; or note that this is an issue they wish to discuss with their doctor. There are also buttons which will display a relevant picture, or jump to a relevant topic not mentioned in the text. At the end of the session the user can obtain a printout of the session and a list of the issues marked for discussion with the doctor.

6 Evaluation

The evaluation of this initial prototype of PIGLET was in three stages, which should be seen as the first part of an ongoing cycle of evaluation and improvement.

Once we had developed a simple working version of the program, we invited two public health students and one general practitioner to test it, and respond to a general questionnaire. Their comments included the suggestions that: the vocabulary be simplified, and with more of the explanations given in terms of generally understood concepts; some way of marking an explanation for further discussion with the doctor be provided; sensitive information (such as potential drug interactions) be minimised or omitted, to be left to the doctor to explain; and that buttons for putting up simple, relevant pictures (such as a picture of the heart in an explanation of heart disease) be provided.

In general, they were all positive about PIGLET's potential for improving patient–doctor communication and general patient awareness. They all commented that the personalisation of the explanations, particularly the inclusion of relevant measurements from the record in problem explanations, would help patients understand how the information provided relates to them.

Since their suggestions were in line with the goals of the project, we modified the program to incorporate them. Once these changes had been implemented, we approached diabetes professionals with the system.

PIGLET was shown to seven diabetes professionals (five diabetes consultants, one diabetes nurse specialist, and one junior doctor) at three clinics in the Glasgow area. At this stage, we had two hopes: that the medical professionals would offer advice on improvements, and that they would allow us to show PIGLET (with their suggested improvements) to some of their patients. Because of constraints on the health professionals' time, PIGLET was shown to groups of two or three people at a time, and their responses to a questionnaire were recorded on audio tape. Each group was asked questions concerning the hypertext interface, the content and style of the explanations, the personalisation of the explanations, and the general usefulness of the system. They were all positive about the system; in particular:

Interface: All thought that most of their patients would be able to use the hypertext interface, provided that a touch screen was used, although one commented that older patients might not want to use a computer.

Content and Style: Although they all agreed that the explanations were clear and fairly easy to understand, two of the consultants thought the vocabulary level should be reduced still further. All said that the explanations were of about the right length and content.

Personalisation: All thought that the personalisation of the information, particularly the way relevant measurements were mentioned, would be useful: "Yes, that's very good — that makes it relevant to themselves, rather than just general information about their condition." Several ways of increasing

the personalisation were suggested, including the generation of charts showing the patient's position (e.g. in weight, cholesterol level, etc.) with a subset
of the rest of the population.

General usefulness: All thought that, despite the large amount of diabetes
patient education material, PIGLET would still be useful: "They tend to
skim over brochures etc., but would probably take in more of this because
it's directed at them." They were all enthusiastic about the idea of PIGLET
as a 'warm-up' for a consultation, allowing patients to mark issues which
they wished to discuss further. A typical comment was "this kind of [system]
would allow patients to make better use of their time with me." Some worries
about confidentiality and security were expressed, but all agreed that some
sort of card system (similar to that used with banking machines) would
provide adequate protection. All said that PIGLET would help correct errors
in their records.

All of the clinics agreed to let us have access to some patient records, so that
PIGLET could be evaluated with the patients themselves. We entered eleven
diabetes patient records in total, and ten of the patients agreed to try out the
system. There were 6 men and 4 women, and ages ranged from 24 to 65. Each
was given approximately half an hour to use the system to look at, and have
explained, their own record. They were then asked questions about the system,
similar to those asked of the doctors, in a structured interview format.

Interface: All found the hypertext interface was easy to use, even though five
had little or no computing experience. Two had problems using the mouse,
but said that they would be able to use a touch screen without problems.
One commented that older patients might be put off by computers.

Style: All said that the explanations were clear and understandable. "I thought
they'd be understood by anyone. Quite clear." Four said there was some terminology that they didn't recognise at first, but all understood after clicking
on the term in question. None found the explanations stilted: "Here it's [ex-
plained] the same way [as the doctor does]: you get the technical name, then
it's broken down for you, and you can follow it." "It's easy enough language,
and the medical terms are explained." One speculated that other patients
might be put off by the medical language, although he wasn't himself.

Content: All said that the contents of the explanations were useful and appropriate ("I've learned more looking at that than by speaking to the doctor"),
although one would have preferred more detailed, "less simplistic" information, and one would have liked to be able to wander further in 'hyperspace'.
All thought the explanations were of about the right length: "You would
want it short and compact, just enough for you to find out about any problems or any blood results or whatever, ready to go into the doctor." "It's
ideal — simplified for the layman."

Personalisation: All said that providing measurements from the record in
problem explanations was useful, and made the explanations more relevant.
Four commented that a history of tests and measurements, perhaps given in

a chart, would be even more helpful. Five thought that providing qualitative information first, before the actual values of the measurements, would make them easier to understand — for example, three commented that a bit more comparison of measurements with others in their demographic group would be helpful.

General Usefulness: All said that PIGLET was not redundant, even given the huge amount of other patient education materials. Seven found PIGLET particularly useful for reminding them of their current state (recent measurements, etc.) right before an appointment. Three said that they didn't use other patient education materials: "I don't actually use the pamphlets and things ... I find the information I get from [the British Diabetes Association etc.] rather limited. [PIGLET] relates to me, rather than a generalisation."

All said that they wanted access to their record, and all thought that an unexplained paper version was inadequate. Two patients would have also liked access to their doctor's notes, and one information about past hospital admissions.

Other comments : Nine thought that the check-box facility would help them remember questions they'd wanted to ask the doctor, although only two actually used it during their session (probably because none had consultations scheduled immediately after the evaluation session). "If you don't know something, you can get the basics there [PIGLET], then go to the doctor and find out what he thinks." Four commented that they often forget questions because they're worried about wasting the doctor's time.

Five said they'd learned something new — "I knew I had the problems, but I didn't know the words" — while five said they were fairly familiar with their problems and treatments and clicked for explanations mainly out of curiosity. Six pointed out minor errors to be corrected in their electronic medical record — all but one of which were also in their original paper record. One commented that it was reassuring to know that the doctor had taken note of their problems: "Sometimes you explain to the doctors that you have a problem, and you're not sure that they take it in. That [PIGLET] shows you that they do know. You're quite happy when you know that they know what your problems are."

All said they would use the system again, although two commented that it might be most useful to new patients. All asked for, and received, printouts of their session for future reference.

7 Improvements and Future Work

The evaluation suggested several changes and additions which might improve PIGLET's performance.

Some of the suggestions from the patients and medical professionals who tried the system can be implemented fairly easily: giving a history of past tests and measurements, for example. It might also be worthwhile generating some simple graphs, showing the patient's position within their demographic group for any given measurement.

We also hope to expand PIGLET so that it can explain issues related to different chronic problems. For example, a cancer PIGLET could generate explanations of records of cancer patients. Extending the system in this way should involve only adding new knowledge to the knowledge base, and adding a few specialised text plans for types of explanation specific to that subarea of medicine (e.g. explanations of chemotherapy programs for cancer).

PIGLET might also be modified for a slightly different purpose: generating documentation to be printed out, as in de Rosis and Grasso (1994). We have found that PIGLET's hypertext chunks translate nicely into a question-answer format printed page, which might be appropriate for providing additional printed documentation for prescriptions.

8 Conclusion

In building and evaluating this prototype system, we hoped to show how simple explanation generation techniques could be used to improve patient access to records and patient education in general. The resulting system, PIGLET, generates short hypertext explanations of medical issues, centred on a patient's medical record, and personalised to that patient's particular needs. The design of PIGLET reflects the needs of the end users, rather than those of artificial intelligence researchers. More complex and subtle methods of explanation generation certainly exist; in this case, however, relatively simple methods were adequate and appropriate.

Our evaluation of the system showed that both patients and medical staff found the system helpful and easy to use, demonstrating that our techniques can produce clear and comprehensible explanations. The personalisation of the explanations was found to make the explanations more relevant to the patients, which has been shown in other studies (Osman et al 1994) to improve patient care in general. In future we hope to expand PIGLET to cover a wider range of medical issues, build a simple interface for medical staff wishing to add information, and to do more rigorous testing of patient reactions.

Acknowledgements

Thanks are due to the University of Glasgow, for funding this project; to Drs. Fisher, Kesson and Semple, for trying out the system, and letting us try it on their patients; to the staff of the diabetes clinics at the Royal Alexandria, the Southern General, and the Royal Infirmary; and to all the patients who gave of their time and opinions.

References

D Berry, T Gillie, and S Banbury. What do patients want to know: an empirical approach to explanation generation and validation. Technical report, Department of Psychology, University of Reading, 1994.

B Buchanan, J Moore, D Forsythe, G Banks, and S Ohlsson. Involving patients in health care: explanation in the clinical setting. In M E Frisse, editor, *Proceedings of the sixteenth annual symposium on computer applications in medical care (SCAMC 92)*, pages 510–514. McGraw Hill, 1992.

B Buchanan, J Moore, D Forsythe, G Carenini, S Ohlsson, and G Banks. Using medical informatics for explanation in a clinical setting. Technical report, Department of Computer Science, University of Pittsburgh, 1994.

A Cawsey. *Explanation and Interaction: the computer generation of explanatory dialogues*. MIT Press, Cambridge, Massachusetts, 1992.

G Carenini and J Moore. Generating explanations in context. In *Proceedings of the International Workshop on Intelligent User Interfaces*, January 1993.

F de Rosis and F Grasso. Mediating between hearer's and speaker's views in the generation of adaptive explanations. Technical report, Dipartimento di Informatica, Universita di Bari, 1994.

R B Jones, A J Hedley, I Peacock, S P Allison, and R B Tattershall. A computer assisted register and information system for diabetes. *Methods of Information in Medicine*, 22:4–14, 1983.

R B Jones, S M McGhee, and D McGhee. Patient on-line access to medical records in general practice. *Health Bulletin*, 50(2):147–54, March 1992.

L Osman, M Abdalla, J Beattie, S Ross, I Russell, J Friend andJ Legge, and J Graham Douglas. Reducing hospital admissions through computer supported education for asthma patients. *BMJ*, 308:568–71, 1994.

E Reiter and C Mellish. Optimising the costs and benefits of natural language generation. In *Proceedings of IJCAI 1993*. Morgan Kaufmann Publishers, 1993.

D Rosner and M Stede. Generating multilingual documents from a knowledge base: The techdoc project. In *Proceedings of COLING-94*, 1994.

Analysis of Medical Jargon: The RECIT System

A.-M. Rassinoux[a], C. Juge[a], P.-A. Michel[a], R.H. Baud[a],
D. Lemaitre[b], F.-C. Jean[b], P. Degoulet[b], J.-R. Scherrer[a]

[a] Medical Informatics Centre, Geneva University Hospital, CH-1211 Geneva 14, Switzerland
[b] Medical Informatics Department, Broussais University Hospital, 75014 Paris, France

Abstract

Medical language constitutes a large subset of human language. However, it presents special features which have to be taken into consideration for natural language analysis. We have observed that natural language texts dealing with a specific medical area (especially discharge summaries and reports) share a common vocabulary and common habits of word usage. Facing these specificities, we describe in this paper an original system called RECIT, designed for medical text analysis and understanding. In particular, we discuss the principles which guide the analysis of medical jargon using semantic considerations, coupled with syntactic information when needed. As a result of the flexibility of the system, a medical application developed in the context of the AIM project HELIOS-2 is presented.

1. Particularities of Medical Language

Medical language as a scientific language, presents typical characteristics of a sublanguage [1]. Sublanguages share a lot of syntactic structures with natural language (NL). Nevertheless, some structures are used in a way which may appear incorrect or which is unusual from a grammatical point of view. Indeed, the usage of natural language by physicians is purely professional: the main objective being the communication of pertinent information, using a limited amount of space and time. Therefore, medical texts (such as discharge summaries or reports) present the particularity of being both technical (using specific medical jargon) and written in a concise and direct style (resembling the telegraphic style). This means that verbs are frequently omitted and there are a lot of coordinated noun phrases, prepositional phrases and adverbs. This direct style sometimes leads to sentences which are grammatically incorrect but nevertheless understandable. Hence, we have to take into consideration some specific "jargon" peculiar to the medical domain. From the semantic point of view, sublanguages are more restricted than the overall natural language. Indeed, the meaning of the sentences is closely connected to the contextual domain [2]. This implies that the basic vocabulary is well defined and the semantic combinatory appears to be sufficiently limited.

Facing these particularities (closed semantic domain and specific writing style), which essentially form the medical language, the use of traditional parsers presents some difficulties of endless unformalised new combinations which arise from the local jargons. On the one hand, formal grammar producing a parse tree for each sentence to be analysed (as in the line of Chomsky [3]) will fail in the presence of irregular structures. On the other hand, strictly semantic approaches aiming at building the semantic representation directly from the surface string, without the help of any syntactic source (as the theory of conceptual dependency of Schank [4]) will fail in the presence of complex sentences. Hence, it is important to devise methods by which the specific expressions encountered in the medical texts can be automatically understood by a computer. The solution implemented in the RECIT system (a French acronym for *REprésentation du Contenu Informationnel des Textes médicaux*) is principally semantically-driven in that semantics is used as early as possible in the analysis process and syntax is only integrated during the processing in order to solve ambiguities and verify syntactic conditions when needed.

2. Developing a Flexible Medical Analyser

The aim of the RECIT system is the automatic understanding of medical texts, allowing to translate information given in natural language into an elaborated conceptual representation. In this paper, we do not want to specify in detail any component of the RECIT system; we just want to emphasise the different linguistic and semantic tasks that have to be processed in order to transform a text from a set of single words to a conceptual representation. A detailed description of the whole system can be found in [5] and different important points are discussed in [6, 7, 8].

2.1 A Two-phase Process

In designing RECIT, we have chosen an original way to deal with the specific features of medical language, leading to an analysis process decomposed in two phases [9].

- n The first phase, which we call *Proximity Processing*, is a deterministic phase which combines the application of non-conventional syntactic procedures with the checking of semantic compatibilities in order to group neighbouring words together.

- n From this set of relevant fragments, the second phase deals with the building of a sound representation of the sentence meaning in *Conceptual Graphs* (CGs), following the formalism which has been defined by J. Sowa [10].

By its semantic-oriented approach, the analyser is able to accept different European languages (currently French, English and German) and to convert all of them into a unique representation [11]. A prototype of this analyser is available on different platforms (SUN, DEC, HP station and also PC). This analyser is written in Quintus Prolog. Prolog is a suitable language for such an approach, in that it allows rules to be expressed directly in the form of clauses [12].

2.2 The Linguistic and Medical Knowledge of the Sub-domain

The process of understanding expressions, i.e. transforming them from the language in which they are expressed into the chosen semantic representation, needs different kinds of information. The information, used by the RECIT system, is referred to as the Medical Linguistic Knowledge Base (MLKB) [13], and is structured as follows (see Figure 1):

1) *The conceptual typologies and the lexicons considered as the basic units of the RECIT system.*
   All the concepts and the relationships which are useful to describe the semantics of the treated medical domain are classified into a *hierarchical*

structure upon which an inheritance mechanism is performed. Concerning the *lexicons*, they constitute an important part of the analysis process in that they contain for each canonical form of a word, both the basic syntactic information (syntactic categories with number and gender variations...) and the semantic argument describing the meaning of the word (when more than one concept is present, they are grouped together by relationships). The latter argument constitutes an initial conceptual representation. All the lexical information is gathered at the beginning of the analysis for each word (simple word or idiom) composing the analysed sentence.

2) *Linguistic knowledge related to the sentence structure.*

This information is described using linguistic rules, the aim of which is to recognise basic syntagmatic expressions. Firstly, the application of *frequent-association rules* allows the recognition of specific expressions frequently encountered in the medical domain and which may be difficult to handle with syntactic grammar (such as temporal expressions, descriptions of laboratory results, composed forms of verbs and so on). Secondly, the application of local *syntactic rules enables many of the syntactic ambiguities to be solved*, by checking the syntactic categories of words found in the immediate neighbourhood of the detected ambiguity. Finally, the specification of *grammatical rules* allowing basic constituents of the sentence structure to be defined (such as a noun plus an adjective or a noun plus a noun complement) constitutes the last typical kind of linguistic knowledge used during the proximity processing phase. These syntactic attachments are performed only if the semantic combination of the corresponding concepts is sensible. This medical sensibility is described using semantic rules as shown in point 3).

3) *Semantic knowledge related to the sentence meaning.*

In our approach, the so-called *syntaxico-semantico compatibility rules* constitute an important part of the semantic description of the treated medical domain. Their functions are twofold. On the one hand, they are a means to define the sensible combination of a pair of concepts linked by a relationship. On the other hand, this semantic description is associated with syntactic structures allowing the concepts and the relationship to be expressed in a specific language. The application of such rules is a key step towards the final semantic representation insofar as the links established during the proximity processing phase are sound and directly used by the following phase devoted to the building of conceptual graphs. The latter process aims at linking all the fragments resulting from the proximity processing in order to establish the conceptual representation of the meaning of the complete sentences. This stage is mainly driven by semantic information, using *conceptual schemata*. A conceptual schema is defined for each concept representing useful information in the text. It allows the description of the different properties which are attached to a concept. Such a structure is completed by an inheritance mechanism using the typology of concepts. Moreover, the conceptual schemata are also linked to the syntaxico-semantico description of verbs in order to treat complex sentences. As a last resort, the default knowledge is used to define the plausible semantic role of a concept.

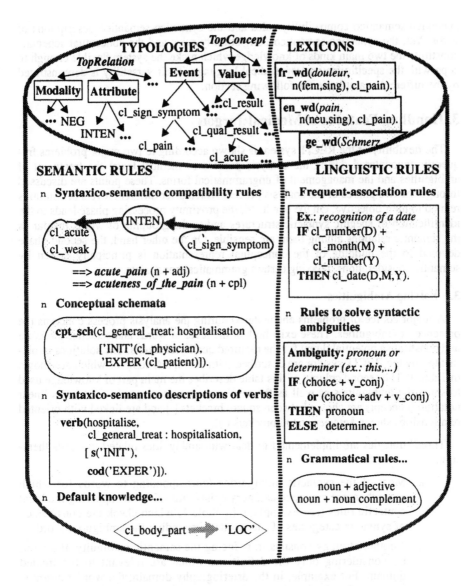

Fig. 1. The Medical Linguistic Knowledge Base (MLKB)

At a first glance, we can consider the MLKB as a recipient for all the declarative knowledge used during the analysis of medical texts. The advantage of working in a closed medical domain is demonstrated by the numerous rules which are domain-dependent. The described syntactic information is principally used by the RECIT analyser both to solve ambiguities and to provide "clues" about where, in a sentence, a conceptual entity might be found. That is, it mediates between conceptual information and the realisation in a specific part of the sentence. In this way, syntax and semantics are strongly integrated and sometimes cannot be described separately, as shown by the syntaxico-semantico compatibility rules and the syntaxico-semantico descriptions of verbs. Yet, the syntaxico-semantico compatibility rules and the conceptual schemata constitute the two main kinds of knowledge which make the system flexible enough to cope with the specific features of medical language. Their versatility is demonstrated by the different examples in the following section.

3. Handling Features of Medical Jargon

The flexibility of the RECIT system is shown according to two major problems frequently encountered during the analysis process of medical texts: the treatment of ambiguities and the management of ungrammatical forms. These points are discussed through the two phases of the analysis process which offer different levels of solutions related to their objective. On the one hand, the proximity processing phase leads to the identification of specific syntactic structures and to the grouping of words insofar as the semantic relations among them are sensible. On the other hand, the second phase devoted to the building of the conceptual representation is principally driven by semantic considerations, relaxing certain grammatical constraints.

3.1 Solving Ambiguities

A major source of problems encountered during the analysis process concerns the presence of ambiguities. These exist both at the level of words and sentence structure and constitute a potential threat for the required quality of results. The solutions implemented by the RECIT system are described with examples, some of which are shown in Table 1. The second column of this table describes the main part of knowledge used to solve the ambiguities as well as the resulting conceptual graph (CG) in its linear notation (concepts are shown between square brackets [ ] and are linked with oriented relationships shown between round brackets ( )).

Three strategies are implemented to deal with ambiguities related to words themselves:

- n Most of the syntactic ambiguities (corresponding to words for which more than one syntactic category exists) are solved by applying the local syntactic rules (see example a. in Table 1) which check the consistency of syntactic categories of the words surrounding the ambiguous word.

- n Another strategy consists in reducing the semantic ambiguity of a word by considering only the definitions which are relevant to the treated domain. For example, in the arteriography domain, the word *rupture* is interpreted as *burst*, whereas the meaning of *quarrel* is discarded.

- n Finally, some ambiguities can be masked with idiomatic expressions. For example, the word *point* becomes unambiguous in the following expressions: *point of view, point of Mc Burney*.

The above solutions are exposed according to their importance in the resolution process. However, this order is inverted during the analysis process which begins by recognising first the idiomatic expressions, then simple words and, in a subsequent step, the rules to solve syntactic ambiguities are applied. Moreover, the last two solutions deal with the semantic restrictions carried out at the level of the lexicons in order to consider only sensible interpretations valid for the treated domain.

Table 1: Treatment of ambiguities

Sentences in NL	Solutions and Results
a. **This** aorta is atheromatous.	- Use of the rule described in Figure 1 which allows to solve the ambiguity about the word *this* - [BODY_ELET:aorta #]- (LOC) <-[LESION:atheroma]\\ .

b.
 	
Infiltration *of* the abdominal aorta;
 	
Normal calibre *of* the right renal artery.
	- Use of the corresponding compatibility rules -
	[LESION:infiltration]-
	   (*LOC*) ->[BODY_ELET:aorta #]-
	        (PART) ->[BODY_AREA:abdomen]\\ .
	
	[MEASURE:calibre]-
	   (RESLT) ->[NORMAL]
	   (*MEASU*) <-[BODY_ELET:renal_artery #]-
	        (LATER) ->[SIDE:right]\\ .

c.
 	
Hospitalisation *for* cholecystitis;
 	
Hospitalisation *for* cholecystectomy.
	- Use of the inferred conceptual schema for hospitalisation -
	[GENERAL_TREAT:hospitalisation]-
	   (*CAUSE*) ->[DISEASE:inflammation]-
	        (LOC) ->[ORGAN:gallbladder]\\ .
	
	[GENERAL_TREAT:hospitalisation]-
	   (*GOAL*) ->[SURGICAL_TREAT:ablation]-
	        (THEME) ->[ORGAN:gallbladder]\\ .

Solving ambiguities at the level of sentences requires the use of more complex semantic structures as shown in the semantic part of the MLKB (see Figure 1). For example, the meaning of the preposition *of* in the expressions shown in example b. of Table 1 is clarified through the application of the syntaxico-semantico compatibility rules. Indeed, the compatibility between a *lesion* and a *body part* is explicit through the location relation ('*LOC*'), while the compatibility between a *body element* and a *measure* is described in the measure relation called '*MEASU*'. Moreover, the prepositional attachments are mainly handled by the conceptual schemata (see example c. in Table 1). In particular, through the conceptual typology, the conceptual schema for *hospitalisation* inherits the relations '*GOAL*' and '*CAUSE*' which come respectively from the schemata of *treatment* and *general treatment*. Finally, also the syntaxico-semantico descriptions of verbs are used to solve ambiguities related to the linking of complements.

3.2 Dealing with Ill-formed Phrases

One important point to be taken into account when analysing these medical texts, is the presence of utterances that are made of constituents without "syntactic glue" or "syntactic correctness" but for which the semantics of each word and the semantic relationships existing between them are well defined. The human ability to understand ill-formed phrases insofar as they are sensible, suggests the reduction of the predominance of syntactic knowledge in order to use heavily semantic information. This is why ill-formed expressions are mainly managed during the second stage of the analysis devoted to the building of conceptual graphs. Indeed, during the first phase of proximity processing, the grouping of words is realised only when all the required syntactic and semantic conditions are verified. Hence, in example a. of Table 2, the adjective *normal* is not linked to the noun *permeability* during the proximity processing phase because of the wrong place of the adjective. Nevertheless, as the underlying concepts are semantically sensible, they are linked with the relationship 'RESLT' (for result) during the subsequent phase of conceptual graph building. Example b. of Table 2 shows that grammatical errors, such as agreement in number between a verb and its subject, are not an obstacle to carry on the building of the semantic representation. A warning is nevertheless displayed for the user. Finally, the last two examples highlight the robustness of the analyser, the final goal of which is to represent most of the sensible medical information, bypassing inconsistencies and unknown words. In example c., a *stenosis* was recognised as a *lesion* during the proximity processing phase. However, the treatment of the passive form of the verb *hospitalise* fails because the syntaxico-semantico description of the verb *hospitalise* does not allow *lesion* as a passive subject. Therefore, only the information about the stenosis is retained in the final semantic representation. In example d., the presence of the unknown word *xyz* prevents the linking of the adjective *narrow* with the noun *stenosis* during the proximity processing phase. However, this link is performed during the conceptual graph building by considering only semantic aspects.

Table 2: Treatment of ill-formed phrases

Sentences in NL	Proximity Processing / Conceptual Graphs

a. Permeability *normal* of the right renal artery.	permeability normal of the_right_renal_artery .
	 [MEASURE:permeability]-
	    (RESLT) ->[NORMAL]
	    (MEASU) <-[BODY_ELET:renal_artery #]-
	           (LATER) ->[SIDE:right]\\ .

b. The sub-renal aorta *seem* infiltrated.	the_sub-renal_aorta seem infiltrated .
	 ERROR: Wrong agreement with the subject: The _sub-renal_aorta and the verb: seem.
	 [(PSBL) ->[BODY_ELET:aorta #]-
	         (PART) ->[AREA:sub-renal]
	         (LOC) <-[LESION:infiltration]\].

Table 2: Treatment of ill-formed phrases

Sentences in NL	Proximity Processing / Conceptual Graphs
c. *The stenosis *is hospitalised.*	the_stenosis is_hospitalised . *ERROR: Analysis not conclusive...* [LESION:stenosis #].

d. *Narrow *xyz* ostial stenosis.	narrow xyz ostial_stenosis .
	 [LESION:stenosis]-
	     (LOC) ->[STRUCTURE:ostium]
	     (CHRC) ->[CHARACTERISTIC:narrow]\ .

4. Integration of the RECIT System in the HELIOS-2 Environment

Since the beginning of 1992, the Geneva University Hospital has been involved in the European AIM project HELIOS-2 [14]. This project deals with the development of Hospital Object Software Tools. The HELIOS-2 environment includes several Medical Oriented Services, among which, a Natural Language Processing (NLP) component has been delivered by the Geneva partner [15]. This component enables the analysis of free texts, together with their storage in the form of conceptual graphs and the subsequent knowledge representation queries. It also provides maintenance tools for the typologies and the lexicons. The RECIT system was first used in the digestive surgery domain with the treatment of echographic reports. Then, in the context of the Artemis demonstrator [16] of the HELIOS-2 project, the analysis of angiographic reports was chosen as an application to be used in parallel with the image processing [17]. These reports describe pathological aspects of renal artery angiograms such as the specification of a stenosis, a dilatation, an infiltration, etc. Such texts present typical features of medical jargon (most of the examples mentioned in Table 1 and Table 2 come directly from these angiographic reports).

Due to the modular structure of the RECIT analyser, no major problem has been encountered when switching from the analysis of echographic reports to the analysis of angiographic reports. Indeed, it is easy to insert new rules without disturbing the general mechanism used to select and apply them. This facility is especially important in the medical domain where professional language is evolving and is also dependent on the jargon of the medical services. Moreover, the considered medical sub-domain is still a closed world with an inherent semantics which lends itself to formal modelling. Since we remain in the medical domain, minor changes have occurred in the typology of concepts. This typology, by its tree forest structure allows principally to modify the event tree related to all events encountered in the considered domain. Finally, due to the powerful inheritance mechanism, most of the new conceptual schemata take advantage of the already implemented semantic information.

There are nevertheless some drawbacks to a domain-dependent approach. In fact, we shift the inherent NLP difficulties from a generalised syntax-driven strategy to a specific semantic-driven solution which is domain-dependent. The result is a strong dependence on a model of the medical domain. The development and the maintenance of such a model is a huge task which should principally be performed by specialists of the domain. This fact slows down the data acquisition process. Nevertheless, sound medical models begin to be available as the one developed in the AIM GALEN project [18] which offers new perspectives in the enhancement of medical knowledge. The adjustment of the RECIT system to the GALEN model is an experiment which is currently conducted at the Geneva Hospital. It aims at evaluating potential solutions to the above mentioned problems. Moreover, the LRE project DOME [19], which deals with medical documents and in which we have been involved since the beginning of 1995, should strengthen the knowledge engineering capabilities.

5. Conclusions

The RECIT system has been presented in this paper as an analysis tool able to understand medical jargon. The two-phase process of the RECIT system offers several advantages with respect to the use of medical language. The proximity processing approach presents a flexible way to recognise specific expressions of the domain under consideration and to solve syntactic ambiguities. Moreover, the confirmation of syntactic attachments by establishing the semantic relation among words, has shown to be relevant in medicine. It constitutes an original way to produce a structure of relevant components which are the starting point for the building of the conceptual representation. This second phase principally makes use of the semantics of the sub-domain. This implies both formalizing the domain dependent semantic information relevant to the task of understanding and developing an appropriate structure to access the information. The specification of conceptual schemata associated with an inheritance mechanism based on the well-defined domain constitutes the principal knowledge used during the elaboration of conceptual graphs. Finally, the fact that our system overcomes many of the weaknesses associated with traditional parsing strategies, is principally achieved by applying an inference-driven semantic analysis based on a limited and well-defined domain, and by considering syntactic rules as a set of heuristics rather than rigid grammar. Experiments with the available prototype show that our approach of handling medical jargon is both flexible and open to other European languages as well as to other medical sub-domains.

6. Acknowledgements

Part of this work was developed within the AIM Project A2015 HELIOS-2 "Hospital Object Software Tools". The specific funding in Switzerland was supported by CERS (Commission d'Encouragement à la Recherche Scientifique).

7. References

1. Grishman R, Kittredge R. *Analyzing Language in Restricted Domains: Sublanguage Description and Processing*. Hillsdale, NJ: Lawrence Erlbaum Associates, 1986.

2. Hirschman L, Sager N. *Automatic Information Formatting of a Medical Sublanguage*. In: Kittredge R, Lehrberger J (Eds.). Sublanguage: Studies of Language in Restricted Semantic Domains. Berlin: Walter de Gruyter, pp. 27-80, 1982.

3. Chomsky N. *Syntactic structures*. La Hague: Mouton, 1957. Traduction française par Michel Braudeau. *Structures syntaxiques*. Paris: Le Seuil, 1959.

4. Schank RC. *Conceptual dependency: a theory of natural language understanding*. Cognitive Psychology; vol. 3, n° 4, pp. 552-631, 1972.

5. Rassinoux A-M. *Extraction et Représentation de la Connaissance tirée de Textes Médicaux*. Ph D Thesis in Computer Science. Faculty of Sciences, University of Geneva, Switzerland. Geneva: Editions systèmes et information, 1994.

6. Morel-Guillemaz (Rassinoux) A-M, Baud RH, Scherrer J-R. *Proximity Processing of Medical Texts*. In: O'Moore R, Bengtsson S, Bryant JR, Bryden JS (Eds.). Medical Informatics Europe '90 (MIE 90). Proceedings, Berlin: Springer-Verlag, pp. 625-630, 1990.

7. Rassinoux A-M, Baud RH, Scherrer J-R. *Conceptual graphs model extension for knowledge representation of medical texts*. In: Lun KC, Degoulet P, Piemme TE, Rienhoff O (Eds.). MEDINFO 92. Proceedings, Amsterdam: Elsevier Science Publishers B. V. (North-Holland), pp. 1368-1374, 1992.

8. Baud RH, Rassinoux A-M, Wagner JC, Lovis C, Juge C, Alpay L, Michel P-A, Scherrer J-R. *Representing Clinical Narratives using Conceptual Graphs*. International Working Conference on Natural Language and Medical Concept Representation. Proceedings, Vevey, Switzerland, June 1994. To be published in: Methods of Information in Medicine, Spring 1995.

9. Rassinoux A-M, Baud RH, Scherrer J-R. *Analysis of Medical Texts using Proximity Processing and Conceptual Graph Representation*. AISB Quarterly 92, Newsletters of the Society for the Study of Artificial Intelligence and Simulation of Behaviour (SSAISB, University of Sussex, Brighton), Special Issue on AI in Medicine; Winter 92/3, pp. 11-15, 1992.

10. Sowa JF. *Conceptual Structures: Information Processing in Mind and Machine*. Reading, MA: Addison-Wesley Publishing Company, 1984.

11. Rassinoux A-M, Baud RH, Scherrer J-R. *A Multilingual Analyser of Medical Texts*. In: Tepfenhart WM, Dick JP, Sowa JF (Eds.). Second International Conference on Conceptual Structures (ICCS 94). Proceedings, Berlin: Springer-Verlag, pp. 84-96, 1994.

12. Geetha TV, Subramanian RK. *Representing Natural Language with Prolog*. IEEE Software; pp. 85-92, March 1990.

13. Baud RH, Lovis C, Rassinoux A-M, Michel P-A, Alpay L, Wagner JC, Juge C, Scherrer J-R. *Towards a Medical Linguistic Knowledge Base*. To appear in: MEDINFO 95, Vancouver, Canada, July 1995.

14. Engelmann U, Jean FC, Degoulet P. *The HELIOS Software Engineering Environment*. Supplement of Computer Methods and Programs in Biomedicine; vol. 45, Suppl, December 1994.

15. Rassinoux A-M, Michel P-A, Juge C, Baud R, Scherrer J-R. *Natural Language Processing of Medical Texts within the HELIOS Environment*. In [14], pp. S79-S96, December 1994.

16. Lemaitre D, Jaulent M-C, Günnel U, Demiris AM, Michel P-A, Rassinoux A-M, Göransson B, Olsson E, Degoulet P. *ARTEMIS-2: An application development experiment with the HELIOS environment.* In [14], pp. S127-S138, December 1994.

17. Engelmann U, Schröter A, Günnel U, Demiris AM, Makabe M, Evers H, Meinzer H-P. *The HELIOS Image Related Services.* In [14], pp. S64-S78, December 1994.

18. Rector AL, Solomon WD, Nowlan WA, Rush TW. *A Terminology Server for Medical Language and Medical Information Systems.* International Working Conference on Natural Language and Medical Concept Representation. Proceedings, Vevey, Switzerland, June 1994. To be published in: Methods of Information in Medicine, Spring 1995.

19. The DOME Consortium for the EC LRE programme. *DOME: Document Management for Health Care Applications.* Technical and Financial Annex, 1994. (Available from the Medical Informatics Centre, University Hospital of Geneva).

Medical Knowledge Representation for Medical Report Analysis

J. F. Smart[1,2], M. Roux[1]

[1] Department of Medical Informatics,
Faculty of Medicine of Marseille,
France

[2] Laboratory of Computer Science of Marseille
URA CNRS 1787
Faculty of Science of Luminy
France

Abstract. We present a knowledge representation formalism designed for medical knowledge-based applications, and more particularly for the analysis of descriptive medical reports. Knowledge is represented at two levels: a definitional level, which describes general medical concepts and the relations between them, and an assertional level, where individual cases are represented. At the definitional level, a concept type hierarchy and a set of schematic graphs define the concepts used and the relations between them, as well as different types of cardinality restrictions on these relations. A compositional hierarchy with a set inclusion relation allows concept composition to be precisely defined. At the assertional level, graphs representing "instances" of this knowledge can be created and manipulated taking into account the knowledge defined at the definitional level.

1 Introduction

In this article we present a knowledge representation formalism designed for medical knowledge-based applications which has been successfully used in the development of a system of automatic analyse and storage of French thyroid pathology reports. This formalism has been developed as part of the Aristotle project [1] under the direction of Professor Michel Roux in the Department of Medical Informatics at the Faculty of Medicine of Marseille.

The model is primarily designed for documentation representation and processing applications; for example, the automatic treatment of medical reports written in natural language. The most immediate application involves the creation of databases of analysed pathology reports, along with any associated images. Applications for this type of database are medical research, education and the exchange of information within and between hospitals. Despite the interest of the information contained in these records, their exploitation in text form is currently extremely difficult. Techniques which provide a normalised computer-usable form for these reports are therefore of great interest.

Medical knowledge representation is a major area of research in medical artificial intelligence, and many different approaches exist. The model presented here draws its inspiration from research in three main areas:

1) Conceptual Graph theory,

2) The various classification-based knowledge representation systems, and

3) The recent research in object-oriented software engineering and database technology.

The theory of Conceptual Graphs [2,3,4,5,6,7], a generalised logic-based knowledge representation formalism, is used as a cornerstone of the model. Although providing solid theoretical bases and a well-defined set of graph operations [8], the Conceptual Graph model lacks certain descriptive capacities at a conceptual level. Notably, cardinality restrictions on relations are poorly supported, and concept composition cannot be adequately represented for our purposes.

The model presented here extends the Conceptual Graph model using notions drawn from other areas of research in knowledge representation (notably, the classification-based knowledge representation systems such as KL-ONE and its descendants [9,10,11] and object-oriented database technology [12]) as well as some original features, to provide a solution particularly well adapted to the problem of descriptive medical discourse analysis.

2   The System Architecture

Following the tradition of the conceptual graph model [2] and the terminology-based classification languages such as KL-ONE [9,10], knowledge is defined at two distinct levels:

1) *The Definitional Level*, which allows the description of the concepts used in the knowledge base, as well as the possible relations between them, and

2) *The Assertional Level*, which allow particular cases or situations to be described and manipulated, "instances" of the concepts and relations described at the definitional level.

In the following sections, we will examine in more detail each of these levels.

3   The Definitional Level

At the Definitional Level, the knowledge base is comprised of three main parts:

1) *The Concept Type Hierarchy*, similar to that of the traditional Conceptual Graph model, such as the one illustrated in Figure 1;

2) *The Schematic Graphs*, which allow the definition of possible relations between concepts as well as cardinality constraints on these relations; and

3) *The Compositional Hierarchy*, which allows a flexible description of concept composition.

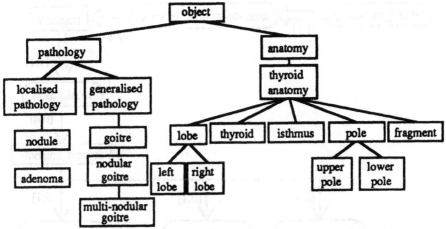

Fig. 1. An extract of a concept type hierarchy: thyroid anatomy and pathology.

3.1 The Schematic Graphs

The schematic graphs and the compositional hierarchy together describe the potential relations between domain concepts. An extension of the canonical graphs in the conceptual graph model, the schematic graphs allow cardinality restrictions to be imposed on the relations. Cardinality restrictions play an important role in many knowledge representation formalisms [9,10,12,13]. The cardinality restrictions used in this formalism follow the lines of those used in certain object oriented knowledge representation systems [12], with the additional possibility of defining certain relations as being "necessary and sufficient" for the description of a concept.

In Figures 2 and 3, the basic graphical notation used for the schematic graphs is illustrated. Concepts are represented by rectangles, and the relations between them by ovals. Each arc between a concept and a relation is labelled by minimal and maximal cardinality restrictions; these represent the number of associations of this type in which a given instance of the concept in question can participate.

For example, in Figure 2, the cardinality restrictions on the relation (anatomical_point_of_view) between [localised_pathology] and [fragment] specify that a localised_pathology must have exactly one fragment as its anatomical point of view, whereas a given fragment may be the anatomical point of view of any number (including 0).

Cardinality restrictions can be divided into three categories:

1) *Weak Cardinalities*, drawn by a simple arrow, where a range of possible values, with a minimal value of 0, is specified. For example, a lobe may have between 0 and 1 specified values of the property laterality:

weak cardinality

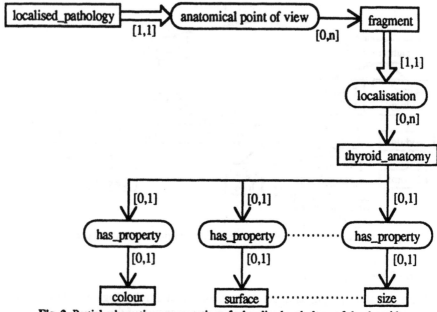

Fig. 2. Partial schematic representation of a localised pathology of the thyroid.

2) *Strong Cardinalities*, drawn by a hollow arrow, are those where the minimal cardinality value is greater than 0. For example, a localised pathology has precisely one thyroid fragment as its anatomical point of view:

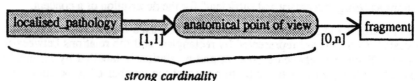

strong cardinality

3) *Definitional Cardinalities*, denoted by a solid black arrow, are strong cardinalities which participate n the set of necessary and sufficient relations which, along with its position in the concept type hierarchy, define a concept. For example, in Figure 4, the concept left_lobe is defined as a lobe with the laterality value [laterality:left]. Furthermore, every lobe with this laterality value is by definition a left_lobe.

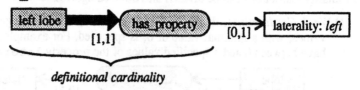

definitional cardinality

As can be seen in Figure 3, these different cardinality strengths allow knowledge to be described in more or less detail at different levels of generalisation.

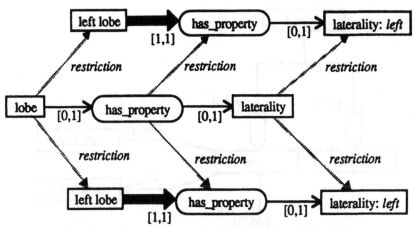

Fig. 3. Definition of the concepts lobe, left_lobe and right_lobe.

3.2 The Compositional Hierarchy

An important part of any medical knowledge representation formalism is the representation of concept composition. In certain application areas, such as natural language understanding and medical report analysis, a complete and flexible representation of this type of knowledge is particularly important.

Furthermore, the classic aggregation relation "has-part", is not in itself sufficient to represent certain important medical knowledge. Although many knowledge representation formalisms provide some support for the representation of concept composition [12], few, if any, allow the use of the set-inclusion relations needed to fully represent this type of knowledge. Thus, for the model presented here, a new formalism for compositional representation was developed.

In the formalism described here, concept composition is described by a set of compositional hierarchies; each compositional hierarchy describes the composition of some concept in the knowledge base using two relations:

1) A "has-part" relation, (K), and

2) An "inclusion" relation ($\supset$)

Figure 4 illustrates the complete compositional hierarchy for the concept thyroid.

The "Has-Part" Relation: K

The "has-part" relation, represented by the symbol K ("Komposition") corresponds to the standard aggregation relation.

It is indicated by a label of the following form:

$$K_{\lambda}^{[\mu, \nu]}$$

where

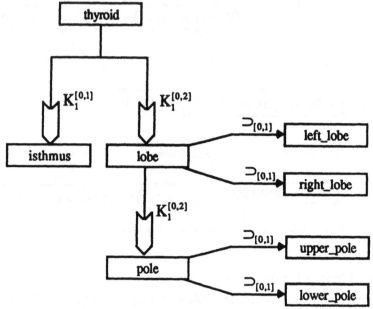

Fig. 4. Composition of the thyroid

$$0 \le \mu \le \nu$$
$$\lambda \in \{0,1\}$$

Following the direction of the arrow, the relation is read "is composed of". The values (μ, ν) correspond to the minimum and maximum number of components of a given type of which a concept may be composed. In this formalism, a given object can be a component of no more than one other object. The value λ thus corresponds to the minimum number of objects of which a given object may be a component. Thus, in Figure 4, a thyroid is composed of between 0 and 2 lobes, and a lobe is always part of exactly one thyroid.

The notation used for the arc is a simplification of the general notion used to represent the different cardinalities described above. In Figure 5 some examples of the different relations possible are illustrated.

The "Inclusion" Relation: ⊃

The other relation used in the hierarchies of composition is the set inclusion relation. Unlike the "has-part" relation described above, this relation cannot be expressed using the notation introduced in the above.

The relation of set inclusion is denoted by the set-inclusion symbol ⊃, and takes the following form

$$\supset_{[\alpha,\beta]}$$

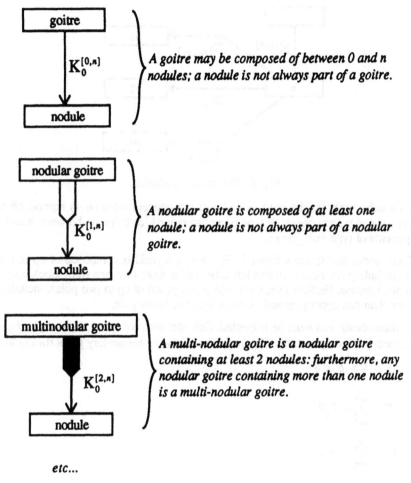

etc...

Fig. 5. Some possible "has-part" relations

This relation has no meaning in an isolated context; it is only meaningful once associated with a concept which is the component in a "has-part" relation, with a maximum outgoing cardinality restriction greater than 1. Consider the following general case (ignoring the cardinality types for simplicity):

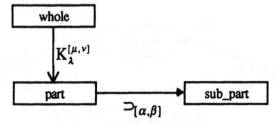

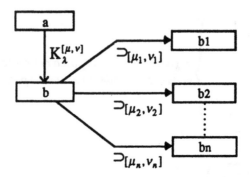

Fig. 6. The inclusion relation.

This sub-graph can be read as follows: the concept whole is composed of between μ and ν components of type part, among which there are between α and β components of type sub_part.

Thus, according to the schema in Figure 4, a thyroid is composed of up to two lobes (including no more than one left lobe and no more than one right lobe), and of up to one isthmus. Furthermore, each lobe is composed of up to two poles, including no more than one upper pole and no more than one lower pole.

Certain constraints must be respected. Consider the concepts $a, b, b_1, b_2, ..., b_n$ in the hierarchy of composition in Figure 6. The following elementary properties hold:

i) $\forall b_i, b_i < b$

ii) $\sum_{i=1}^{n} \mu_i \leq \mu$

iii) $\sum_{i=1}^{n} v_i \leq v$

iv) $\forall b_i, a \xrightarrow{K_\lambda^{[\mu_i, v_i]}} b_i$ holds.

At the assertional level, the relation $K_\lambda^{[\mu, v]}$ corresponds to the "has-part" relation in one direction, and to the "localisation" relation in the other. Thus, if the relation $c_1 \xrightarrow{K_\lambda^{[\mu, v]}} c_2$ holds, than the following schematic graph can be established:

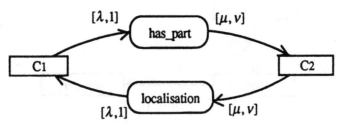

4. The Assertional Level

At the assertional level, conceptual graphs are used to represent and manipulate specific "instances" of the knowledge described at the definitional level. However, the conventional graph construction and unification operations must be extended to take into account the totality of the information represented at the definitional level.

4.1 Graph Construction

During graph construction, non-explicit information may have to be added to a graph in order to respect the cardinality constraints at the definitional level:

1) Relations and concepts may have to be introduced. For example, the creation of a graph containing the isolated concept [left_lobe:l_1] automatically results in the introduction of its associated necessary property [laterality:*left*], resulting in:

   [left_lobe:l_1]->(has_property)->[laterality:*left*]

2) Newly added information may allow the class of a concept to be redefined. Thus, if the property [laterality:*left*] is attributed to an isolated concept [lobe:l_2], the lobe is redefined as a left lobe, giving [left_lobe:l_2].

The cardinality constraints also prevent certain additions or modifications to a graph. For example, consider the following graph:

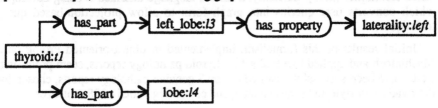

The addition of the property [laterality:*left*] to the concept [lobe:l_4] would cause this concept to become [left_lobe:l_4]. However, this modification would violate the conceptual constraint defined in the compositional hierarchy which specifies that, though a thyroid may be composed of up to two lobes, it can never have more than one left lobe. Thus, this addition is not permitted.

4.2 Graph Unification

The graph unification operation is equally extended to respect the full scope of the knowledge represented at the definitional level. Any unification or join producing a graph which violates the cardinality constraints of the definitional level is therefore forbidden. For descriptive text analysis, this extended form of the unification operation is a surprisingly flexible tool for descriptive discourse analysis. Once certain elementary syntactic considerations are taken into account (agreement on number etc...), these graph unification techniques can be used as an effective tool for the construction of discourse representations out of individual sentence representations in graph form.

The originality of this approach resides in its highly semantic nature - roughly speaking, syntactic information is used to resolve semantic ambiguities, rather than the inverse. The basic strategy for the unification of the graphs representing successive sentences involves correctly identifying the sub graphs in each graph which describe concepts common to both sentences. Only when this is done can the sentences be unified using conventional graph unification. If no common concepts exist, there may exist potential relations between some of the concepts in the graphs which allow the construction of a graph representing the two sentences. This technique will be detailed in a forthcoming article.

5. Discussion and Conclusion

This formalism has been used to implement a system of analysis and representation of French thyroid pathology reports. The system uses a combined syntactical/semantic analysis to treat the sentences of a report, producing a representation of the meaning of the sentence in graph form [1,14]. These graphs are then combined to build a complete representation of the report, using a process of discourse analysis based on the technique of semantic graph unification described above. The system thus allows the creation of a database of analysed reports. Interrogations and searches may be performed using conventional graph projection-based techniques [2,15]. Currently, a natural-language interface is being developed which converts user queries into graph form to allow projection-based query techniques.

Initial results of this formalism, implemented in object-oriented Prolog on a Machintosh and applied to a trial set of thyroid pathology reports, seem promising. Tests have been successfully carried out on 3 genuine pathology reports, chosen for their variety in style and subject matter, and containing a total of 28 sentences.

The formalism is particularly well suited for descriptive discourse analysis. In fact, in technical medical texts such as pathology or X-ray reports, the formalism seems capable of resolving a considerable part of the discourse analysis task with little or no recourse to linguistic or syntactically-based techniques.

Other studies [4,5,14] would indicate that conceptual graph-based models can be applied to report analysis involving simple temporal data, such as in hospitalisation reports. However, the treatment of time in a medical context is a complex problem [16,17], and the formalism described here does not claim to provide any more than elementary support for temporal reasoning.

Current work primarily involves the extension of the knowledge base and the syntactical/semantic analyser to allow a greater number of texts to be treated. Work is also being carried out to establish a formalised representation of set inclusion at the assertional level, in order to represent expressions such as "the thyroid is affected by several nodules, among which three are benign".

The representation of negation is limited to a relatively simple level, such as the absence or presence of certain properties. The representation of more complicated negations remains as yet a problem to be resolved.

A graphical knowledge base tool designed to facilitate the management of the knowledge base is also being developed. Such a tool will allow the base to be modified and updated without the risk of inadvertently introducing incoherent information. For larger knowledge bases, such a tool is invaluable.

References

[1]   Ledoray V, Giusiano B, Roux M. A system for understanding medical reports: Architecture and knowledge required, *MEDINFO 92*, 1992:1389-1394.

[2]   Sowa J. *Conceptual Structures : Information Processing in Mind and Machine*, Reading: Addison-Wesley Publishing Company, 1984.

[3]   Sowa J. Towards the Expressive Power of Natural Language. In: J. Sowa, ed. *Principles of Semantic Networks*, Morgan Kaufmann Publishers, 1991:57-189.

[4]   Rassinoux A et al. Conceptual graphs model extention for knowledge representation of medical texts, *MEDINFO 92*, 1992:1368-1374.

[5]   Baud R et al. Natural Language Processing and Semantical Representation of Medical Text, Meth Inform Med 1992; 2:117-25.

[6]   Schröder M. Knowledge Based Analysis of Radiology Reports Using Conceptual Graphs. In: Pfeiffer HD. ed, Proceedings of the Seventh Annual Workshop on Conceptual Graphs, New Mexico State University, Las Cruces, New Mexico, 1992, 213-222.

[7]   Alpay L et al. Interfacing Conceptual Graphs and the Galen Master Notation for medical knowledge representation and modelling. In: Andreassan S et al. eds, Artificial Intelligence in Medicine. Amsterdam: IOS Press, 1993: 337-347.

[8]   Chein M, Mugnier M. Conceptual Graphs: Fundamental Notions, Revue d'intelligence artificielle 1992; 6:4:365-406.

[9]   Brachman R, Schmolze J. An Overview of the KL-ONE Knowledge Representation System. Cognitive Science, 1985, 9:171-216.

[10]   Mac Gregor R. The Evolving Technology of Classification-Based Knowledge Representation Systems, in: J. Sowa, ed. *Principles of Semantic Networks*, Morgan Kaufmann Publishers, 1991:385-400.

[11]   Bernauer J, Goldberg H. Compositional Classification Based on Conceptual Graphs. In: Andreassan S et al. eds, Artificial Intelligence in Medicine. Amsterdam: IOS Press, 1993: 337-347.

[12]   Bouzeghoub M. *Du C++ à Merise Objet: Objets*, Paris: Editions Eyerolles, 1994.

[13]   Godbert E, Pasero R, Sabatier P. Natural Language Interfaces: Specifying and Using Conceptual Constraints, in: G. Salvedny and M. Smith, eds., *Human-Computer Interaction* (Elsevier, Amsterdam, 1993). 385-390.

[14] Ledoray V. "La sémantique dans l'analyse de comptes rendus médico-techniques rédigés en Français: application à l'anatomie pathologie de la thyroïde" PhD thesis, Université Paris 7 (1991).

[15] Fargues J et al., Conceptual graphs for semantic and knowledge processing, *IBM J. RES.DEVELOP.* 1986*30*:1: 70-79

[16] Kahn M G. Modelling time in medical decision-support programs. Medical Decision Making 1991:*11*; 249-64.

[17] Kahn M G. The Pursuit of Time's Arrow: Temporal Reasoning in Medical Decision Support In: Andreassan S et al. eds, Artificial Intelligence in Medicine. Amsterdam: IOS Press, 1993: 3-6

Temporal Reasoning and Simulation

Modelling Medical Concepts as Time-Objects

E.T. Keravnou

Department of Computer Science
University of Cyprus
Kallipoleos 75, P.O.Box 537, CY-1678 Nicosia, Cyprus
email: elpida@turing.cc.ucy.ac.cy

Abstract

Time is intrinsically related to medical problem-solving in general. Modelling time from the perspective of computer-based, competent, solution derivation of real-life medical problems is a challenging undertaking. Starting from the premise that temporal reasoning is an integral aspect of medical problem-solving, necessary requirements for medical temporal reasoning are listed, the notion of a time-object is proposed as an appropriate ontological primitive for modelling medical concepts, and generic models for patient data, disorders, and actions based on time-objects are discussed.

Keywords: Medical temporal reasoning, time-object, temporal granularity, temporal uncertainty, medical concepts, knowledge engineering.

1 Requirements for Medical Temporal Reasoning

Time is important in medical problem-solving. For many medical problems, it is the pattern of changes which is significant in solving the problem rather than a snapshot description of the state of affairs that prevailed at a particular time. Durations, temporal uncertainty and multiple temporal granularities are inherently relevant both to medical knowledge and patient data. Most general theories of time proposed in the literature do not adequately cover these temporal requirements. For example Allen's theory of time and action [1] does not address durations, absolute temporal uncertainty or multiple time scales; Kowalski and Sergot's event calculus [12] does not allow vagueness in the occurrence of events, nor does it address multiple granularities; and Dean and McDermott's time map manager [5] covers durations and absolute temporal uncertainty but does not cover multiple granularities. Most of these theories were not in fact developed to support problem-solving but rather to support natural language understanding or temporal database management.

Time constitutes an integral and important aspect of medical concepts and thus warrants explicit representation. This is why logic-based approaches to temporal reasoning (temporal-arguments, modal temporal logic, reified logic) [19] are not adequate for medical problem-solving. In temporal-arguments methods, eg situational calculus [14], time is represented as just another parameter in the relevant predicates which does not give it explicit status. In reified logics general temporal knowledge such as "the effects never precede their causes" cannot be expressed [19]; researchers have in fact argued against reification [6]. In medical problem-solving the entities of importance are compound objects (disorders, treatments, patient states) which have temporal existences and can interact with each other in complex ways, usually through mechanisms that are incompletely understood. For a temporal ontology to be adequate for medical problem-solving it must enable the representation of objects with vague existences and the representation of various temporal interactions between such objects (eg causal interactions). This suggests some kind of a *time-object* as a central primitive for such an ontology.

The general requirements for medical temporal reasoning are:

- *The occurrences of domain objects can be expressed in both absolute and relative terms.* For either case an occurrence can be considered a point or a period, under some granularity. Absolute occurrence means that the given point or period is explicitly specified, in relation to some fixed origin. Relative occurrence means that the occurrence of some object is specified in relation to the occurrences of other relevant time objects.

- *The occurrences of domain objects can be vague.* An absolute occurrence, at some granularity, is vague if it is not possible to specify precisely, at the particular granularity, the end points of the given period. A relative occurrence is vague if it is expressed as a disjunction of possible primitive temporal relations. For example the occurrence of object A either precedes or follows the occurrence of object B.

- *The occurrences of domain objects can be represented at multiple temporal granularities.* Occurrences of different domain objects can refer to different temporal scales (eg hours, days, months, years, etc), and in fact the occurrence of the same object can be expressed at different granularities.

- *The causality relation can have explicit temporal constraints.* Causality is an important relation in medical problem-solving [2-3]. Atemporal causal models have been used, and in some cases quite successfully. However, generally speaking, causality void of time is a contradiction of terms, since it prevents the representation of the generic, commonsense, constraint, that an effect cannot precede its cause. The temporal constraints can be absolute or relative, and can involve both quantitative and qualitative expressions [2].

- *An ability to perform temporal data abstraction efficiently and effectively.* Data abstraction is a necessary reasoning process for most medical problem-solvers. The requirement here is to extend the normal functionality of this process, which is roughly summarised in the question "What significant (from the perspective of the particular problem-solving) abstractions can be drawn from the current set of patient data?", to be able to abstract from *temporal* data [17].

- *An ability to perform temporal abduction, deduction and induction.* Problem solving in general involves these three basic forms of inference [15]. The requirement here is again to extend from an atemporal, to a temporal, type for these inferences, eg [4].

The above requirements can be refined into *static* and *dynamic* requirements (see Table 1). The former specify ontological requirements, what the ontology of the domain objects must provide, and the latter specify reasoning functionalities which need to be built on top of the ontology. Statically, the overall requirement is for considerable freedom, and hence power, of expression regarding the occurrences of individual objects (multiple time scales, absoluteness, relativeness, vagueness, incompleteness) and for time to figure explicitly and prominently in relations between objects (causality in specific, and abductive, deductive and inductive relations, more generally). Thus patient data that constitute the starting point for any medical problem-solving activity are primarily temporal entities. Dynamically, the overall requirement is problem-solving flexibility; nothing is gained by having a rich syntax but weak semantics; the expressiveness of a language is wasted if not accompanied with sufficiently powerful reasoning mechanisms. For example there is no point in representing vagueness, if the system cannot reason flexibly with vague temporal expressions. Thus the overall dynamic requirement is for a system to be able to reason flexibly both with the occurrences of individual objects as well as with inter-object relations involving temporal constraints. The paper addresses only some of the static requirements. Other of the requirements are discussed in [8-10].

A general architecture for medical knowledge-based systems needs to address all these requirements and to provide means for tailoring these to the specific needs of particular systems, which could be to eliminate temporal reasoning all together. Temporal uncertainty can be easily converted to temporal certainty but not the other way round, multiple granularities can be easily mapped to a single granularity but not the other way round, etc.

Static Requirements	Dynamic Requirements
Multiple granularities Point and interval duration	Mapping objects across temporal scales
Absolute occurrence Relative occurrence	Mapping absolute occurrences to relative occurrences and vice versa
Vague occurrence	Reasoning flexibly with vagueness
Temporal causal relation	Dynamically instantiating potential: - causal antecedents of some object - causal consequents of some object - causal links between objects
Temporal abductive relation Temporal deductive relation Temporal inductive relation	- Triggering temporal hypotheses based on the extent of the satisfaction of temporal constraints - Logically deducing temporal expectations of hypotheses - Inducing temporal conclusions (eg disorder D operative from time t)
Temporal patient data	Temporal data abstraction

Table 1: Ontological and Functional Temporal Reasoning Requirements

2 Time-Object as the Ontological Primitive

The appropriateness of the chosen ontological primitives depends on whether they provide the right abstraction for representing the models of the relevant medical concepts such as disorders, patient data, information-acquisition actions, therapeutic-actions, etc. All these concepts denote compound objects with rich internal structures and rich inter-object relationships. In this paper the notion of a *time-object* is proposed as an appropriate ontological primitive for modelling medical concepts. This proposal fully complies with the view that temporal reasoning is an integral aspect of medical problem-solving. The argument is that we want to reason about time-objects as integral objects. Disembodying the time attribute from the other atemporal attributes, and reasoning separately with each category of attributes implies that the two are completely independent of each other. If that were the case then indeed it would have been possible to initially concentrate on the atemporal aspects of these objects and simply to consider time at a later stage in the development.

Viewing a compound entity, the time-object, as the ontological primitive is quite unorthodox. The approach normally adopted in designing some temporal calculus is to choose, for example, between a time-point or a time-interval as the (atomic) ontological primitive [19]. Such approaches treat time as a separate, rather than as an integral, aspect of whatever concepts are to be modelled; time is modelled as a dimension per se, and largely independently of the entities which have an existence in time. Deciding between time-points alone, time-intervals alone, or both time-points and time-intervals, is seen as a technical, lower level, decision. The true conceptual issue is the adequate representation of objects which have (vague, unknown, or relative) existences expressed at different temporal scales, and this necessitates primitives that provide the appropriate level of abstraction.

2.1 The Notion of a Time-Object

A period of time of relevance to some application is modelled as a named time-axis with its appropriate time scale. The relevant time-axes collectively provide the context for the creation of time-objects and thus a time-axis constitutes an essential primitive for the definition of time-objects. Examples of time-axes could be lifetime, infancy, childhood, 1990, cancer-monitoring-period,, etc.

A time-object is an entity which has a temporal existence; this existence can be expressed in relation to different time-axes and thus time scales. A time-object is a coupling between a *property* and a *temporal existence*. Properties can be complex entities and their semantics are domain specific. Formally a time-object, τ, is denoted by the pair $< \rho, \varepsilon_\tau >$ where ρ is the property and ε_τ is the existence function for τ defined as follows:

$\varepsilon_\tau : A\chi \to Eexp$
where $A\chi$ is the set of discrete time-axes and
      Eexp is the set of existence expressions (see below)

The general existence function, ε, has two formal parameters, a time-object and a time-axis. The value returned by a specific call of the existence function is either a specification of some region on the argument time-axis, over which the argument time-object is said to exist (an existence expression), or an indication that the given time-axis is irrelevant to the given time-object. An existence function for the specific time-object τ, ε_τ, is obtained by performing partial parameterisation on the general function, ie fixing the first parameter to the specific time-object.

Time is treated as a linear, discrete, entity. A time-axis α denotes an ordered set of discrete time-values, Times(α), referring to a specific time-unit. A time-axis can be infinite or bounded, and can have an origin. The entire set of time-units, $M\chi$, specifies the different temporal granularities. Relation ma $\subset A\chi \times M\chi$ gives the time-unit for each time-axis. Explicit links between time-axes can be given through relation t-link which associates a time-value on one time-axis with a time-value on another time-axis. Other relations between time-axes (eg concurrent, includes, intersects-with, etc) can be derived from relation t-link.

Properties are associated with the time-units meaningful to them, and are specified as infinitely or finitely persistent. A finitely persistent property is further specified as reoccurring or not, and also, if possible, maximum and minimum durations for any of its occurrences under a specific time-unit are defined. The earliest and latest time-values, relative to the origin of some time-axis, for the initiation of any occurrence of a property, on the particular time-axis, can also be specified. An important relation between properties (and thus time-objects) is the *causality* relation; other relations are mutual exclusion and co-occurrence [8].

The same time-object can have an existence on different time-axes; for a time-object to have an existence on some time-axis it is required that (a) the time-period over which the given object has existed/is existing/will exist is (at least partly) covered by the time-period denoted by the time-axis, and (b) it is meaningful to talk about the given object at the temporal granularity associated with the given time-axis. If either of these requirements is not satisfied then the existence of the particular object on the particular time-axis is undefined, denoted by the symbol $\perp$. Neither of the above requirements is independent of the property of the object; the actual existence of an object depends on its property (and the external environment interacting with that property) and it is the property that defines which temporal scales are meaningful or not.

Thus the existence function for some time-object maps a time-axis to the (absolute) existence expression for the particular object on the particular time-axis. Let $\exp_\alpha$ be the set of absolute existence expressions for time-axis α. The set of all absolute existence expressions, Eexp, is the union of the $\exp_\alpha$ sets for all $\alpha \in A\chi$. Each $\exp_\alpha$ set is defined as follows:

$$\exp_\alpha = \{\perp\} \cup \{<l_\beta, u_\beta, l_\lambda, u_\lambda, \varsigma> \mid l_\beta, u_\beta, l_\lambda, u_\lambda \in \text{Times}(\alpha); l_\beta \leq u_\beta; l_\lambda \leq u_\lambda;$$
$$\varsigma \in \{\texttt{closed, open, open-from-left, open-from-right, moving}\}\}$$

An absolute existence expression for some time-axis, α, defines a period on the time-axis; it is given as a quintuple where the first four elements are time-values and the fifth element denotes a status (for the relevant time-object). The semantics of these elements are given below:

- l_β and u_β respectively give lower and upper bounds on the actual value for the initiation of the given period

- l_λ and u_λ respectively give lower and upper bounds on the actual value for the termination of the given period

- ς indicates whether any of the ends of the given period are precisely known or not; more specifically:

   — If ς = closed then $l_\beta = u_\beta$ and $l_\lambda = u_\lambda$.

   — If ς = open-from-left then $l_\beta \neq u_\beta$ and $l_\lambda = u_\lambda$.

   — If ς = open-from-right then $l_\beta = u_\beta$ and $l_\lambda \neq u_\lambda$.

   — If ς = open then $l_\beta \neq u_\beta$ and $l_\lambda \neq u_\lambda$.

Status value moving is reserved for special time-object *now* (see below).

Thus if the status regarding the existence of some time-object on some time-axis is not closed then there is vagueness about its actual existence at the given granularity. If some of the above values are unknown then the existence of the time-object at the given granularity is incomplete. The maximal and minimal durations for a time-object at a given granularity, are respectively given by $(l_\lambda - u_\beta)$ and $(u_\lambda - l_\beta)$. In addition if the exact duration of a time-object on a given time-axis is known then it can be expressed through relation Dur $\subset$ TO $\times$ Aχ $\times$ N where TO is the set of time-objects.

The existence expressions as described above constitute internal representations. Externally, information is more likely to be expressed through qualitative sentences such as "the throat has been sore from about midday yesterday", "the headache started about a week ago and it lasted for nearly 2 days", etc. Such sentences which assert the occurrences of properties (sore throat, headache) in absolute but not precise terms need to be translated into the corresponding internal existence expressions. Once translated at some granularity then a mechanism is required for mapping such expressions across different temporal granularities [8]. A vague existence expression at some granularity may be mapped to a precise expression at a coarser granularity. It could therefore be meaningful to talk about the absolute existence of some time-object at different granularities. At some granularity the existence could be just a time-point while at some other finer granularity the existence could be a time-interval. Relative temporal relations between time-objects, in the context of some time-axis, may be directly specified or derived on the basis of the relevant existence expressions for the time-objects [8]. A relative relation between the same pair of time-objects could be different on different time-axes. For example on one time-axis the two objects overlap while at a less refined time-axis they simply coincide.

Apart from the temporal relations, two structural relations, isa-component-of and its inverse contains, enable the definition of hierarchical relationships between time-objects. Thus a time-object can be refined into a set of temporally-related sub-time-objects and similarly a set of temporally-related time-objects can be collated into a super time-object. Informally a time-object must exist on any time-axis on which at least one of its component time-objects exists and no time-object can 'outlive' the time-object that contains it.

Finally, there is a global time-object, *now*, which is marked as a time-point on every time-axis to which it applies, ie every time-axis which has an ongoing time-value; now is a moving time-object and its rate of movement on some time-axis depends on the granularity of the time-axis. If now is not marked on some time-axis then the given time-axis refers entirely to the past or entirely to the future. The time-objects on some time-axis can therefore be partitioned into past, ongoing and future time-objects.

3 Generic Models for Patient Data, Disorders, and Actions in Terms of Time-Objects

In medical problem-solving patient data constitute *findings*, and disorders and actions constitute *hypotheses*. The solution to any problem is made up of concluded hypotheses. Thus a diagnostic problem-solver aims to conclude a number of disorder hypotheses as the explanation of the patient's condition, while a therapy planner aims to construct a therapy plan consisting of concluded therapeutic-action hypotheses. All these entities (data, disorders, actions) are naturally represented as time-objects.

3.1 Patient Data

Patient data are facts about the patient; each patient datum asserts the occurrence of some property (with respect to the patient) over some time-period. This is the basic definition of a time-object. Patient data can be classified in different ways; each classification is either based on the asserted property, or on the temporal existence, or both. For example, based on their properties, data can be divided into symptoms, signs, laboratory results, etc; based on their temporal existencies they can be divided into past and ongoing; and based on both their properties and temporal existences, into hard and soft findings. Patient data can be hierarchically or causally related; both types of relations are included in the time-object ontology. Collectively data can refer to different temporal granularities, which in turn refer to different spans of time relative to the lifetime of the given individual. This information is naturally translated into the set of time-axes for the given patient-model. Some of the time-axes have as their origin the birth of the particular patient (although their time-values could span a period prior to the birth of the patient).

3.2 Disorders

Two basic types of models for a disorder, are the *associational* (or heuristic) model and the more detailed *causal-associational* model. A causal-associational model can be converted to an associational model but not vice versa. An associational model for a disorder is simply a listing of the ´observable´ manifestations of the disorder, together with some indication of the strength of association between the disorder and each of its manifestations. A causal-associational model gives the observable manifestations of the disorder but in addition it gives the internal (non-observable) patterns of malfunctioning (causal sequences of internal states) which give rise to the particular observable manifestations; this is the causal part of the model.

A disorder, its internal causal states, and external, observable, manifestations, are again naturally expressed in terms of time-objects. Let Δ be the time-object denoting some disorder; $\{o_1, o_2, ..., o_n\}$ be the time-objects denoting its observable manifestations; and $\{s_1, s_2, ..., s_m\}$ be the time-objects denoting its internal states, its non-observable manifestations. Time-objects $o_1, ..., o_n$ are normally assumed to be mutually independent. The existences of Δ, $o_1, ..., o_n$, and $s_1, ..., s_m$ do not necessarily refer to the same time-axis. The existence of a generic disorder is expressed relative to the lifetime (using birth as origin) of any potential patient. For example the dysmorphology syndrome Morquio is presented from about the age of 1 year and persists forever, while the skeletal dysplasia Brachyolmia presents from about the age of 5 years and persists forever. In contrast, the common cold can present at any time, for any number of times, and thus any presentation of it has a finite duration. This knowledge can be expressed as follows, where π is the selector function for the property part of a time-object, ie $\pi : \text{TO} \rightarrow \Phi$, where TO is the domain of time-objects and Φ is the domain of properties:

$\pi(\Delta 1) = \text{Morquio-present}; \quad \pi(\Delta 2) = \text{Brachyolmia-present}; \quad \pi(\Delta 3) = \text{cold-present}$

$\varepsilon_{\Delta 1}(\alpha_1) = \text{<9, 14, +\infty, +\infty, open-from-left>}$

$\varepsilon_{\Delta 2}(\alpha_1) = \text{<54, 66, +\infty, +\infty, open-from-left>}$

$\varepsilon_{\Delta 3}(\alpha_2) = \text{< x, \_, \_, x+n, \varsigma >}$

where the time-unit for time-axis α_1 is months and for time-axis α_2 is days. For any potential patient the generic time-objects $\Delta 1$ and $\Delta 2$ can be instantiated once; a given instantiation can differ on the actual initiation time but not on the termination time which is fixed to $+\infty$. Properties Morquio-present and Brachyolmia-present are specified as infinitely persistent. The generic time-object $\Delta 3$ can be instantiated many times for any potential patient (property cold-present is finitely persisting); in the above generic existence expression, x is a time-value variable and n is a time constant specifying the maximal duration of the given property at the given granularity. For a given instantiation of $\Delta 3$, x will be bound to the relevant´ initiation time while the status variable's, ς's, value will depend on whether the end points of the given time-period are accurately known or not.

The existence of a generic disorder Δ is expressed relative to the lifetime of a potential patient, while the existences of the manifestations of the disorder are expressed either relative to the initiation time of a potential instantiation of Δ or relative to the lifetime of a potential patient. The existences of all the time-objects are expressed under the assumption of no external interference, such as the undertaking of therapeutic-actions, which could result in the premature termination of the temporal persistencies of time-objects. The following relations hold: manifests(Δ,o_i) and manifests(Δ,s_i). By adding another argument to relation manifests it is possible to express a quantitative or qualitative strength of association. Alternatively manifests can be refined to a set of more specific binary relations denoting different strengths of association such as necessarily-manifests, only-explanation-for, commonly-manifests, occasionally-manifests, etc. The following axioms are used:

Axiom 1: manifests(Δ,m) $\Rightarrow$ contains(Δ,m)

Axiom 2: (contains(Δ,m) $\wedge$ $\varepsilon_m(\alpha) \neq \bot \wedge \varepsilon_\Delta(\alpha) \neq \bot$) $\Rightarrow$ (m $\sqsubset=_\alpha \Delta$)

The time-objects denoting the manifestations of some disorder are considered to be components of the time-object denoting the given disorder (axiom 1). Through axiom 2, which is a generic axiom of the temporal ontology used, it is dictated that no manifestation can precede or outlive the disorder that 'created' it. A disorder is therefore refined in terms of the temporal patterns of occurrence of the abnormal properties manifested by it. Similarly some manifestation (o_i or s_i) can be expressed, at a more detailed level, as a collection of more refined time-objects and their interrelationships, thus enabling the description of a disorder at multiple levels of abstraction.

An associational model for a disorder consists of the time-object, Δ, denoting the disorder, the time-objects, o_i, i = 1, ..., n, denoting its external manifestations (their properties and existence expressions), and the manifests relation giving the relative strengths of association. A causal-associational model consists of an associational model as specified above, plus the time-objects s_i, i = 1, ..., m (their properties and existences), the relevant extension to the manifests relation to include the s_i time-objects and the causes relation: causes(x,y,c), where the antecedent x $\in$ {s_i | i = 1, ..., m}; the consequent y $\in$ ({s_i | i = 1, ..., m} $\cup$ {o_i | i = 1, ..., n}); and c is a quantitative or qualitative indication of the strength of causality:

Axiom 3: (causes(x,y,_) $\wedge$ $\varepsilon_x(\alpha) = <\_,v\_,\_,\_,\_> \wedge \varepsilon_y(\alpha) = <u\_,\_,\_,\_,\_>$) $\Rightarrow$ (u)= v)

Axiom 4: $\forall$ o_i $\exists$ s_j causes(s_j, o_i, c)

Axiom 5: $\forall$ s_i $\exists$ x $\in$ ({o_j} $\cup$ {s_j}) causes(s_i, x, c)

No effect can precede its cause (axiom 3); every observable manifestation must have at least one (immediate) cause from amongst the internal manifestations (axiom 4); every internal manifestation must have at least one causal effect which is either an internal manifestation or an observable manifestation (axiom 5). If it is required for the graph defined through relation causes to be connected then the relevant axiom must also be specified. An internal manifestation s_i constitutes an *ultimate-cause* iff it has no causal antecedents: ultimate-cause(s_i) $\Leftrightarrow$ $\neg$($\exists$ s_j s.t. causes(s_j, s_i, _))

In addition to what is specified above, relative, a priori, likelihoods of occurrence for the given disorders in the given population of patients can be specified. Furthermore if disorders are represented in terms of causal-associational models, the ultimate causes (causal antecedents) for each disorder can be associated with a priori likelihoods of occurrence.

3.3 Actions

In medical problem-solving actions can be divided into information-acquisition procedures and therapeutic procedures. The former simply aim to ´observe´ the current status of the patient without bringing about any changes, while the latter aim to change, hopefully for the better, the status of the patient.

An action in the context of medical problem-solving, like any action [16], has a set of preconditions justifying the application of the action, and, if executed, it is expected to cause various effects. An action, its preconditions and effects are all time-objects. The property associated with an action time-object gives the textual description of the particular undertaking which may be do-nothing. A given action time-object can be decomposed, through relation contains, into a temporal plan of sub-action time-objects.

Let A be the time-object denoting some generic therapeutic action; $\{p_1, p_2, ..., p_n\}$ be the generic time-objects denoting the action's preconditions; and $\{e_1, e_2, ..., e_m\}$ be the generic time-objects denoting the action's effects. The following relations hold: requires(A, p_i, n) and causes(A, e_i, c). Relation requires associates an action with its preconditions, the third argument of the relation indicating how necessary a given precondition is to a given action. Relation causes introduced above in the context of generic disorders, is also used for associating an action with its effects. The existence expression for a generic action gives the expected duration for any instantiation of the particular action, if such information can be given, otherwise it is given entirely in terms of relevant type variables. Most likely, the actual initiation and termination of some instantiation of the action would be expressed in absolute calendar terms. The property associated with an action time-object indicates whether more than one instantiation for the given action can apply to any potential patient and/or whether any constraints apply on the time-interval between successive instantiations of some action. Other temporal constraints, such as a given operation, y, can not be performed on a patient of less than x years of age, should best be expressed through preconditions:

$\pi(y)$ = perform-operation-y; $\pi(x)$ = the-patient-is-at-least-10-years-old
ε_x(lifetime) = <10, 10, +∞, +∞, closed>; requires(y, x, definitely)

Time-axis lifetime is expressed at temporal granularity years. For a patient who is less than 10 years of age, time-object x constitutes a future time-object and thus y cannot be instantiated for such a patient since the particular precondition is specified as a definite requirement for the action. In general the preconditions for an action instantiation must constitute past (or almost past) time-objects at the point of commencing the execution of the action instantiation; unless a precondition involves a time-invariant property and thus would constitute an ongoing time-object at any time. The existence expressions for the effects of some action are given relative to the initiation of any instantiation of the action. The existence expressions for the preconditions of an action are either given relative to the lifetime of any potential patient or relative to the initiation of any instantiation of the action. This enables the expression of statements like "continuous chest pain for 3 days requires a chest radiograph", "severe chest pain requires an ECG immediately", etc:

$\pi(A_1)$ = take-a-chest-radiograph; $\pi(A_2)$ = do-an-ECG
$\pi(p_1)$ = continuous-chest-pain; $\pi(p_2)$ = severe-chest-pain
requires(A_1,p_1_); requires(A_2,p_2_)
$\varepsilon_{p1}(\alpha)$ = <-3,-3,0,0,closed>; $\varepsilon_{p2}(\alpha)$ = <0,0,0,0,closed>

Time-axis α is expressed at the granularity of days. Special time-value 0 denotes the initiation of the execution of an action instantiation. More complicated preconditions such as "more than two distinct incidents of severe tonsillitis in a year requires the removal of the tonsils" can also be expressed:

$\pi(P)$ = more-than-two-severe-incidents-of-tonsillitis
$\pi(A)$ = remove-tonsils; requires(A,P,definitely)
$\pi(p_1)$ = severe-tonsillitis; $\pi(p_2)$ = severe-tonsillitis;
$\pi(p_3)$ = severe-tonsillitis
contains(P,p_1); contains(P,p_2); contains(P,p_3)
disjoint(p_1,p_2); disjoint(p_1,p_3); disjoint(p_2,p_3)
$\varepsilon_p(\alpha)$ = <-12,-12,0,0,closed>

where time-axis α is at the granularity of months. Lastly, the occurence of some events may signal the premature termination of the action instantiation. Such events can also be associated with generic actions.

Patient data, disorders and therapeutic-actions are therefore naturally modelled as time-objects. Triggered disorder or action hypotheses need to be temporally and contextually screened, and the evaluation of hypotheses involves the 'matching' of time-objects [10].

4 Conclusion

Time is intrinsically related to medical problem-solving in general. The first generation of medical knowledge-based systems were, with the odd exception, atemporal systems; time did not figure at all or it figured in an implicit way. Some of the more recent systems, however, do have explicit temporal models and associated temporal reasoning, eg [2-4,7,10,13,17]. Temporal reasoning, if relevant to some application, should be an integral aspect of the engineering of the given application from the beginning. For temporal reasoning to be properly and explicitly modelled in some application, time must be given special status with regard to the relevant medical concepts; the required temporal reasoning functionalities can then be explicitly modelled, based on the chosen ontology. In the paper the notion of a time-object has been proposed as an appropriate ontological primitive and generic representations for patient data, disorders, and actions, in terms of time-objects, have been presented. The seeds of the proposed temporal ontology were sown during the development of the SDD system [10], the temporal reasoner of which is based on a layered, extensible, architecture. The next step is to implement the theoretical framework as a generic, reusable, tool for medical temporal reasoning.

Acknowledgements: I am greatly indebted to John Washbrook, Eleni Christodoulou, and the two anonymous referees for their constructive criticism on earlier drafts of the paper. The work reported here was developed in the context of CEC AIM Project A2034 GAMES II and Project "Temporal Diagnostic Reasoning" which is partly funded by the University of Cyprus.

References

[1]   Allen J.F. (1984), "Towards a general theory of action and time", *Artificial Intelligence*, Vol.23, pp.123-154.

[2]   Barahona P. (1994), "A causal and temporal reasoning model and its use in drug therapy applications", *Artificial Intelligence in Medicine*, Vol.6, No.1, pp.1-27.

[3]   Console L., Dupre D.T. and Torasso P. (1990), "Fuzzy temporal reasoning on causal models", *Internat. J. Intelligent Systems*.

[4]   Console L. and Torasso P. (1991), "On the co-operation between abductive and temporal reasoning in medical diagnosis", *Artificial Intelligence in Medicine*, Vol.3, pp.291-311.

[5]   Dean T.L. and McDermott D.V. (1987), "Temporal data base management", *Artificial Intelligence*, Vol.32, pp.1-55.

[6]   Galton A. (1991), "Reified temporal theories and how to unreify them", *Proceedings IJCAI'91*, pp.1177-1182.

[7]   Kahn M.G., Fagan L.M., and Tu S. (1991), "Extensions to the time-oriented database model to support temporal reasoning in medical expert systems", *Meth Inform Med*, Vol.30, pp.4-14.

[8]   Keravnou E.T. (1994), "A formal ontology for temporal reasoning: open intervals and multiple granularities", *submitted to Artificial Intelligence*.

[9]   Keravnou E.T. (1994), "Temporal vagueness in medical reasoning", to appear *Int J. Systems Research and Info Science*.

[10]   Keravnou E.T. (1994), "Temporal reasoning tool", AIM Project A2034 GAMES-II Deliverable 34.

[11]   Keravnou E.T. and Washbrook J. (1990), "A temporal reasoning framework used in the domain of skeletal dysplasias", *Artificial Intelligence in Medicine*, Vol.2, pp.239-265.

[12]   Kowalski R. and Sergot M. (1986), "A logic-based calculus of events", *New Generation Comput.*, Vol.4, pp.67-95.

[13]   Larizza C., Moglia A. and Stefanelli M. (1992), "M-HTP: A system for monitoring heart transplant patients", *Artificial Intelligence in Medicine*, Vol.4, pp.111-126.

[14]   McDermott D. (1982), "A temporal logic for reasoning about processes and plans", *Cognitive Science*, Vol.6, pp.101-155.

[15]   Peirce C.S. (1878), "Illustrations of the logic of science, sixth paper - deduction, induction, hypothesis", *The Popular Science Monthly*, Vol.1, pp.470-482.

[16]   Rutten E. (1993), "An imperative planning language: from temporal representation to real-time execution", ERCIM Research Report ERCIM-93-R004 GMD.

[17]   Shahar Y. (1994), "A knowledge-based method for temporal abstraction of clinical data", PhD Thesis, Department of Computer Science, Stanford University (Report No. KSL-94-64 or STAN-CS-TR-94-1529).

[18]   Tong D.A. and Widman L.E. (1993), "Model-based interpretation of the ECG: a methodology for temporal and spatial reasoning", *Computers and Biomedical Research*, Vol.26, pp.206-219.

[19]   Vila L. (1994), "A survey on temporal reasoning in artificial intelligence", *AICOM*, Vol.7, No.1, pp.4-28.

Modeling Medical Reasoning with the Event Calculus: an Application to the Management of Mechanical Ventilation

Luca Chittaro, Marco Del Rosso
Dipartimento di Matematica ed Informatica,
Università di Udine,
Via Zanon, 6
33100 Udine - ITALY
chittaro@dimi.uniud.it

Michel Dojat
INSERM U.296
Faculté de Médecine
8, Rue du Général-Sarrail
94010 Créteil Cedex - FRANCE
dojat@laforia.ibp.fr

Abstract

Explicit representation of time and change is an essential feature for building systems that are supposed to interact with real-world dynamic environments. In this paper, we propose to use the Cached Event Calculus (CEC), an improved version of the Event Calculus [Ko86], to represent temporal aspects in intelligent medical monitoring systems. In particular, we explore the application of CEC to the management of mechanical ventilation, using it to interpret change in data over time, assess patient status and its evolution, and choose the proper level of mechanical assistance.

1 Introduction and Motivation

In data-rich clinical environments such as Intensive Care Units (ICUs) or operating rooms, there is a crucial need for intelligent monitoring systems that can help the clinician to deal with the massive flux of information. These systems should be able: (i) to acquire and analyse the mass of data available to propose a diagnosis of the current state of the patient, (ii) to filter the numerous alarms from monitors to indicate only those that require a human intervention and (iii) to propose specific therapeutic strategies depending on the evolution of the patient's state. Explicit time and change representation is essential for building such intelligent monitoring systems. A typical clinical application of intelligent monitoring systems is the management of the mechanical respiratory assistance provided to patients who suffer from a lung disease and are hospitalised in ICUs.

1.1 Intelligent Systems for Mechanical Ventilation Management

Recent physiological studies [Br87, Va91] have convinced physicians to mechanically ventilate patients as soon as possible with partial assistance modalities: a variable level of mechanical assistance is added to the spontaneous respiratory activity of the patient. Although partial mechanical support such as *Pressure Support ventilation* (PS) [Ma86] is simple in its principle, its use generally requires the presence of a trained and experienced physician who adapts the level of assistance to the evolution of the patient's state. This is emphasised when the physician, applying specific strategies, tries to decrease gradually the assistance and appreciates the patient's capacity to breathe alone. This procedure (called *weaning procedure*) must be performed carefully to improve the quality and the success rate of such a difficult process.

Many decisions and adjustments performed on the ventilator settings are based on objective data and can be formalised and modeled with appropriate knowledge representation techniques. Ideally, the advantages of a knowledge-based system for the management of ventilator therapy are: (i) to function on a 24 hours per day basis, allowing a continuous adaptation of the level of the assistance and a reduction of total duration of ventilation, and (ii) to develop specific weaning strategies, including a gradual decrease of the mechanical support, difficult to obtain in clinical practice without the assistance of a computerised system. Such a system must work in a closed-loop to be useful to the clinical staff. Recently, clinical studies have validated this approach [Do92].

Since the precursor work of Fagan [Fa80], several systems have been designed to assist respiratory management (see [Uc93] for a broad survey on recent intelligent patient monitoring projects). Briefly, systems can be divided in two categories:

- *Systems solving a practical clinical problem.* For example, [Si88] deals with the complex task of ventilating patients with Acute Respiratory Distress Syndrome, while [To91] tackles the problem of assisting therapists and nurses in weaning post-operative cardiovascular patients from mechanical ventilation. These systems are hardly adaptable to other clinical contexts.

- *General architectures for intelligent monitoring.* These are long term research projects, such as [Ha92], which proposes a general architecture for intelligent agents and is applied to intensive care monitoring problems, and [Ru93], which mixes qualitative and quantitative computation in a ventilator-management advisor. The first objective of these researches is not to design a prototype working at the patient's bedside, but to explore novel AI techniques potentially useful to solve medical problems.

Our project is intermediate. On one hand, our goal is to solve a clinical problem (i.e. the management of PS ventilation) and to test a closed-loop prototype at the patient's bedside. On the other hand, we aim at an extendible system for ventilation management based on generic mechanisms in which we gradually incorporate new capabilities validated in the clinical environment. We are especially concerned with modeling temporal aspects of medical reasoning for interpreting clinical data in real-time.

Existing systems often use an embedded implicit temporal representation (e.g. [Fa80], [Co90]). Unlike these systems, we adopt an explicit representation of temporal knowledge and reasoning to monitor and control in real-time ventilator therapy.

1.2 The Temporal Dimension

Time is a central factor in intelligent monitoring systems that are supposed to interact with real dynamic environments. The need for time representation covers two major aspects:

- *Modelling of temporal concepts and inferences used by the physician*: the physician i) builds dynamically an interpretation of the evolution of the patient's ventilation, ii) predicts the patient's evolution with regard to previous states and iii) constructs and executes a plan of actions to drive the patient to an expected state. The physician adapts his/her strategy to the history of the patient's ventilation and to the time the patient spent in a given respiratory state. To build a global dynamic interpretation of the patient's behaviour, he/she must recognise only relevant changes and forget non-relevant information (such as short instabilities) [Do95].

- *Respect of real-time constraints*: the system should be able to i) acquire physiological data provided by several monitors, ii) plan the sequencing of the three fundamental tasks in medical reasoning [St92] - observation, diagnosis and then therapy - each task being in turn decomposable in several sub-tasks and iii) have a prompt reaction in alarming situations, which impose to short-cut some sub-tasks.

The need for an explicit representation of time in medical diagnosis has been advocated by several researchers, but very few clinical decision support systems incorporate clear formalisms for temporal reasoning [Sh90]. Moreover, the integration of a time map manager into real-time systems (such as intelligent monitoring systems) is difficult due to the high complexity of temporal constraint propagation algorithms [De91, Va92]. In patient monitoring we need to reason about disease evolution (e.g., "Has the patient's ventilation been stable enough to envisage a decrease in mechanical assistance?") or to judge a patient's response to therapy (e.g., "Is the increase of mechanical assistance effective?"). A model with events, states and cause-effect relationships seems well adapted [Do95]. In this paper, we explore the application of the Event Calculus (EC) proposed by [Ko86] to interpret change in data over time in the context of mechanical ventilation management.

2 Temporal Ontology and Inference in EC

Kowalski and Sergot's Event Calculus (EC) is a general approach to representing and reasoning about events and their effects in a logic programming framework [Ko86]. From a description of events which occur in the real world and properties they initiate or terminate, EC derives the maximal validity intervals (MVIs) over which properties hold. It takes the notions of event, property, time-point and time-interval as primitives and defines a model of change in which *events* happen at *time-points* and initiate and/or terminate *time-intervals* over which some *property* holds. It embodies a notion of *default persistence* according to which properties are assumed to persist until an event occurs that interrupts them.

A model of the world based on events, whose occurrence modifies the state of the world and properties that have a tendency to persist during time, is well adapted to our applications [Do95].

2.1 The Basic Event Calculus

Formally[1], we represent an event occurrence by means of the happens_at(event, timePoint) clause[2]. The relation between events and properties is defined by means of initiates_at and terminates_at clauses:

```
initiates_at(event1,property,T):-        terminates_at(event2,property,T):-
   happens_at(event1,T),                     happens_at(event2,T),
   holds_at(prop1,T), ...,                   holds_at(prop1,T), ...,
   holds_at(propN,T).                        holds_at(propN,T).
```

The initiates_at (terminates_at) clause states that each event of type event1 (event2) initiates (terminates) a period of time during which property holds, provided that n (possibly zero) given conditions hold at instant T. In EC, both initiates_at and terminates_at are context-independent predicates: they do not admit preconditions. However, even when modeling very simple real-world example, the context is essential to decide which properties are initiated or terminated by the occurrence of an event. For example, the event "turn on the switch" in a simple light bulb circuit, initiates the property "the light is on" at a given instant, only if the property "electrical power is supplied" holds at that instant [Ch96]. Thus, we added preconditions to initiates_at and terminates_at in order to model complex domains, such as mechanical ventilation management.

A particular initiates_at clause is used to deal with initial conditions. Initial conditions describe a possibly partial initial state of the world and are specified by

---

[1]We follow PROLOG's convention of beginning variables' names with an uppercase letter and constants with a lowercase letter.

[2]In general, the requirement of knowing the exact time point when each event happened can be relaxed (we dealt with the case of partially ordered events in [Ch94]) at the expense of efficiency. Fortunately, all the events in the considered application can be time-stamped.

means of a number of events of type `initially(prop)`. Their validity from the beginning of time can be derived by means of the clause:

```
initiates_at(initially(Prop),Prop,0):-
   happens_at(initially(Prop),0).
```

The EC model of time and change is defined by means of the axioms:

```
mholds_for(P,[Start,End]):-
   initiates_at(Ei,P,Start),
   terminates_at(Et,P,End),
   End gt Start,
   \+broken_during(P,
      [Start,End]).
```

```
mholds_for(P,[Start,infPlus]):-
   initiates_at(Ei,P,Start),
   \+broken_during(P,[Start,infPlus]).

broken_during(P,[Start,End]):-
   terminates_at(E,P,T),
   Start lt T, End gt T.
```

where the predicate `gt` extends the ordinary ordering relationship > to include the cases involving infinite arguments (positive infinite is syntactically denoted by `infPlus`). Analogously, we defined the predicates `ge`, `lt` and `le` which extend $\geq$, < and $\leq$, respectively. The `mholds_for` axiom states that a property P maximally holds between events `Ei` and `Et`, if `Ei` initiates P and occurs before `Et` that terminates P, provided there is no known interruption in between (the negation involving the `broken_during` predicate is indeed interpreted using negation-as-failure). The interval `[Start,End]` is thus a Maximal Validity Interval (MVI) for property P. The `broken_during` axiom states that a given property P is interrupted between `Start` and `End` if there is an event `E` that happens between them and terminates P. This axiom provides a so-called *weak interpretation* [Ch96] of `initiates_at` clauses (in a *strong interpretation* [Ch96], the `broken_during` axiom would consider also initiating events for P and not just terminating ones as possible interruptions for the interval `[Start,End]`).

Finally, the holds_at axiom relates a property to a time-point rather than to a time-interval:

```
holds_at(P,T):-
   mholds_for(P,[Start,End]), T gt Start, T le End.
```

The `holds_at` predicate conventionally assumes that a property is not valid at the starting point of the MVI, while it is valid at the ending point.

The axioms presented above already allow to model a temporal knowledge base and run it PROLOG. Unfortunately, the obtained computational performance is not acceptable. This is the reason why we proposed the Cached Event Calculus (CEC), an efficient implementation of EC with preconditions.

2.2 The Cached Event Calculus (CEC)

EC records any input event without processing it (the cost of update processing in EC is thus constant), and temporal reasoning is performed at query time. More specifically, in order to determine the set of MVIs for a given property p, `mholds_for` generates all ordered pairs of initiating and terminating events for such a property, and then checks the absence of known interrupting events in between. Thanks to the negation as failure rule, MVIs no longer supported are not derived anymore. This generate-and-test strategy makes query processing in EC very inefficient. Moreover, results of computations are not recorded for later use, further deteriorating performance.

To face these efficiency problems, we extended EC with a caching mechanism that records MVIs of properties in the form of `mholds_for` facts for later use in query processing and updates them as soon as a new event occurrence is entered in the database. Unlike general caching mechanisms, the Cached Event Calculus (CEC) does not only add and/or remove assertions, but also clips and/or extends existing MVIs according to domain and temporal knowledge. Moreover, it replaces the generate-and-test strategy with a more focused one.

[Ch96] describes in detail the architecture and the implementation of CEC and formally proves the computational complexity of EC and CEC. We only summarize here the complexity results, shown in Table 1. The query columns give the cost of deriving the full set of MVIs for a given property, the update columns give the cost of adding a new event to the database. We assumed that the database contains n initiating events and n terminating events for any property (with $n \geq 1$). Since a precondition in a context-dependent `initiates_at` or `terminates_at` may itself be context-dependent, we consider in general an arbitrary nesting level of preconditions: the parameter Lbk is the maximum level of nesting from a single property and can be statically determined for a given knowledge base [Ch96]. The complexity is measured in terms of the number of accesses to facts in the database.

	Ec update	EC query	CEC update	CEC query
$L_{bk}=0$	constant	$O(n^3)$	$O(n^2)$	$O(n)$
$L_{bk}=1$	constant	$O(n^6)$	$O(n^4)$	$O(n)$
$L_{bk}=2$	constant	$O(n^9)$	$O(n^5)$	$O(n)$
...	...	...	...	...
$L_{bk}=k$	constant	$O(n^{(k+1) \cdot 3})$	$O(n^{(k+1)+2})$	$O(n)$

Table 1. Computational complexity of EC and CEC [Ch96].

In summary, CEC moves computational complexity from query to update processing, and features an absolute improvement of performance, because its update processing costs less than EC query processing. Note that these results refer to the worst-case complexity analysis. Experimental results show that CEC is even better on the average case and that the chronological acquisition of events (as it is the case of the mechanical ventilation domain) further increases performance.

3 Using CEC to Reason about Time and Change in Mechanical Ventilation Management

In this section, we describe the use of CEC to model time and change in an intelligent monitoring system for ventilator therapy management. The medical expertise was obtained from clinicians at the Henri Mondor Hospital, in France. Clinicians provided about 50 rules covering both interpretation of data, prediction of patient's evolution and therapeutic decisions. This domain knowledge has been modeled using events and properties. Temporal reasoning is performed by CEC to interpret the status of the patient, the evolution of patient's ventilation and the evolution of the therapy. CEC relies on an additional module for interface and data acquisition purposes.

3.1 Architecture of the Prototype

The implemented prototype is organized into two different parts, as depicted in Figure 1. The Interface allows clinical staff to initialize the system according to information about the patient's (weight, type of pathology, ...). It receives incoming data from a gas analyzer (for expiratory partial pressure of CO_2: PCO2) and from the ventilator (for expired volume VT and respiratory rate RR) and provides them to CEC in a proper format. CEC performs all the inferential activity needed to interpret the data received by the Interface and take the proper therapeutic decisions. The interface modifies the ventilator settings (level of PS, mode of ventilation) according to CEC decisions, carries out the therapeutic decisions taken by CEC, and presents the inferences done by CEC to the clinical staff.

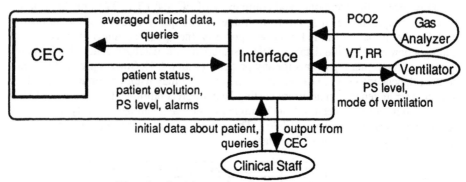

Fig. 1. Architecture of the prototype.

The Interface reads the values of the clinical parameters with a frequency of 0.1 Hz, for a period of time whose duration can be changed (by CEC) according to the current clinical situation. At the end of this observation period, the Interface sends to CEC the averages of the clinical parameters read all over the period. As a result, CEC performs its inferential activity to update its database in response to the new information available about the patient. When the database update ends, the Interface reads the current value of those properties that prescribe the current level of assistance and the duration of next observation period, respectively. The former are sent to the ventilator, and the new observation period is started. In some cases, CEC may also determine that an alarm should be raised to the clinical staff, because some specific situation is occurring (for example, patient is hypoventilated and the mode of ventilation should be changed to Control Mechanical Ventilation, or despite successive increases in assistance, the ventilation remains incorrect).

At a more detailed level, the task of determining the adequate therapy after observation is divided into two ordered main activities: diagnosis and therapy. Diagnosis is further divided into two steps. First, CEC has to classify the patient ventilation according to the observed values of the clinical parameters received from the Interface. These values are compared with the threshold values set (as a function of patient's pathology and morphology) at the initialisation of the system, to determine tolerance ranges of variations for physiological parameters, duration of observation before decreasing the assistance, etc. Second, CEC has to assess the evolution of the patient status by considering its history over previous periods of observation and judge patient's response to therapy. Moreover, this second step tests the patient's capacity to breathe without assistance by starting or assessing the progress of a specific strategy (weaning procedure). In the therapeutic activity, CEC chooses the adequate ventilator therapy depending upon the results of the diagnosis activity.

3.2 Using Events and Properties to Represent Ventilation Management Expertise

In this section, we present how the considered domain, especially the temporal part of the expertise, was modelled in terms of events and properties. Some examples, extracted from the prototype knowledge base, are used to illustrate specific points.

Events.

The sequencing of the three main tasks: observation, diagnosis and therapy is represented using three types of events that control the interactions between CEC and the Interface. The first type of event models the reception of new values for the observed clinical parameters and has the form `receive(VT,PCO2,RR)`, where VT, PCO2 and RR are numerical values for the total inspired volume (in ml), expiratory partial pressure of CO_2 (in mmHg) and respiratory rate (in cycles/min), respectively. The second and the third type of event model requests for the assessment of the evolution of the patient status (`evolutionRequest` event), and for the determination of therapy (`therapyRequest` event).

The reasoning steps illustrated in the previous section take place as follows. When an observation period ends (*observation task*), the Interface sends to CEC a `receive(VT,PCO2,RR)` event containing the averages of the observed clinical parameters. This results in the derivation of the patient classification (*first step of diagnosis task*). Then, the Interface sends an `evolutionRequest` event (*second step of diagnosis task*), and finally it sends a `therapyRequest` event (*therapy task*). The time of occurrence of each of these three subsequent events is read from a system clock and sent to CEC together with the event (this allows to keep track of the precise times at which each activity started). After the third event is processed, the Interface reads the current suggested therapy from CEC and is ready to send commands to the ventilator and to start a new observation period, or to warn the clinical staff that some special case is taking place.

Property	Description
vt(X)	average value of total inspired volume observations taken over the last observation period
pcO2(X)	average value of expiratory partial pressure of CO_2 observations taken over the last observation period
rr(X)	average value of respiratory rate observations taken over the last observation period
patientStatus(X)	classification of patient's status
evolutionStatus(X)	classification of the evolution of patient status over previous observations
weaningStatus(X)	classification of the progress of the weaning procedure
weaningDuration(X)	planned duration for the weaning procedure
assistanceLevel(X)	prescribed level of mechanical ventilatory assistance
minimumAssistance(X)	minimum allowed level of assistance
dx(X)	planned change in the level of assistance
obsDuration(X)	planned duration of the next observation period
abnormalStateCount(X)	number of subsequent observations periods for which the patient was abnormal
alarmingState(X)	classification of the situation aimed at possibly requiring staff intervention
normalSinceChange(X)	how long the patient has been normal since the last change in assistance
normalDuringWean(X)	how long the patient has been normal during the weaning procedure

Table 2. Listing of the main properties.

Properties.

Table 2 summarizes the properties that have been defined to model the domain. As it can be noticed, all these properties have one parameter (denoted as x in the Table). Moreover, the values each parameter can take are *mutually exclusive*. This allows us to assume that when a property with a certain parameter value starts to hold, the previously holding property with *the same name* but with a different value in the parameter is terminated. Thus, we expressed a termination condition valid for all the properties of this specific domain in a very compact and practical way, by using just one terminates_at clause. This clause states that a generic property Property (characterized by a value value for its parameter) terminates at instant T whenever an event Event, that initiates the same property but with a different parameter value (NewValue), happens at instant T. All the other clauses in the knowledge base will be initiates_at clauses. The weak interpretation (Section 2.1) considers by default all the properties to be concatenable [Sh87]: if whenever they hold over two consecutive intervals (i.e. two consecutive period of observation) they hold over their union.

We will now provide examples of properties characterizing patient classification, evolution of patient status, and therapy.

Depending on the values of the three clinical parameters (VT, PCO2, and RR), the patient's ventilatory state can be classified as normal or abnormal. In particular, we distinguish six different abnormal states: insufficientVentilation, tachypnea, severeTachypnea, hyperventilation, hypoventilation, inexplicableHypoventilation. The patient's ventilatory state is classified by the patientStatus(X) property, where the parameter x can assume the value normal or can be one of the six abnormal states. The regions of parameter values characterizing ventilatory states are not unique, but depend on patient's morphology. As an example, if the patient's weight is greater than or equal to 55 kg, the lower bound (vtmin) for his/her inspired volume in a normal ventilatory state is 300 ml, otherwise 250 ml can be sufficient. Analogously, upper bounds for the expiratory pressure of CO_2 (pCO2max) and the respiratory rate (rrmax) depend on patient's pathology. For example, pCO2max is 55 mmHg or 65 mmHg for BPCO patients, and rrmax is 28 or 32 cycles/min for patients with neurologic disorders. These upper or lower bounds for parameters are set during system initialization by means of initially events. The acquisition of new information about parameter values is modeled by means of the receive event type. For example, the update receive(420,36,17), happening at a given time, reports that VT is 420 ml, PCO2 is 36 mmHg and RR is 17 cycles/min. In response to receive events, a set of initiates_at clauses updates recorded parameter values, while another set handles the patient ventilatory state (the state is normal in this example). A sample clause taken from the former set, and one from the latter are listed in the following:

```
initiates_at(receive(VT,PCO2,RR),vt(VT),T):-
   happens_at(receive(VT,PCO2,RR),T).

initiates_at(receive(VT,PCO2,RR),patientStatus(normal),T):-
   happens_at(receive(VT,PCO2,RR),T),
   holds_at(vtmin(Vm),T), VT>=Vm,
   holds_at(pCO2max(Pm),T), PCO2=<Pm,
   holds_at(rrmax(Rm),T), RR>12, RR<Rm.
```

The first clause states that when a new value VT for the average of the inspired volume parameter is received, property vt(VT) initiates. The terminates_at clause previously discussed and the weak interpretation guarantee that the already holding MVI for vt is clipped only if the value in it is different from VT.

The second clause states that whenever an event of type `receive(VT,PCO2,RR)` happens, and property `vtmin(Vm)` holds at that time (indicating the minimum allowed value Vm for the inspired volume), and the observed value VT is greater or equal to Vm, and property `pCO2max(Pm)` holds (indicating the maximum allowed value Pm for the pressure parameter), and the observed value PCO2 is lower or equal to Pm, and property `rrmax(Rm)` holds (indicating the maximum allowed value Rm for the respiratory rate parameter), and the observed value RR is greater than 12 and lower than Rm, then the property `patientStatus(normal)` starts to hold (if it was not already holding). If a `patientStatus(X)` property with X not equal to `normal` was holding previously, it terminates in the same instant, as an effect of the general `terminates_at` clause previously discussed.

Properties concerning the evolution of the patient are initiated by events of the form `evolutionRequest`. The main property in this group is the `evolutionStatus(ES)` property, which gives an assessment of the evolution of the patient state over previous periods of observation. For example, `evolutionStatus(s0)` indicates that the patient is actually normal and he/she has not been abnormally ventilated (tachypnea or insufficient ventilation) since `evolutionStatus(s0)` was initiated.

An example of a clause describing a change in `evolutionStatus` is the following:

```
initiates_at(evolutionRequest,evolutionStatus(s1),T):-
    happens_at(evolutionRequest,T),
    holds_at(evolutionStatus(s0),T),
    holds_at(patientStatus(PtnSts),T),
    (PtnSts==tachypnea; PtnSts==insufficientVentilation).
```

This clause states that whenever an `evolutionRequest` event happens, and `evolutionStatus(s0)` holds, and `patientStatus(PtnSts)` holds with PtnSts equal to `tachypnea` or `insufficientVentilation`, then the property `evolutionStatus(s1)` is initiated. The property `evolutionStatus(s1)` indicates that the patient is in tachypnea or insufficient ventilation for the first time since the instant when s0 was started.

The `evolutionStatus` property is crucial in the therapy activity, because it is often necessary to consider more than one single observation period to determine the appropriate assistance level. For example, in case of tachypnea, the level of assistance is raised as soon as the tachypneic status is observed. But in order to tolerate short instabilities without modifying the global therapy, the initial level of assistance is restored if in the next two observations the patient's status returns normal. On the contrary, the new level is maintained if another tachypneic status is observed in one of the next two observation periods.

The `assistanceLevel(X)` property indicates which level of assistance the ventilator has to deliver to the patient during the next observation period. As a response to `therapyRequest` events, the previous value of the assistance can be incremented or decremented by a specified step. Consider, for example, the clause:

```
initiates_at(therapyRequest,assistanceLevel(NewX),T):-
    happens_at(therapyRequest,T),
    holds_at(evolutionStatus(s1),T),
    previous(T,evolutionStatus(s0)),
    holds_at(dx(D),T),  holds_at(assistanceLevel(X),T),
    Xn is X+D, min(40,Xn,NewX).
```

This clause states that if the parameter of the `evolutionStatus` property at time T is s1, while it was s0 in the previous MVI (a simple `previous` predicate is used to derive which was the value of a given property in the MVI immediately preceding the

MVI containing a given time instant), and the planned amount (which depends on the level of assistance already delivered and is represented by the dx property) for the next variation (if any) in the level of assistance is D, and the value of the assistance is X, then assistanceLevel(NewX) initiates, with NewX being the minimum between 40 and X+D (a simple min predicate finds the minimum among two given numbers).

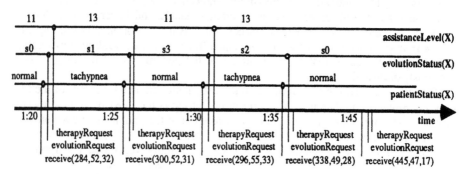

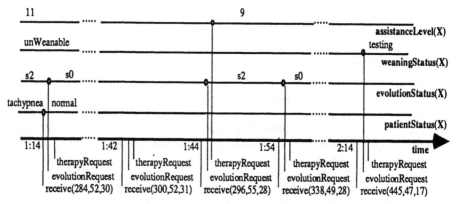

Fig. 2. Two examples of system's operation.

3.3 Examples

This section provides two examples (Figure 2) taken from prototype operation over real clinical situations. For clarity purposes, the figure depicts only the properties in the database which are most relevant to the considered cases. All initiations and terminations of properties are graphically linked with their causing events.

The first part of Figure 2 illustrates a situation where a patient with neurologic disorders (rrmax equals 32) is initially in a normal state, then it becomes tachypneic leading to a temporary increase in assistance. Assistance is later decremented because the patient is again normal, but then it is incremented permanently, because the patient becomes tachypneic a second time. That is, a short ventilation instability does not lead to modification in the global therapy, while a successive inadaptation does.

The second part of Figure 2 takes into account a larger time span (dotted lines highlight periods for which events sent to CEC are not shown, because they did not cause changes in the illustrated properties). In the considered situation, after 30 minutes of normal ventilation the level of assistance is decremented from 11 to 9, and after 30 more minutes, the system starts the weaning procedure. That is, the assistance is gradually decreased in function of the stability of ventilation, and when the patient tolerates a low level of assistance (9 cmH2O), a specific strategy is adopted to test the patient's ability to breath without assistance.

4 Evaluation of Results and Future Work

The application of CEC to the patient monitoring problem presented in this paper has been extensively tested on patient's data reflecting several real clinical situations. In the following, we provide some remarks on the obtained results, and sketch further research activity.

4.1 Respect of Real-Time Constraints

CEC allows to introduce an explicit time and change representation into real-time patient monitoring systems while ensuring computational tractability. The response time of the prototype is within the bounds required by the application and it is equal to few seconds on a portable 80386 computer and fractions of a second on a Sun SparcStation 2. As shown in the paper, the execution time for a CEC update increases polynomially with the number of recorded events. Nevertheless, response within a given specified time can be guaranteed in these applications, by regularly moving the oldest, no more necessary events and MVIs, from the active CEC database to a separate database when new events arrive (forgetting mechanism); the separate database is used for historical purposes, and is queried with CEC.

We pursue our work to integrate the management of all the alarms provided by the ventilator. The occurrence of alarms (apnea, disconnection of the patient, ...) can lead to the modification of the natural sequencing of the reasoning. The second step of the diagnosis phase (evaluation of the evolution of the ventilation) has to be short cut in order to envisage as soon as possible a therapy and to send information to the staff.

4.2 Modeling of temporal concepts and inferences

We modelled the patient's evolution as a state transition problem and change as a discrete process. Events introduce an abrupt change in state. States are bounded by events and we exploited instant-based (events) and interval-based (states) entities. Similar considerations have been adopted in other medical contexts (for example [La93] uses state transition to model hemodynamics evolution over time after cardiac surgery), but instead of adopting custom-tailored temporal-reasoning methods, our approach is based upon the Event Calculus, a sound formalism that can be extended and adapted to the purposes of the application.

An interesting feature, which comes at no additional cost when using CEC, is the ability of not only to monitor and control in real-time the therapy, but also to reason non-monotonically about the data that were recorded in its database. For example, it is possible to change the initial conditions or add events referring to the past, by simply inserting them into the database. CEC will non monotonically revise the contents of the database in response to new information. For example, episodic information obtained from laboratory tests reflecting the quality of blood gas exchange, can lead to a new interpretation of the expired CO_2 pressure and to revise the assumption about the quality of the patient's ventilation and of the adequacy of the past and current therapy.

CEC is also an interesting tool to maintain and query the temporal database dynamically built. Besides existing query capabilities that allow to retrieve important information about the history of a patient's case, the addition of new functions to allow the comparison of histories of different patients can contribute to the improvement of the actual system. We also pursue our effort to integrate CEC into a prototype working at the patient's bedside [Do92] and to test it in a clinical environment.

References

[Br87] Brochard L., Pluskwa F., Lemaire F. *Improved efficacy of spontaneous breathing with inspiratory pressure support*, American Review of Respiratory Disease, vol.136, 411-415, 1987.

[Ch94] Chittaro L., Montanari A., Provetti A. *Skeptical and Credulous Event Calculi for Supporting Modal Queries*, Proc. ECAI '94: 11th European Conference on Artificial Intelligence, Amsterdam, The Netherlands, John Wiley and Sons, 361-365, 1994.

[Ch96] Chittaro L., Montanari A. *Efficient Temporal Reasoning in the Cached Event Calculus*, to appear in Computational Intelligence Journal, 1996.

[Co90] Cohn A.I., Rosenbaum S., Factor M. and Miller P.L., *DYNASCENE: An approach to computer-based intelligent cardiovascular monitoring using sequential clinical "scenes"*, Methods of Information in Medicine, vol. 29, 122-131, 1990.

[De91] Dechter R., Meiri I. and Judea P. *Temporal constraints networks*, Artificial Intelligence, vol. 4, 61-95, 1991.

[Do92] Dojat M., Brochard L., Lemaire F. and Harf A. *A knowledge-based system for assisted ventilation of patients in intensive care*, International Journal of Clinical Monitoring and Computing, vol. 9, 239-250, 1992.

[Do95] Dojat M. and Sayettat C. *A realistic model for temporal reasoning in real-time patient monitoring*, to appear in Applied Artificial Intelligence Journal, 1995.

[Fa80] Fagan L.M. *Representing time dependant relations in a medical setting*, Ph.D. thesis, Stanford University, 1980.

[Ha92] Hayes-Roth B., et al. *Guardian: a prototype intelligent agent for intensive-care monitoring*, Artificial Intelligence in Medicine, vol. 4, 165-185, 1992.

[Ko86] Kowalski R.A. and Sergot M.J. *A logic-based calculus of events*, New generation computing, vol. 4, 67-95, 1986.

[La93] Lau F., and Vincent D. D. *Formalized decision support for cardiovascular intensive care*, Computers and Biomedical Research, vol. 26, 294-309, 1993.

[Ma86] MacIntyre N.R. *Respiratory function during pressure support ventilation*, Chest, vol. 89, 677-683, 1986.

[Ra93] Rasmussen J., *Diagnostic reasoning in action*, IEEE Transactions on Systems, Man and Cybernetics, vol. 23, 981-992, 1993.

[Ru93] Rutledge G.W. Thomsen G.E., Farr B.R., Tovar M.A., et al., *The design and implementation of a ventilator-management advisor*, Artificial Intelligence in Medicine, vol. 5, 67-82, 1993.

[Si88] Sittig D.F. *A computerized patient advice system to direct ventilatory care*, Ph.D. thesis, University of Utah, 1988.

[Sh87] Shoham Y. *Temporal logic in AI: semantical and onthological considerations*, Artificial Intelligence, vol. 33, 89-104, 1987.

[Sh90] Shortliffe, E.H. *Clinical decision support systems in medical informatics*, in Computer applications in health care, eds. E. H. Shortliffe and A. Perrault, Reading: Addison Wesley, 466-501, 1990.

[St92] Stefanelli M., Lanzola G., and Ramoni M. *Knowledge acquisition based on an epistemological model of medical reasoning*, Proc. IEEE-EMBS, vol 3, Paris, 880-882, 1992.

[To91] Tong, D.A. *Weaning patients from mechanical ventilation. A knowledge-based approach* . Computer methods and programs in biomedicine, 35, 2677-278, 1991.

[Uc93] Uckun S. *Intelligent systems in patient monitoring and therapy management*, Technical Report KSL 93-32, Stanford University, 1993.

[Va92] Van Beek P. *Reasoning about qualitative temporal information*, Artificial Intelligence, vol. 58, 297-326, 1992.

[Va91] Van de Graaff W.B., et al. *Pressure support. Changes in ventilatory pattern and components of the work of breathing*, Chest, vol. 100 , 1082-89, 1991.

A General Framework for Building Patient Monitoring Systems

Cristiana Larizza, Gabriele Bernuzzi and Mario Stefanelli

Dipartimento di Informatica e Sistemistica, Università di Pavia (Italy)
e-mail cri@ipvlim1.unipv.it

Abstract. *This paper describes a general framework designed to assist the physician in the management of patients long-term monitoring. It exploits a temporal model based on the temporal primitives time-point and interval and provides powerful mechanisms performing temporal abstractions and temporal reasoning that can be used to assess the patient clinical evolution in various medical domains. The framework is integrated into a clinical workstation providing several tools designed to assist the clinical staff in the management of the patients records and in the definition of the domain-specific knowledge.*

1 Introduction

The diffusion of information technologies, and especially of AI applications, in clinical environments still presents problems. However, it is acknowledged that an appropriate use of these technologies in patient care could be very effective to improve the efficiency and the quality of the care. In particular, the use of a *decision support system* in the management of the patients monitoring, provided that it is *integrated* into a hospital information system, could provide real benefits in clinical practice. However, a knowledge based system must be capable of handling *time-oriented data* in order to be very effective and useful to qualify and use efficiently the clinical information for monitoring purposes.

There have been many attempts to develop systems integrating hetherogeneous tools in an advanced workstation to use in daily medical practice and to propose medical knowledge based systems able to perform temporal reasoning as required by the monitoring task [4], [5], [7], [9]. Nevertheless, until now the efforts in this field did'nt produce systems to be used in routine.

In 1989 we tackled the problem of managing the patients long-term surveillance by developing the system M-HTP (Monitoring system of Heart Transplanted Patients) [6]. M-HTP was purposely designed to assist the infectious diseases specialists in monitoring the onset of infectious complications in heart transplanted patients. Our successive efforts were devoted to the design of a more powerful and domain-independent system supporting the physicians in clinical monitoring. The purpose of the present paper is to describe the framework we developed for building monitoring systems to be exploited in any medical domain requiring the evaluation of the clinical patient evolution through the analysis of longitudinal clinical records. It is composed by a knowledge based system able to *interpret* time-stamped data, stored into a valid-time relational database [10], [11], and to suggest diagnostic hypotheses on the basis of the occurrence of specific patterns detected in the courses of the clinical data. The patterns to be searched in the time series can be defined by the physician through a knowledge acquisition tool.

In this paper first we describe the framework architecture, then we describe the temporal abstraction mechanisms and the inferential mechanisms incorporated into the system. Finally, we present a monitoring application developed with the framework.

2 System architecture

The goal of an intelligent monitoring system is to analyze and interpret clinical data for diagnostic and therapeutic purposes. This paper describes the implementation of a framework designed to develop intelligent monitoring systems in specific medical domains. The principal goals of this project were: (1) to develop a general architecture integrating heterogeneous modules to build effective monitoring applications that could also exploit already existing databases; (2) to adopt methods of detecting efficiently relevant trends from multiple time series; (3) to exploit the trends detected in the time series to support clinical decision making. Figure 1 shows the software architecture. The system was developed with standard environments that could guarantee the portability on most hardware platforms (UNIX systems, DOS PCs, AppleMacintosh) and the extension with external applications.

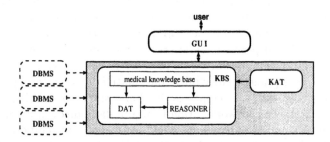

Fig. 1. The framework *architecture*. Its components are: the Knowledge Based System (KBS) and the Knowledge Acquisition Tool (KAT). The KBS contains the medical knowledge base, the Data Abstraction Tool (DAT) and the Reasoner.

The scheme shows also the DBMSs representing the data sources of all the longitudinal data to be interpreted by the system. They must be valid-time databases [3], [10] storing each information with its time-stamp. For the applications developed until now, we adopted the *relational model* and the *client/server* architecture. We developed also a Graphical User Interface (GUI) to accede to all the modules composing the intelligent monitoring system in uniform way.

The framework components are the Knowledge Based System (KBS) and the Knowledge Acquisition Tool (KAT):

The KBS is the module that analyzes the patient clinical parameters courses to intepret their progress and make hypotheses about the evolution of a pathological process in the patient. It is composed by two sub-modules: the Data Abstractions Tool (DAT) and the Reasoner. The most innovative aspect the KBS is represented by the temporal abstraction mechanisms incorporated into the DAT for finding patterns in the time courses of the variables under monitoring. The Reasoner performs the inferential process aimed to provide the physician with diagnostic and therapeutic suggestions based on the interpretation of relevant pathophysiological states detected by the DAT. The KBS was developed using a shell for developing knowledge based applications providing rules and frames as knowledge representation formalisms.

The KAT allows the definition of the *medical knowledge base* according to the data abstractions ontology and the inferential model underlying the KBS. The entities composing the *application knowledge* (*findings* over which the system must perform the reasoning, *abstractions* relevant to the specific task, *diseases* to be diagnosed and monitored and corresponding *therapeutic protocols*) must be defined using the KAT through a suitable graphical interface providing the user with an on-line help.

An example of the GUI will be shown in Figure 3. In the following two sections the mechanisms embodied into the DAT and the Reasoner modules are extensively described.

3 The abstraction mechanisms

To interpret efficiently long time series, a monitoring support system must perform abstractions on the raw data. This task constitutes a fundamental phase of the reasoning arising from the need of generating an effective patient clinical history representation.

Our system was designed to carry out two different types of abstractions. A first type of abstraction is *qualitative abstraction* involving the evaluation of single events at a time. It can be defined as *time-dependent mapping* of quantitative values into symbolic categories representing abstract concepts like low, normal, high [2]. The classification depends on the observation time of the datum [6].

A further very important kind of abstraction is *temporal abstraction*. The *time* is a fundamental component in clinical monitoring and a suitable representation and interpretation of the patient evolution is essential. The framework described here adopts the same model of time based on *time-points* and *intervals* used in M-HTP project [6]. Nevertheless, while M-HTP carried out just one type of temporal abstraction based on the evaluation of the persistency during time of a finding categorical values, the framework described in this paper contains a more general and extended set of temporal abstraction mechanisms.

We define *temporal abstraction* a mechanism allowing extracting patterns of different types from time series by aggregating several events into intervals. The usefulness of this kind of abstractions stands in the capability of representing a large set of raw data with a small number of meaningful abstract entities (*relevant episodes*) sufficient to obtain a complete synthesis of the patient clinical condition. In addition, a patient history can be interpreted more efficiently if it is represented through short sequences of clinical episodes rather than through long sequences of events.

Concerning this basic subject, we made a great deal of efforts to define a temporal abstraction ontology taking into account the abstractions more frequently involved in the physicians reasoning on time-dependent clinical data. Moreover, we pursued the task of defining domain-independent mechanisms in order to exploit them in various monitoring applications.

Our temporal abstraction ontology consists of two main classes of temporal abstractions concerning: (1) the finding of temporal trends in a single time series, called *simple temporal abstractions*, (2) the finding of temporal relationships between intervals, called *complex temporal abstractions*. Simple temporal abstractions detect patterns like `increase` or `stationarety` in a single parameter course, while complex temporal abstractions detect more complicated patterns involving the analysis of one or more time series, like `"increase of the finding V after decrease of the finding W"`. In the following sub-sections the two temporal abstraction mechanisms are described.

Simple temporal abstractions. *Simple abstractions* allow detecting patterns of simple shape from time series. They are similar to temporal abstractions performed in [7]. In Figure 2 examples of the four different types of simple abstractions (`state, stationarety, increase, decrease`) performed by the system on a time series are shown. The points represent observations of a finding stored into the patients database (*events*) and the thick segments correspond to *intervals* aggregating the instantaneous events into episodes. Each one originates from a specific abstraction and is detected throughout a different mechanism.

A `state` abstraction is obtained by aggregating adjacent observations of a finding whose *symbolic* values belong to a pre-defined set. It is described in detail in [6]. Later on a `state` abstraction will be denoted with the expression `state(V,symbolic values list)`.

The remaining three simple abstractions were introduced in the present version of the framework. They involve a process of *trend detection* applied to observations of *numerical findings* and allow extracting patterns, like a monotonic change (increase or decrease) of a finding or a stationary course, that can be completely defined through set of attributes.

The abstractions `increase` and `decrease` are specified in terms of minimal speed of change of the finding (`rate threshold` property), maximum distance (in days) between two observations to be aggregated into an episode (`granularity` property), and minimal duration of an episode to be considered significant (`minimal span` property). These mechanisms can detect different

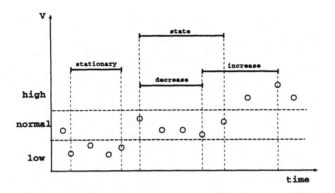

Fig. 2. The *simple temporal abstractions* recognised by the system

trends (**fast** or **low** changes of clinical parameters) by tuning opportunely the values of the properties characterizing the abstraction. In particular appropriate values for granularity and rate threshold properties can allow include also small fluctuations of the course into an episode. The algorithm used to perform these abstractions was designed to extract patterns from the data by disregarding little or local diversions from the trend to detect, in a similar way as a human expert does. The **stationarety** pattern aggregates observations from a time series whose trend is stationary with small fluctuations. The algorithms for detecting increase, decrease and stationarety abstractions are based on linear regression methods.

In Figure 3 a working session of a patient history abstraction phase performed by the DAT module is reported. It shows how the episodes generated by the system are displayed by the system interface (GUI). In the upper part of the figure the window showing the occurrences of all patient episodes detected by the DAT is reported. In the lower right part of the figure the definition window of the **WBC decrease** abstraction is reported. The **WBC decrease** abstraction is defined as a WBC decrease of at least 200 *units/day* (**rate threshold** property). The **minimal duration** property (2 days) establishes the minimal span of an episode to be considered significant. In the lower left part of the figure the patient **WBC** observations are plotted. The dark bands appearing in the plot corresponds to the episodes detected. In particular, the **WBC decrease** episode, spanning the period [100, 112] (days since monitoring onset), aggregates also the datum at day 107 causing a small non-monotonicity into the global trend. This obtains thanks to the value of the **rate threshold** property, set up to 200 *units/day*, and to the mechanism for trends detection employing linear regressions on different length sub-intervals.

Complex temporal abstractions. The search of patterns similar to those described above is an essential task of the monitoring systems, but this only kind of temporal abstraction can become insufficient in a lot of applications. To meet the need, frequently required in clinical monitoring, of investigating also

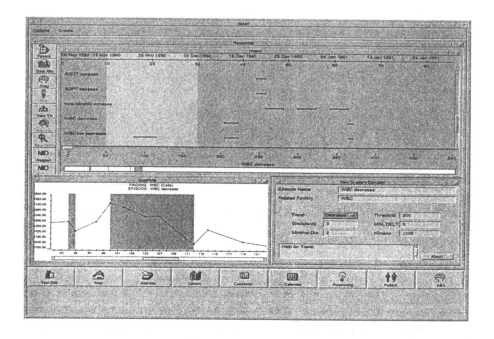

Fig. 3. The graphical interface during browsing of a patient history.

temporal relationships between episodes, we introduced in the temporal abstractions ontology also *complex* abstractions. The task of a *complex abstraction* is just finding out specific relationships between intervals. A first aspect of these abstractions is that they are not built directly on the raw data, but on abstractions (simple or complex), called **sub-episodes**; moreover, the number of sub-episodes is bound to two. However, this is not a limit to the generality of the mechanism since complex abstractions can be defined iteratively on the basis of others previously defined.

The definition of a complex abstraction is given through the name of the two sub-episodes classes (**operands**) and a temporal connective (**operator**). The operator specifies the temporal relation between the two operands to find out. The set of relationships includes the temporal operators derived from the Allen algebra [1] and two further operators, **OR** and **CONTEMPORANEOUS**. The **OR** operator is used to test the presence of at least one instance of the two classes of sub-episodes, while the **CONTEMPORANEOUS** operator is defined to synthetize a whatever overlapping between the instances of the sub-episodes. Each temporal operator is binary and, in general, is not simmetric, to say it is important the order of the two operands.

The representational power of this type of abstraction goes beyond the simple temporal relationship detection and can supply a very flexible mechanism to search compound patterns in a single time series. In fact, since the two complex abstraction sub-episodes can be associated to the same finding, one can search patterns more complicated than those detected with simple abstractions. For example, the complex abstraction `increase(V) MEETS stationarety(V)` detects the compound pattern **"increase with saturation of V"**. As well the more complicated pattern **"increase with saturation at high values of V"** can be detected by defining a complex abstraction as in the following expression: `increase(V) MEETS (stationarety(V) CONTEMPORANEOUS state(V,high))`. in which a simple and a complex abstraction are combined. In Figure 4 some other examples of compound patterns that can be detected by defining complex abstractions are reported.

A difficult problem to tackle in finding temporal relationships between intervals is the management of the inaccuracy on the location or the span of an episode. We don't treat this subject in explicit way, but we supply the opportunity of managing the problem through the **temporal shift** property of complex abstractions. This attribute represents a tolerance value on the location of the intervals that frequently is not known accurately.

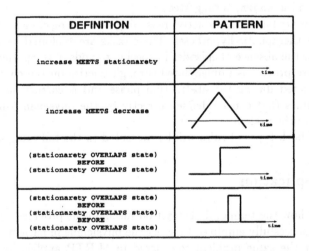

Fig. 4. Example of different levels complex abstractions performed on a single time series.

4 The inferential process

The importance of the temporal abstraction task by means of the various mechanisms described above is determined by two different reasons. First, it is fundamental to obtain a complete and significant description of the patient history. Second, it is an essential step to arrive at performing efficiently the final task of the monitoring, that is the assessment of the patient clinical condition in order to establish if a pathologic process is developing in the patient or a therapeutic action should be recommended. An important aspect of our system is that the efficiency of the inferential module is strictly linked to the quality and the completeness of the patient history representation provided by the DAT. In fact, the reasoning develops taking into consideration the set of episodes detected in the patient history, implying that the more powerful are the mechanisms generating the temporal abstractions, the more effective will be the inferential process. A collateral, but not less important, consequence is also that the knowledge base definition of the inferential module is more smooth and agile if the temporal abstractions represent high-level clinical concepts. A fundamental step in building a monitoring system with our framework is, therefore, the definition of a wide set of appropriate abstractions representing all the clinical episodes important for the specific monitoring task.

The inferential process is performed through two steps. At the first step the system assigns to the pathologies defined in the knowledge base one of the certainty levels: **ruled out**, **unknown**, **suspected** or **definite**. Then, on the basis of the patient pathological status, the timeliness of undertaking a therapeutic action is evaluated and one of the following therapeutic actions, **start**, **stop**, **no change** and **unknown**, is suggested.

Both steps are based on abductive reasoning associating clinical episodes with diagnostic or therapeutic hypotheses. These tasks are exploited by verifying the occurrence or the absence of episodes in specific periods of the patient follow-up. The abductive process is implemented through queries on the collection of the episodes detected during the abstraction phase and is carried out throughout production rules that can be defined directly by the physician using the KAT embodied into the system.

Examples of both types of rules will be presented in the following section.

5 An application

To illustrate how the temporal abstraction mechanisms provided by our framework can be succesfully emploied to assist the physicians in patients monitoring, let's consider the same problem considered in M-HTP application concerning the infectious surveillance of heart transplanted patients. The main goal of the infectious specialist is preventing the onset of dangerous deteriorations in the patient by detecting some critical situations and undertaking a therapy protocol as early as possible. Since the physician should really take advantage of using an integrated framework to highlight the occurrence of significant episodes during the patient follow-up, we employed our framework to develop an infectious monitoring workstation prototype that will be tested in a Department of Infectous Diseases of the University Hospital Policlinico S. Matteo of Pavia (Italy). The workstation is composed by the modules reported in Figure 1. The database is composed by 22 sections (about 400 fields). We calculated that during the first year of monitoring since the transplantation about 400 records, corresponding to about 200.000 bytes per patient, are collected.

The KBS. To describe this module we restrict to the problem of monitoring Cytomegalovirus (CMV), an infectious complication frequently occurring in immunosuppressed patients. In a little fraction of patients the pathologic process remains at a subclinical stage and recovers without treatment. More frequently, the infection manifests itself with various symptoms like fever, tiredness, gastroenteritis, platelets (PLT) count fast decrease, low level white blood cells (WBC) count and can progress to cause very dangerous complications. A recent approach in treatment is *preemptive therapy* using quantitative methods, called CMV viremia and CMV antigenemia, to identify patients with sub-clinical infection. Because low levels of CMV antigenemia and viremia are clinically irrelevant, whereas high levels are associated with clinical symptoms, the monitoring of these quantitative methods can allow the physician assessing the timeliness of the treatment. Nevertheless, the trends of WBC and PLT counts and some symptoms must be monitored in order to make an early diagnosis of the infection, when the quantitative tests are'nt available.

We defined an appropriate medical knowledge base containing the domain entities (findings, abstractions, pathologies, therapies), the rules useful to help the physician to diagnose the infection at the sub-clinical stage and the rules, conforming to the preemptive therapeutic strategy, to advise the treatment.

The findings to be considered in the reasoning are: PLT, WBC, CMV viremia, CMV antigenemia, body temperature and symptoms. For each numerical finding the categorical values needed for the qualitative abstraction were specified. Moreover, for each finding, several temporal abstractions useful to represent all the episodes relevant to take decisions were defined. Important abstractions are for example:

simple temporal abstractions	
PLT fast decrease	decrease(PLT, 15%, 7 days)
normal WBC	state(WBC, normal)
PLT stationarety	stationarety(PLT, 5)
leukopenia	state(WBC, low, very low)
fever	state(body temperature, high, very high)
CMV symptoms	state(symptoms, abdominal pain, nausea, hypoxemia)
CMV antigenemia	state(CMV antigenemia, low, high, very high)
CMV viremia	state(CMV viremia, low, high, very high)
low CMV antigenemia	state(CMV antigenemia, low)
complex temporal abstractions	
CMV alarms	CMV symptoms OR leukopenia OR PLT fast decrease OR fever
sub-clinical CMV infection	low CMV antigenemia CONTEMPORARY normal WBC AND PLT stationarety
CMV infection	CMV viremia OR CMV antigenemia
CMV symptomatic infection	(CMV infection) CONTEMPORARY (CMV symptoms OR PLT fast decrease OR leukopenia)

By looking at the tables above, it is immediately evident what are the advantages from the representational power point of view of being able to define high-level clinical concepts by means of complex episodes. In fact, once the set of abstractions is complete, then also the task of assessing the patient clinical condition is straightforward. This task requires the only association of the combined presence and/or absence of specific episodes with diagnostic or therapeutic clinical decisions. Below two rules defined for this application representing the physicians heuristic knowledge are reported.

Rule 1
```
IF      DURING last 10 days ARE PRESENT
            low CMV antigenemia OF TIME SPAN at least 7 days
                            AND
            leukopenia OF TIME SPAN at least 5 days
        AND
            DURING last 15 days IS NOT PRESENT
                CMV infection OF TIME SPAN at least 1 day
        THEN CMV infection IS highly suspected
```

Rule 2
```
IF      DURING last 10 days IS PRESENT
            low CMV antigenemia OF TIME SPAN at least 1 day
        AND
            DURING last 5 days IS NOT PRESENT
                CMV alarms OF TIME SPAN at least 1 day
        THEN no change the therapy
```

The first rule establishes an high suspect of the infection based on the occurrence of **low CMV antigenemia** and **leukopenia** and on the absence of specific quantitative tests that can confirm definitely the infection. In analogous way, the second one allows representing the medical knowledge concerning the preemptive therapy strategy. These two examples make even more evident the great importance of defining appropriate abstractions by means of which the specialists heuristic knowledge can be represented with a limited number of simple rules.

6   Discussion

The framework presented in this paper represents an effort to provide an effective environment to be used by the clinical staff in a wide class of monitoring applications and to guarantee a sensible improvement of the quality of the care. The integration of a hospital information system with a decision support system able to perform temporal abstractions could, in fact, assist the physicians to *exploit* the whole clinical information collected over long periods without run the risk of missing or ignoring potentially meaningful data. Multiple are the possible uses of this framework.

First to help the clinical staff in focusing the attention on the most important clinical episodes that can be associated to a patient critical condition without consulting a very bulky medical record. From this point of view, we believe that the mechanisms defined in our system to perform temporal abstractions are enough flexible to be used in the majority of long-term monitoring applications in which some specific courses of the clinical parameters assume a great relevance. The capability of making and combining both these kind of abstractions, trend detection and temporal relationships finding, to define high-level clinical concepts, increases greatly the possibility of employment and the representational power of our mechanisms. Moreover, we believe that the temporal abstractions, as they are defined in our sysem, can well represent the temporal knowledge emploied by the human expert in monitoring, so that it can be specified directly on behalf of the physician in easy way.

Other recent approaches to this problem are interesting, but some of them appear less general than our.

Staimann et al. [8] exploits fuzzy sets theory to detect pre-defined trends in a time series. This mechanism used for on-line intelligent monitoring of ICU patients is appealing because it can detect various patterns. However, it has two limits: first, trends like increase and decrease are not detected, second the template defining the trend to be searched must be specified completely in the time-value space.

An other system, described by Wade et al. [11], exploits a set-based method to discover temporal patterns in data. This approach takes into consideration the only detection of temporal relationships among events stored into a database, without processing the data to find temporal trends in a time series.

The approach of Shahar [7] is the more close to our regarding the kind of temporal abstractions performed. However, the methods adopted to implement the abstraction mechanisms are different. In particular, in our system the mechanisms to find increase, decrease and stationary trends are based on linear interpolation of numerical values, while in [7] they are derived from abstracted data through temporal inference tables. This implies that no compound patterns, except NON-MONOTONIC abstractions, can be detected easily.

Besides temporal abstractions, our system has the ambitious goal of assisting the physician in applying diagnostic and therapeutic protocols. In this case the inferential model plays an important role and the capability of representing the medical knowledge in a compact way with short and readable rules, thanks to the combined use of simple and complex abstractions, is an important aspect.

Finally, our framework could be very useful for the physician to *refine* his/her patient management policy or to tackle problems analogous to that described in [11]. The possibility of analyzing past patient histories highlighting the only relevant situations allows, in fact, to investigate and validate new diagnostic or therapeutic strategies as the experience increases.

While the mechanisms performing temporal abstractions were defined with the valid contribution of the physicians, we think that will be still necessary an effort to evaluate the performance of the system in the routine activity. The assessment should be devoted, in particular, to verify whether the system is able to detect all the relevant patterns as the specialist does and to check if the knowledge base definition and update are sufficiently agile for the physician as appears from our early experience. Moreover, we whish go on to make the system more powerful and general by incorporating several techniques for trends detection depending on the kind of time series to be considered and on the kind of knowledge elicited from the physician. A promising approach could regard the use of bayesian methods to improve the trend detection task and also to pursue the more ambitious goal of forecasting the evolution of the patient status on the basis of the time course of clinical parameters and of the prior knowledge about the process under monitoring.

Acknowledgments. This work has been supported by *Progetto Finalizzato: Sistemi Informatici e Calcolo Parallelo* of C.N.R. under grant n. 92.01617.PF 69. The authors thank Dr. Faravelli for his contribution in the development of the system interface and Dr. Grossi providing his clinical experience in the development of the Medical Knowledge Base.

References

1. J. F. Allen. Towards a general theory of action and time. *Artificial Intelligence*, 23:123-154, 1984.
2. W. J. Clancey. Heuristic classification. *Artificial Intelligence*, 27:289–350, 1985.
3. C.S. Jensen et al. (editors). A consensus glossary of temporal database concepts *Report* R-93-2035, Aalborg University, November 1993
4. E.T. Keravnou and J. Washbrook. A temporal reasoning framework used in the diagnosis of skeletal dysplasias *Artificial Intelligence in Medicine*, 2:239–265, 1990
5. C.D. Lane, Joan D. Walton, E.H. Shortliffe. Graphical Access to Medical Expert Systems: Design of an Interface for Physicians. *Methods of Information in Medicine*, 25:143-150, 1986.
6. C. Larizza, A. Moglia, and M. Stefanelli. M-HTP: A system for monitoring heart transplant patients. *Artificial Intelligence in Medicine*, 4:111-126, 1992.
7. Y. Shahar and M. Musen. RESUMÉ: a temporal–abstraction system for patient monitoring *Computer and Biomedical Research* 26, 255 – 273 1993.
8. F. Steimann, K.P. Adlassing. Two–stage interpretation of ICU data based on fuzzy sets. Proceedings of AAAI 94 Spring Symposium, Stanford, 152–156, 1994
9. Paul C. Tang, Jurgen Annevelink, Henri J. Suermondt, Charles Y. Young. Semantic integration of information in a physician's workstation. *International Journal of Bio-Medical Computing*, 35:47-60, 1994.
10. A. Tansel, J. Clifford, S. Gadia, S. Jajodia, A. Segev, R. Snodgrass (eds.). Temporal Databases': theory, design, and implementation Database Systems and Applications Series. Benjamin/Cummings, Redwood City, CA, 1993
11. T.D. Wade, P.J. Byrns, J.F. Steiner, J. Bondy. Finding temporal patterns – A set–based approach. *Artificial Intelligence in Medicine*, 6:263-271, 1994.

Semi-Qualitative Models and Simulation for Biomedical Applications

Pedro Barahona

Universidade Nova de Lisboa, Departamento de Informática
2825 Monte da Caparica, Portugal

Abstract. Simulation of biomedical systems (e.g. drug metabolism) is an important component of decision support in many medical applications. To cope with the uncertainty of the parameters that are used in the models, one may rely on either applying stochastic methods to quantitative simulation or using qualitative simulation instead, where a number of techniques infer as much as possible in the lack of quantitative information. Many cases, however, rather than lack of quantitative information, are cases of incomplete information, in that numerical bounds can be assigned to the unprecisely known parameters of a model. In such cases, constraint solving techniques might be used to cope with such source of uncertainty and a balance must be sought between the expressive power of the models and the constraint solving capabilities that can be effective applied to these models. This paper presents semi-qualitative modeling and simulation, a new approach aimed at reaching such appropriate balance. This new formalism is introduced with a simple example and by means of a more formal presentation, and its application to a multicompartmental model for drug metabolism is discussed.

1. Introduction

The simulation of biomedical systems is an important component of decision support in many medical applications, namely the simulation of drug metabolism to assess the quality of a therapy plan, the risks involved and possible outcomes. Notwithstanding more unconventional approaches (e.g. Petri nets [1]) the most common approach to simulation consists of modeling a physical system by means of differential equations and solving them. This has been done extensively for drug metabolism, where multi-compartmental models are often used [2].

In practice, the parameters of the equations are only known approximately and this poses a number of problems. When differential equations are solved analytically, the influence on the solutions from the variation of the parameter values can be assessed and possible behaviours of the system can be predicted according to the bounds imposed on these parameters. More often the differential equations describing the physical system can only be solved numerically, which requires exact values of the equations parameters to be known. To cope with the uncertainty of these parameters one has to rely on Monte Carlo simulation techniques that use samples from the (infinite) set of possible values of the parameters. Nevertheless the cost of Monte Carlo simulation, as well as the risk of missing important behaviours, increases rapidly with the dimension of the parameters space. Other alternatives might be considered, namely to transform the original equations into more convenient models, as belief networks [3], but this seems to be problem dependent and hard to generalise.

A different alternative has been followed in qualitative approaches such as QSIM and QPT [4, 5]. Here, the system is modeled rather "coarsely" (e.g. numerical values and derivatives are only considered qualitatively as being positive, negative or null), and special techniques are used to predict behaviour on the basis of these coarse models. Despite its use in some medical applications and tools [6, 7], the lack of quantitative information in these approaches imposes limitations to their applicability, in that an explosive number of possible behaviours can be generated.

Hence, extensions were proposed to introduce, if only partially, quantitative information in order to increase the modeling power of these formalisms. These include the introduction of fuzzy linguistic qualifications for the variables of the qualitative models (as done in a revised proposal of QPT [8]) and refining the qualitative models by assuming numerical ranges within which qualitative variables can take values (as in QSIM+Q2 [9] and its later extension Q3 [10]).

This semi-quantitative approach imposes constraints on the values of the variables and hence solving sets of constraints. This is in general quite hard and dependent on the type of constraints used. For example in Q2, arbitrary functions are accepted but users must supply envelopes to constrain the upper and lower bounds of these functions. As such a trade-off must be considered between the expressive power of the model and the constraining capabilities that can be extracted from them.

The new semi-qualitative formalism presented in this paper aims at an adequate balance in this trade-off, namely by only requiring linear constraints on real numbers, that can be efficiently handled. The modeling is similar to that in [11], but the modeling constructs are formalised and the handling of uncertainty is introduced.

The paper is organised as follows. In section 2, the semi-qualitative approach is informally introduced by means of an example. Sections 3 and 4 include a more formal presentation of the semi-qualitative approach, namely the definition of semi-qualitative models and simulation. Section 5, shows the application of the semi-qualitative approach to a multi-compartment model of drug metabolism. Section 6 summarises the main conclusions.

2. An Introductory Example

Consider an organ A that produces substance S and delivers it into the blood stream, from where S is consumed by organ B. Given a number of regulatory mechanisms that control the production of S, A is not able to yield S if the concentration of S in the blood stream is too high. On the other hand, the rate at which S is consumed by B also depends on its concentration: if the concentration of B rises above a certain threshold, then B is consumed at a maximum rate, which decreases for lower concentrations of B. In a quantitative approach this can be modeled through differential equations (1) and (2) below (a, m, c and d are all positive constants)

$$(1) \quad \frac{dS}{dt} = a \ (m-S) \qquad\qquad (2) \quad \frac{dS}{dt} = -\frac{c \ S}{d+S}$$

The behaviour of this system can be intuitively predicted. Starting with S=0, equation (1) and (2) impose a positive and negative rate on S, respectively. Because S cannot be negative then the positive rate must initially be greater (in absolute value) then the negative, resulting in an overall positive rate that forces S to increase. After some time, the positive rate (that decreases with S) is equal to the negative rate (which increases in absolute value with S) and the system reaches a steady state.

In a quantitative approach, the behaviour of the system can be predicted by solving the above equations, either analytically or numerically. In this and many other cases, finding an analytical solution is not possible (or at least easy). The behaviour of the system must be obtained by numerical simulation, requiring that parameters a, m, c and d are known exactly. Otherwise, samples from the parameters can be taken as done in Monte Carlo simulation. However, the cost of Monte Carlo simulation and the risk of missing important behaviours, increases rapidly with the dimension of the parameters space.

In this particular case, it is possible to find analytically S_f, the final value of S, by equating the rates given by equations (1) and (2). Its value depends of course on the value of parameters a, m, c and d. If no exact values of these are known, then neither can be computed an exact value for S_f, but if bounds for the parameters are known, then bounds can be computed for S_f. As to the time when S reaches its final value (or rather a percentage, say 99%, given its asymptotic nature), the computation of bounds for it requires the use of Monte Carlo (or similar) simulation techniques.

Alternatively, qualitative simulation (e.g. as provided by QSIM) implements the intuitive reasoning above and predicts that S increases and eventually reaches a final and steady value. The fact that a single qualitative behaviour is obtained, regardless of the actual value of the parameters of the differential equations is clearly one important advantage of this approach. Instead of verifying, a posteriori, that all the quantitative simulations lead to the same qualitative behaviour, QSIM predicts, in one simulation alone, that there is only one qualitatively distinct behaviour of the physical system.

However, no quantitative information is provided to bound the final value of S, nor the time it takes to reach it, and this can be quite important. For example, the concentration of S in the blood stream above some level might be unacceptable.

This drawback motivated the development of quantitative extensions to qualitative simulation, such as QSIM+Q2 [9] and Q3 [10], that are able to deal with incomplete quantitative information. They rely on the ability to express constraints over the variables (namely bounds on their values) and their relationships, and use techniques (e.g. constraint propagation) for solving such constraints. However, arbitrary relationships lead to inefficient constraint solving, and a balance must be found between expressive and solving power.

The semi-qualitative approach presented in this paper aims at achieving this adequate balance. The example above will be used to illustrate the approach. Informally, a physical system is modeled by a set of variables, and a set of processes that specify the changes to which the variables are subject to. These changes are not merely qualitative values (positive, negative or null) as in qualitative approaches, nor involve them arbitrarily complex calculations on real numbers (as in differential calculus). Instead, they will be allowed to take a finite set of real values, as can be seen in the following examples. The production of substance S produced by organ A is modeled by process production.

```
process production (in: S, out: S)
      M   < S          ->    rate(S)  =   0
    M/2 < S < M        ->    rate(S)  =   3
        S < M/2        ->    rate(S)  =   5
    where 9 < M < 11
```

In the differential equation the rate of production of S is proportional to the difference between the value of S and the value of parameter m. In the semi-qualitative model above, the different rates of S are obtained by linear interpolation of this quantitative rate. If the concentration of S exceeds a certain value M, than its rate becomes null. Otherwise two interpolation regions (or *qualitative states* of process production) were considered, one above and the other below M/2. For each of these regions a constant contribution for the rate of S is considered, that should lie between the upper and lower ranges defined by equation (1). For the sake of this example, these rates will be equated to 3 and 5, respectively.

Notice that the rate of production of substance S is not specified in the *threshold points*, that delimit the qualitative states of the process, namely when S = M/2 and

S = M. This is an important point, discussed below, but for the time being, one may assume that in these threshold points a process produces a rate that lies somewhere between the values in the neighbouring qualitative states (e.g. when S = M/2 the rate produced by the process is somewhere between 3 and 5).

Of course, this semi-qualitative modeling can be made "more quantitative" and accurate if more interpolation regions are considered (i.e. more than 2 regions between S=0 and S=M). In particular, this would achieve the improved results that step-size refinement introduced in Q3 achieves over Q2 [10].

The consumption of substance S by organ B can be similarly modeled, by means of the process below (it is assumed that the Michaelis-Menten equation (2) imposes a maximum value of the consumption rate when S >> d)

```
process consumption (in: S, out: S)
          C < S          ->    rate(S) = -6
    C/2 < S < C          ->    rate(S) = -4
      0 < S < C/2        ->    rate(S) = -2
          S < 0          ->    rate(S) =  0
    where 6 <  C < 10
```

The semi-qualitative simulation of this model can be briefly presented. At S=0, process production imposes a rate of +5, whereas the consumption imposes a rate somewhere between 0 and -2. Regardless of the exact value of the consumption rate, the balance is positive and hence an episode starts in which S increases at a rate of +3, the sum of the production and consumption rates, respectively +5 and -2.

This episode lasts until either one of the processes, production and consumption, leaves their current qualitative states, i.e. until the value of S violates the constraint that defines that state. That depends on the exact value of C and S, which, given the underlying uncertainty are only known approximately (C in]6..10[and M in]9..11[). This leads to two different possibilities: either C < M in which case S reaches C/2 first, or M < C in which case S reaches M/2 first.

Let us assume the first scenario, i.e. C < M. In this case, S reaches C/2 in a time t_1 equal to C/6 which is bound by]1 .. 1.33[. This time point ends episode E_1 and starts episode E_2. In this episode, S increases at rate +1, the result of positive rate +5 and negative rate -4. Hence, episode E_2 lasts until S reaches C or M/2. But, since C is greater than M/2, the latter is reached first, in time t_2 equal to C/6+(M/2-C/2)/1. Taking into account that C<M, the value t_2 is bound by]1.5 .. 3.5[.

At this time point a stable situation is reached. If S were to increase above M/2 than the combination of a positive and negative rate of +3 and -4 would force S to decrease. On the other hand, if S was to decrease below M/2 than the combination of a positive and negative rate of +5 and -4 would force S to increase.

The alternative scenario, when M<C would also have two episodes. In the first, S reaches M/2, in time t_1', equal to M/6. In the next one it would reach C/2 in time t_2'= M/6+(C/2-M/2)/1. Taking into account that M<C, t_2' is bound by]1.5 .. 2[.

Although these scenarios correspond to similar qualitative behaviours, an important difference between the two alternative scenarios is that the rates of S that are imposed by the processes in the last episode. Although the overall rate is null in both cases, in the first scenario this is achieved by a negative rate of -4 (S is greater than C/2) and a positive rate of +4 (S is equal to M/2 and the rate must be somewhere between +3 and +5). In the second scenario the null rate is however achieved by balancing a positive rate of +3 and a negative rate of -3.

When compared to a qualitative approach, the semi-qualitative approach has therefore the advantage of producing quantitative information for the value of the variables, as well as to the timing in which these values are reached. In this case, the scenarios discussed would make S to stabilise at either M/2 or C/2, i.e. somewhere between 4.5 and 5.5 and 4.5 and 5 (remind that is this last scenario M<C), after a time bound by, respectively]1.5 .. 3.5[and]1.5 .. 2[. When compared to the quantitative approach, semi-qualitative modeling has the important advantage of being able to predict the possible scenarios, (two scenarios in the above example) and not to rely on an unpredictable set of simulations (as required by Monte Carlo techniques).

3. Semi-Qualitative Models

This section, presents a somewhat more formal definition of a semi-qualitative model of a physical system (semi-qualitative simulation is presented in the next section). A physical system is modeled by means of a set of variables and a set of processes. Variables range over the reals, and denote entities whose value varies continuously throughout time. Their changes are due to processes of which they are output variables. These changes are specified by means of rates, which are kept constant within a set of (an usually small number of) qualitative states of the input variables of the processes (in a way the semi-qualitative approach degenerates into a quantitative one if an arbitrarily large number of qualitative states are defined for each process).

Qualitative states correspond to ranges on the value of variables. These are open intervals, ending on threshold points, that together completely cover the possible values of the variables. Changes are not defined at threshold points, but must lie somewhere between those imposed in the adjacent ranges.

For example, process production above is defined for 3 qualitative states of the input variable S, respectively]-∞..M/2[,]M/2..M[and]M..+∞[, respectively imposing a rate of 5, 3 and 0 on the same variable S. In threshold points M/2 and M, the rates imposed on S should lie on intervals [3..5] and [0..3].

Qualitative states of a process can be expressed as a conjunction of terms of the form Lower<V_i<Upper, where Vi are input variables and Lower and Upper denote the threshold points delimiting the qualitative states of the variables (in fact, one might allow conjunctions of linear combination of variables of the form Lower < $\sum a_i V_i$ < Upper, but these are not used in the paper). Moreover than one output variable might be defined for each process. This is exemplified in process transformation shown below (later used in section 5) to specify that substance A is metabolised into B.

```
process transformation2 (in: A, B, out: A, B)
        B > B0                    ->   rate(A) =  0, rate(B) = 0
        B < B0, A1 < A            ->   rate(A) = -2, rate(B) = 3
        B < B0,  0 < A < A1       ->   rate(A) = -1, rate(B) = 1.5
        B < B0,      A < 0        ->   rate(A) =  0, rate(B) = 0
    where ... < A1 < ...,    ... < B0 < ...
```

For large values of A this transformation is faster (i.e. A is consumed at rate -2 and B produced at a rate +3), then for lower values of A (where A is consumed at rate -1 and B produced at a rate +1.5). Of course, no changes happen when A <0. In a way this process can be considered to model the differential equations below (it could model them more closely if more qualitative states where used for A).

$$\frac{dA}{dt} = - k_1 A \qquad\qquad \frac{dB}{dt} = k_2 A$$

There are however two important differences in the process specification that take into account the usual saturation phenomena, which are not considered in the equations. When A becomes larger the rate at which it can be consumed cannot exceed a certain value, and this is specified by means of range $]A_1..+\infty[$. On the other hand, B will not be produced if its value is already too high, and expressed by null rates when $B_0 < B$.

3.1 Modeling of Uncertainty

In the above processes, and to denote variations in different instances of a physical system, uncertainty may appear in two different ways. On the one hand, threshold points may vary, usually within certain bounding intervals. On the other hand, rates imposed on qualitative states could also vary from individual to individual.

In principle, both these sources of uncertainty should be modeled. Nevertheless, it is arguable that this would imply a certain degree of redundancy, and therefore it should be sufficient to model only one of these sources. To illustrate this argument, consider the situation where r_b the rate of B is proportional to A (i.e. $r_b = k\,A$), but this constant factor k is not known exactly. For example, let us assume that k is close to 2, and we need to define a value of r_b when A lies in the range $]4..5[$. If k=2, we could consider a "central" value for r_b, say 9, and specify

```
....  4 < A < 5       ->   rate(B) = 9 .....
```

Now, let us assume further that, although not knowing k exactly, it is known that its value lies between 1.8 and 2.3. In this case, instead of specifying that the rate of B should change to a value between $9*1.8/2 = 8.1$ and $9*2.3/2 = 10.35$, one might use the bounds of k to change the threshold points that delimit the qualitative state. Since the threshold points corresponded to values of A where r_b would be 8 and 10 if k were equal to 2 now, given the uncertainty on the value of k, all that can be said is that the threshold points should lie, respectively, in the ranges $]8/2.3 .. 8/1.8[$ and $]10/2.3 .. 10/1.8[$. This leads to changing the above specification into

```
    ....  M < A < N      ->   rate(B) = 9 .....
where M in [3.478 .. 4.444], N in [4.347 .. 5.556]
```

This justifies that, to specify the underlying uncertainty on the semi-qualitative models threshold points may vary, but rates can be assumed as constant.

4. Semi-Qualitative Simulation

A semi-qualitative simulation corresponds to determine, from an initial state of the variables, a sequence of episodes, $E_1, E_2, ...E_n$. An episode corresponds to a time interval where the same set of conditions of the processes are satisfied and the rates of all variables are constant. The simulation produces the time evolution of the variables V_i of the system, starting with value V_i^0 at time T_0. Given the constraints imposed on the specifications of processes, the rates of each variable are obtained from the sum of the rates that each of the process individually contributes to that variable. Therefore, the value of a variable V_i along the time is given by

$$V_i(T) = V_i^0 + r_{i1}*(T_1 - T_0) + r_{i2}*(T_2 - T_1) + ... + r_{in}*(T_n - T_{n-1}) + r_{in+1}*(T - T_n)$$

where r_{ij} is the rate of variable V_i throughout episode E_j. The simulation of a system composed by a set of variables and processes thus involves the determination of a) the rate of each variable for every episode E_j; and b) the duration of each episode E_j (i.e. the value of $(T_j - T_{j-1})$).

4.1 Episodes Duration

As observed before, every condition for every process is a conjunction of relations of the form Lower $< V_i <$ Upper. Since the conditions in a process are mutually exclusive, an episode is defined by a conjunction of conditions of the form above, collected from all the processes. Since the rate of all variables is constant throughout an whole episode, the duration of the episode is determined by the minimum time in which one of the conditions satisfied throughout the episode is violated.

Denoting by r_{in} the (constant) rate of variable V_i during episode E_n, an episode ends when one of the conditions Lower $< V_i <$ Upper is violated which occurs, of course, when either $V_i =$ Lower or $V_i =$ Upper. Since rates r_{in} are known, one may determine whether V_i is increasing or decreasing and only consider the appropriate equality. Moreover, the value of variable V_i in episode E_n is given by

$$V_i(t) = V_i^{n-1} + r_{in}*(t-T_{n-1}),$$

where V_i^{n-1} is the value of the variable at T_{n-1}, the beginning of episode E_n. Hence the violation of a boundary condition occurs at time t such that

$$a_i*V_i^{n-1} + a_i*r_{in}*(t-T_{n-1}) = K$$

Considering a different variable t_j for each of the conditions that define the episode, then T_n, the end of episode E_n, will be the lowest of these t_j. The simulation may then proceed by computing, for each variable V_i, its value at $t=T_n$ and iterating the above procedure from this time point.

4.2 Uncertainty and Alternative Scenarios

If the upper and lower bounds of the conditions in the processes are constant, then the determination of the sequence T_1, T_2, ..., T_n is trivial. Nevertheless, uncertainty on a semi-qualitative model enables these bounds to range over intervals. Nevertheless, the linearity of the constraints still allows the simulation to be performed with a controlled, if increased, complexity. In fact, the set of constraints shown above for the case where K was a constant, is now replaced by a set of constraints of the form

$$a_i*V_{in-1} + a_i*r_{in}*(t_j-T_{n-1}) = K \quad \text{and} \quad K < k_{up} \quad \text{and} \quad K > k_{lo}$$

where k_{up} and k_{lo} are the (constant) bounds of K. Since a_i and r_{in} are constants, all these constraints on the t_j are linear. Moreover, the determination on which of the t_j is the lowest, must be replaced by the imposition of constraints that enforce that one particular t_j is indeed the lowest. For example, if an episode E_n is defined by conditions on t_1, t_2, and t_3, then one might impose the additional constraints $t_1 \leq t_2$ and $t_1 \leq t_3$ to enforce that the episode under consideration ends at t_1 and make $T_n = t_1$. Given constraints on t_1, .., t_m, the number of possible alternatives is thus equal to m.

The complexity of this simulation has therefore to consider two aspects: a) solving of a set of linear constraints, and b) exploiting a potentially exponential number of alternatives. The solution of a set of constraints is in general too complex to be of any practical use. However, inn some cases there are efficient algorithms that verify whether a set of constraints is satisfiable, namely linear constraints, and languages such as CLP(R) [12] include these efficient constraint solving techniques.

Assuming that each of the alternatives can be efficiently handled, the exponential number of alternatives raises the problem that already plagued qualitative simulation: the explosion of possible scenarios. When compared with the qualitative approach, the semi-qualitative may take advantage on the quantitative information to further reduce the number of alternatives. Moreover, if two episodes from different scenarios start with the variables in the same state than they can be safely merged (which is not the case with qualitative simulation).

4.3 Determination of the Rates

When all the processes have their input variables within a qualitative state (i.e. different from threshold values), the rates applicable to the variables are simply the sum of the rates imposed by each of the processes. The case where some of the variables are in threshold points is more complex, since the rates are not defined in these points. This section only addresses the case of one single variable in a threshold point, those with two or more variables are simply exemplified in section 5.

When in both neighbouring qualitative states of a threshold point, say V=K the rate of variable V is of the same sign, positive or negative, a new episode starts in the qualitative state with V>K or V>K, respectively. To explain the case where the sign of the rate in the neighbouring qualitative states are different, let us consider the following process, with initial values A=0, and B=0, and where A_0 is positive.

```
process P (in: A, out: A, B)
   A < A0    ->      rate(A) =  3, rate(B) = 1
   A > A0    ->      rate(A) = -2, rate(B) = 5
```

Initially, process P clearly imposes rate(A) = 3, rate(B) = 1, corresponding to the state of the process where A < A_0. Once the threshold point A = A_0 is reached there are no explicitly defined values for the rates of A and B.

In this case it is easy to verify that A should remain at A_0: if the value were to increase, than a negative rate (i.e. -2) would bring its value back to A_0; if the value were to decrease, than a positive rate (i.e. +3) would bring its value back to A_0. Hence the only possible value for A is A_0 and the system infers rate(A) = 0.

The situation is however less clear with respect to the rate of B. Again, it seems reasonable to assume that it should be between 1 and 5. Applying an argument similar to that used for A, if A is subject to infinitesimal variations that increase its value by ε, then A is a negative rate of -2 is imposed on A and after time $\tau_1 = \varepsilon/2$, it will return to A=A_0. On the other hand, if the A decrease by ε, the positive rate of +3 makes A to reach A_0 in time $\tau_1 = \varepsilon/3$.

Assuming random perturbations, then the ratio between the time where A>A_0 and A<A_0 is τ_1/τ_2 and the rate of B should be the average of its rates for A>A_0 and A<A_0 weighted by the time that they remain there, respectively $\tau_1/(\tau_1+\tau_2)$ and $\tau_1/(\tau_1+\tau_2)$. In a system with n variables, $V_1...V_n$, in threshold point $V_k = C$ then, denoting the rates of variable V_k when $V_k > C$, $V_k < C$ and $V_k = C$ by r_i^+, r_i^- and $r_i^=$, respectively, it is

$$r_i^= = \frac{r_i^+ * |r_k^-| + r_i^- * |r_k^+|}{|r_k^-| + |r_k^+|}$$

In the example above the rate of B is 3.4 = (1*2+5*3) / (2+3). Of course, when i=k, the above formula imposes $r_k^=$ = 0 (i.e. rate (A) = 0 and A=A_0).

5. Simulation of a Multi-compartment Model

In this section the semi-qualitative approach is applied to a multi-compartment model for insulin metabolism [3], with two compartments specified by equations

$$\frac{dA}{dt} = k_1 A + k_2 B \qquad\qquad \frac{dB}{dt} = k_3 A + k_4 B$$

The corresponding semi-qualitative model, shown in Figure 1, include two processes (p_2 and p_4) similar to process transformation shown in section 3, and takes into consideration the modeling of saturation, as explained in that section. Additionally, the semi-qualitative model includes two other transformation processes (p_1 and p_3) that specify the metabolisation of substance A into C and substance B into D. The complete set of processes is shown in Figure 1.

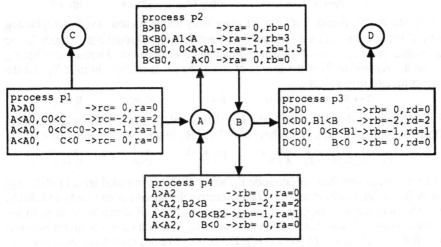

Figure 1- A multi-compartment model

5.1 Complete Saturation

Assuming that, initially, the amount of C is very large (i.e. C >> C_0) and the concentration of A, B and D are all null, then at time T_0, which starts episode E_1, all processes but p_1 are in threshold points (A=0, B=0). In this case it is clear that a positive rate for A must be considered since the rate at which it is produced in process p_1 is greater then any negative contribution that process p_2 might impose on it. As to B, it is r(B)=-0.5 for B>0 and r(B)=1 for B<0 (or rather B=0, as B cannot be negative). Applying equation (3) from the previous section, one finds, for the first episode

$$\text{Episode 1:} \quad r(A) = 1.75 \quad r(B) = 0 \quad r(C) = -2 \quad r(D) = 0.75$$

E_1 ends when either A reaches one of {A_0, A_1, A_2} (in which case a threshold point of processes of processes p_1, p_2 or p_4 is reached), or C reaches C_0 (a threshold in p_1), or D reaches D_0 (in p_3). Different scenarios can thus be considered, depending on the value of these parameters. Assuming that A_0 is reached first, then episode E_1 ends at time T_1 when the concentration of the substances are the following

$$A_0/1.75 = T_1 \quad A = A_0 \quad B = 0 \quad C >> C_0 \quad D = 0.75\, A_0/1.75$$

At this time point, there are two variables lying in threshold points, A=A_0 and B=0, and the system is in the transition between 2^2 =4 qualitative states. In this case, one must rely on the study of the effect that infinitesimal variations of A and B around these threshold points might have, generalising the case of one single variable discussed in section 4.3. This is outside the scope of this paper, but the end result shows that both A and B remain steady, while C decreases (albeit at a lower rate since A is above A_0 for some periods) and D increases with rates

$$\text{Episode 2:} \quad r(A) = 0 \quad r(B) = 0 \quad r(C) = -0.34 \quad r(D) = 0.74$$

Episode E_2 lasts until either C reaches C_0, or D reaches D_0. Assuming the latter is reached first, at time T_2 it is

$$T_1+(D_0-0.75A_0/1.75)/0.74 = T_2 \qquad A = A_0 \qquad B = 0 \qquad C>>C_0 \qquad D = D_0$$

T_2 begins episode E_3, where three variables (A, B and D) lie in threshold points. This makes it necessary to check $2^3 = 8$ neighbouring qualitative states, to assess the effect of infinitesimal variations of A, B and D around the threshold points. This leads to

$$\text{Episode 3:} \quad r(A) = 0 \qquad r(B) = 0.5 \qquad r(C) = 0 \qquad r(D) = 0$$

which means that A and D are always marginally above A_0 and D_0 (thus explaining why C is not consumed nor is D produced any longer). Throughout episode E_3 the concentration of A, C and D remain constant and B increases until it reaches either B_0, B_1 or B_2. Assuming B_1 is the lowest of them, then it is reached in time T_3 and the concentrations have values

$$T_2 + B_1/0.5 = T3 \qquad A = A_0 \qquad B = B_1 \qquad C>>C_0 \qquad D = D_0$$

Episode E_4 starts then and similar analysis shows that the rates applicable are equal to those in episode E_3, until B reaches either B_0 or B_2, in time T_4. Assuming $B_0 < B_2$,

$$\text{Episode 4:} \quad r(A) = 0 \qquad r(B) = 0.5 \qquad r(C) = 0 \qquad r(D) = 0$$
$$T_3 + (-B_0-B_1)/0.5 = T_4 \quad A = A_0 \qquad B = B_0 \qquad C>>C_0 \qquad D = D_0$$

The simulation will then start episode E_5, where the rates imposed are all null except for A (in fact processes p_1 and p_3 become inactive due to saturation ($A>A_0$ and $D>D_0$) and the only activity happens in processes p_2 and p_4; all B produced in p_2 in the periods where B is less than B_0 is transformed into A, which cannot be all transformed into B due to saturation of process p_2). Hence, A increases until it reaches either A_1 or A_2. Assuming $A_2 < A_1$, then

$$\text{Episode 5:} \quad r(A) = 0.33 \qquad r(B) = 0 \qquad r(C) = 0 \qquad r(D) = 0$$
$$T_4 + (A_2-A_0)/0.33 = T5 \quad A = A_2 \qquad B = B_0 \qquad C>>C_0 \qquad D = D_0$$

Here, all the activity comes to an halt due to saturation of the processes (activity would resume if this saturation could be ended, say by a process decreasing the concentration of D). Figure 2 shows this behaviour of the system that occurs when the following constraints are imposed

$$A_0 < A_2 < A_1 \qquad A_0/1.75 < D_0/1.75 \qquad B_0 < B_2 < B_1 \qquad C_{init} >> C_0$$

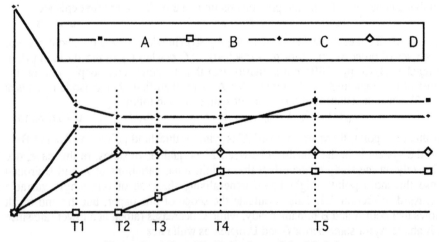

Figure 2 - A Possible Behaviour of the Model

5.2 Complete Metabolisation

In a way, the stability of the system was caused by the fact that D could not be produced any longer. In the alternative scenario of this section C eventually becomes null , as shown in figure 3, and the full metabolisation of A and B takes place.

In the first three episodes, the concentrations of A and B increase and reach stable values, C decreases at the maximum rate of -2. When $A = A_0$ and $B = B_0$, A and B stabilise due to saturation and this decreases the rate at which C is consumed to -0.61. Eventually C reaches C_0, but the situation remains similar (in fact the consumption rate of C decreases to -0.49) until C reaches 0. At this stage, A can no longer be maintained at A_0 and the system decreases A first, then B and finally A again until both A, B and C reach 0. Throughout these episodes D (represented in a different scale in the figure) increases at its maximum rate of 1, except in the first episode where the rate is slightly less, and in the episode starting in T_7. In this latter, the decrease of A while B is already null is insufficient to keep D increasing at its maximum rate. Finally, when A becomes null, no more changes are produced.

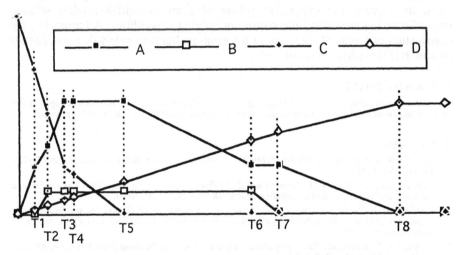

Figure 3 - An Alternative Behaviour of the Model

The set of episodes generated and the constraints considered are shown below

Episode 1:	$r(A) = 1.75$	$r(B) = 0$	$r(C) = -2$	$r(D) = 0.75$
Episode 2:	$r(A) = 1$	$r(B) = 1$	$r(C) = -2$	$r(D) = 1$
Episode 3:	$r(A) = 1.67$	$r(B) = 0$	$r(C) = -2$	$r(D) = 1$
Episode 4:	$r(A) = 0$	$r(B) = 0$	$r(C) = -0.61$	$r(D) = 1$
Episode 5:	$r(A) = 0$	$r(B) = 0$	$r(C) = -0.49$	$r(D) = 1$
Episode 6:	$r(A) = -0.33$	$r(B) = 0$	$r(C) = 0$	$r(D) = 1$
Episode 7:	$r(A) = 0$	$r(B) = -0.5$	$r(C) = 0$	$r(D) = 1$
Episode 8:	$r(A) = -0.25$	$r(B) = 0$	$r(C) = 0$	$r(D) = 0.75$
Episode 9:	$r(A) = 0$	$r(B) = 0$	$r(C) = 0$	$r(D) = 0$

$A_1 < A_0 < A_2$ $B_0 < B_2 < B_1$ $D_0 \gg$ $A_1/1.75 < (C_{init}-C_0)/2$ $B_0 < (A_0-A_1)/1.67$

6. Conclusions and Further Work

This paper presented semi-qualitative models and simulation, a new formalism to predict the behaviour of physical systems that lies between the pure quantitative and qualitative approaches. It extends previous work, presented in [11], by allowing uncertainty to be specified in threshold points of process specificatons.

The objective of introducing this formalism is common to a number of similar proposals, such as QSIM+Q2 and Q3, that take advantage of quantitative, if incomplete information, to control the usual explosion of behaviours generated by qualitative simulation. All these proposals rely on constraint solving to handle numerically bound values that model the underlying uncertainty. The main difference between the above mentioned approaches and that presented in this paper is that this latter imposes some restrictions on the expressive power of the models, namely by imposing constant rates (albeit in intervals whose bounds are not fixed), in order to take advantage from efficient constraint solvers. In particular, semi-qualitative simulation only requires linear constraints over the reals, for which efficient constraint solvers exist in programming languages such as CLP(R).

Despite the limitations of the current modeling, the examples discussed show the ability of this approach to model a number of physical systems with relevance in the medical domain (e.g. compartmental models often used in the biomedical modeling) and justifies the current effort to develop an efficient tool for semi-qualitative simulation. A number of improvements on the work described are under consideration, namely the inclusion of weaker restrictions on the specification of rates, to improve the expressive power of semi-qualitative models without jeopardising (significantly) the constraint solving capabilities. Additionally, the introduction of temporal operators in the semi-qualitative models that was already present in [13] is currently under study.

Acknowledgments

This research was carried out in the Artificial Intelligence Centre of UNINOVA, and was supported by the Commission of the European Communities (Project Dilemma A2005 from the AIM Programme).

References

1. R. Hofestadt, *Petri Nets to Model Metabolic Processes*, in Proceedings of MIE'94, pp.115-120, Lisbon, May 1994.

2. E. Ackerman, L. Gatewood, J. Rosevear and G. Mollnar, *Blood Glucose Regulation and Diabetes*, in Concepts and Models in Biomathematics, Marcel Dekker, New York-Basel, 1969.

3. U.G Oppel, A. Hierle, L. Janke and W. Moser, *Transformation of Compartmental Models into Sequences of Causal Probabilistic Networks*, in Proceedings of AIME'93, IOS Press, pp. 75-83, Munich, October 1993.

4. B. Kuipers, *Commonsense Reasoning about Causality: Deriving Behavior from Structure*, Artificial Intelligence 24, pp.169-203, 1984.

5. K.D. Forbus, *Qualitative Process Theory*, Artificial Intelligence 24, pp. 85-168, 1984.

6. B. Kuipers and J. Kassirer, *Causal Reasoning in Medicine: Analysis of a Protocol*, Cognitive Science 8, pp. 363-385, 1984.

7. L. Ironi, A. Catanneo and M. Stefanelli, *A tool for pathophysiological knowledge acquisition*, in Procceedings of AIME'93, IOS Press, pp. 13-31, Munich, October 1993.

8. B. D'Ambrosio, *Qualitative Process Theory Using Linguistic Variables*, Springer-Verlag, 1989.

9. B. Kuipers and D. Berleant, *Using Incomplete Quantitative Knowledge into Qualitative Simulation*, Technical Report AI90-122, Artificial Intelligence Laboratory, Dept. Of Computer Sciences, University of Texas at Austin, 1990.

10. B. Kuipers and D. Berleant, *Combined Qualitative and Numerical Simulation with Q3*, in Recent Advances in Qualitative Physics (B. Faltings, P. Struss eds.), MIT Press, 1992.

11. P. Barahona, *A Causal and Temporal Reasoning Model and its Use in Drug Therapy Applications*, Artifificial Intelligence in Medicine vol. 6, no. 1, pp. 1-28, 1994.

12. J. Jaffar and S. Michaylov, *Methodology and Implementation of a CLP System*, in Proceedings Fourth International Conference of Logic Programming, Melbourne, 1987.

13. F. Azevedo, P. Barahona and F. Ferreira da Silva, *Modeling causal and temporal knowledge to support therapy planning*, , in Proceedings of MIE'94, pp.109-114, Lisbon, May 1994.

Generating explanations of pathophysiological system behaviors from qualitative simulation of compartmental models

Liliana Ironi[1] and Mario Stefanelli[2]

[1] Istituto di Analisi Numerica del C.N.R., Pavia, Italy
[2] Dipartimento di Informatica e Sistemistica dell'Università di Pavia, Pavia, Italy

Abstract. This paper describes a method for generating, in a language that is comprehensible by the user, explanations of the behaviors of a pathophysiological system from compartmental models. Such behaviors are obtained through the qualitative simulation of a model of the pathophysiological system, which has been built within the QCMF (Qualitative Compartmental Modeling Framework) framework and directly coded into the QSIM language. The main advantage offered by this explanation facility over the simulation outcome plots consists in its capability of explaining *how* and *why* the predicted behavior arises from the structure of the modeled system and physical laws. The basic idea underlying the explanation algorithm consists in determining a causal concatenation of the changes from a state to its successor: the algorithm when comparing two successive states will single out those changes that are a direct consequence of the direction of change in the state and will propagate the effects of these changes following the causal order between variables which is implicitly defined by the compartmental structure.

1 Introduction

In this paper we describe a method for generating explanations of the simulated behavior of a compartmental system. We consider explanations as presentations of information that offer a meaningful interpretation of simulation results, which describes, in a language that is comprehensible by the user, *how* and *why* a behavior occurs. Two different aspects must be dealt with when explanations are automatically generated: on one side giving a causal interpretation of the behavior and on the other presenting this interpretation in natural language. Therefore, the core of the paper deals with a method for generating a causal interpretation of behaviors; more precisely, given a succession of states which defines a simulated behavior, the cause-effect links between variables in successive states are identified. One of the basic distinctions between the proposed different approaches to the generation of causal explanations of the system behaviors is whether the relations or constraints used in the model capture the causal dependencies between variables, i.e. they are directional or not. In the former case, a causal account is simpler to be obtained because the values are propagated through the directional constraints supplied by the model builder

[3]. Alternatively, bidirectional constraints require causal ordering techniques [6, 7, 8] because they lack an explicit representation of causality. As far as our explanation algorithm is concerned, it has been designed to generate a causal account of the behavior of a compartmental model, whose structure implicitely defines the causal dependencies between variables.

The work here described (EXPLAIN module) is part of the implemented system QCMF (Qualitative Compartmental Modeling Framework) [5], which aims at integrating tools to reason about a system behavior: model building, model formulation, and results explanations. QCMF includes knowledge of the theory of compartmental systems [1, 2], which is the adopted modeling ontology; whereas the domain ontology, that is the compartmental structure, is entered by the user through an iconic language. QCMF, through an automatic analysis of the given structure and the acquisition of information about functional relationships between variables, generates the behavior model, which is described by a set of Qualitative Differential Equations (QDE) directly coded into the QSIM language [9]. The system behavior can be predicted by simulating the model starting from an initial state which describes the perturbations acting on the system. The QSIM code defining the initial state is automatically built by QCMF as well. Another feature of QCMF is its capability to generate and maintain a library of models which can be efficiently retrieved and used by a MKBS (Medical Knowledge Based System) for a given goal. As regards QCMF's possible use, many issues can be considered. First, it can work as a stand-alone system resulting in a powerful didactic tool for reasoning about the pathophysiological behaviors of a system. Then, QCMF can be fully integrated within larger knowledge-based systems that use different formalisms. In fact, it may generate knowledge sources that can be properly exploited in the deductive inference of medical reasoning in the execution of both diagnostic and therapeutic tasks.

The behaviors predicted by the simulation are a mathematical consequence of the model equations and of the initial state. Therefore, the interpretation of the predicted behaviors requires the knowledge of both the modeling assumptions and the simulation algorithm. If the user looking at the simulation output is not the person who built the model, or if the model is complex and contains hidden assumptions, then it can be difficult for her to make sense of the obtained results. QCMF's explanation facilities are intended to address this problem by relating predicted behaviors to the underlying modeling choices. Therefore, the EXPLAIN module interprets the system's behavior obtained through the simulation of the QSIM model by exploiting the underlying causal direction in the compartmental structure. In the transition from a state (s_i) to its successor (s_{i+1}) one or more variables will change qualitative state. This transition from state s_i to s_{i+1} will be necessarily influenced by the sign of the time derivatives ($qdirs$) in state s_i which describes the evolution occurring in the system at time t_i. This principle is the basis of the algorithm underlying the EXPLAIN module: the algorithm when comparing two successive states will single out those changes that are a direct consequence of the directions of change in s_i and will propagate the effects of these changes through the directional constraints to account for the remaining changes. Applied to a behavior chosen by the user, this procedure is repeated from the initial state to the whole sequence of qualitative states which describe the behavior.

The presentation of explanations is provided at different level of details: first, the relevant events are presented, and then, on the user request, more detailed causal explanations for any change in the system variables are produced. The user is allowed to ask for two different kinds of explanations: (1) *what* happened either to the system or to a variable at a given significant time-point or time-interval, and (2) *why* a behavior of a state variable occurs at a given significant time-point or time-interval. In addition to the request of explanations of a specific behavior of the system, another option offered by the EXPLAIN module allows the user to ask for a description about the structure and behavior models stored in the library.

2 The *explanation* algorithm

The *explanation* algorithm exploits both the methodological knowledge underlying the compartmental modeling technique and that concerning the QSIM algorithm. It has access to all the information about the system model and to the predicted behavior to be explained and generates a causal chain describing the system transition from one state to another. To this end, it exploits (1) the structure model as a causal dependency structure, (2) the behavior model to detect both the causes and the effects of a change, and (3) knowledge of the QSIM algorithm, namely the transition rules, to identify which changes occur from one system state to its successor.

Given a system behavior B_i, which is defined by a sequence $S = \{s_0, s_1, ..., s_n\}$ of qualitative states and its corresponding temporal sequence $T = \{t_0, t_1, ..., t_n\}$, the basic steps of the algorithm are the following:

1. analysis of the structure model for defining the causal ordering between variables. It results in a graph, whose nodes and arcs denote the model variables and their causal links, respectively;
2. construction of the subsets $S^* \subset S$ and its corresponding $T^* \subset T$, respectively called set of *significant events* and *significant instants*, in order to explain the significant events, and, therefore, to avoid the generation of redundant explanations;
3. definition of the causes and effects related to the significant events by exploiting the behavior model and QSIM transition rules.

2.1 Causal ordering in a compartmental model

Causal ordering [6, 8] is a method for inferring a causal dependency among the variables in a set of equations, given only a list of variables that are perturbed and produce a change in other variables. As far as compartmental models are concerned, we do not need to exploit such technique for establishing a causal order between variables, since the equations are built from a structure model where causal directions are explicitly defined. As regards the variables in the equations, the first distinction we can make is between *state variables* (x_i), which represent the concentration of substances in the compartments, and *flow variables* (f_{ij}) which express the transfer rate of substance either from a compartment to another or with the environment. Let us recall that the dynamics of each x_i is described by the mass balance equation:

$$\dot{x}_i = f_{i0} + \sum_{\substack{j=1 \\ j\neq i}}^{n} f_{ij}(x_j) - \sum_{\substack{j=1 \\ j\neq i}}^{n} f_{ji}(x_i) - f_{0i}(x_i) \tag{1}$$

Therefore, between the system variables the following causal dependencies may be stated:

- each change in x_i is caused by a change in $\dot{x}_i$;
- each change in $\dot{x}_i$ is caused by changes in the flows that appear in (1), i.e. either in the leaving (f_{ji}) or entering (f_{ij}) flows associated with the $i-th$ compartment (the compartment 0 denotes the external environment);
- each change in f_{ij} is caused by a change in x_j;
- each change in f_{ij} may also be caused by a change in x_k, where $k \neq j$, when the flow is controlled by the quantity or concentration of substance in compartments other than the source one.

If we denote by X, $\dot{X}$, F the vectors of state variables x_i, of their time derivatives $\dot{x}_i$, and of all the flows f_{ij}, respectively, the cause-effect relationships between variables are summarized by the graph given in Fig. 1:

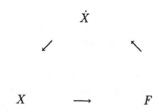

Fig. 1. Possible cause-effect relationships between the variables

Therefore, in our case, the automatic construction of the causal ordering is based on the diagram in Fig. 1: each variable, which corresponds to a node in the graph, is linked by directed arcs to both its influencing and influenced variables. To this end, through an automatic analysis of the structure model:

1. $\forall \dot{x}_i$, the link to x_i is stated;
2. $\forall x_i$, the subset $\mathcal{F}_i \subset F$ is built, where $\mathcal{F}_i := \{f_{kl} \in F | f_{kl} \text{ depends on } x_i\}$, i.e. the elements of $\mathcal{F}_i$ are both the flows which depend directly on x_i and those which are controlled by x_i. Then, x_i is causally linked to each elements of $\mathcal{F}_i$;
3. $\forall i$, $i = 1, .., n$, where n is the number of state variables, the set of all flows entering and leaving the $i - th$ compartment which appear in the equation (1), $\mathcal{F}^i := \bigcup_{j=0,n}(\{f_{ij}\} \bigcup \{f_{ji}\})$, is built. Each element of $\mathcal{F}^i$, taken with its proper sign, is linked to $\dot{x}_i$.

Let us observe that the set of variables which assume a constant value, either possible incoming flows from the environment or fractional transfer rate coefficients are grouped into a set K. The elements of K are exogenous to the system: their values are time-invariant, and then can only be perturbed at the initial time, and therefore can be considered the causes of a change only in the initial state of $\dot{X}$. On the mathematical level, since each equation is structural, that is it describes the dynamics of a unique compartment of the system, the simple method we proposed for building the causal order graph of a compartmental model produces results which do not differ from those obtained by the causal ordering technique [6, 8] but, in our case, much information is already available at a physical level. To generate a causal explanation, the causal order graph is traversed starting from the variables whose values have been initially perturbed. Feedback loops may be handled by exploiting heuristic rules which state "what comes first" in the propagation of causality.

As an example, let us refer to the glucose-insulin regulatory system described in [5], and whose structure model is shown in Fig. 2. At first, the sets $\mathcal{F}_1 = \{f_{01}, f_{10}, f_{20}\}$; $\mathcal{F}_2 = \{f_{02}, f_{10}, f_{01}\}$; $\mathcal{F}^1 = \{f_{10}, f_{01}\}$; $\mathcal{F}^2 = \{f_{20}, f_{02}\}$ are built, and then the directed graph which describes the causal dependencies between the whole set of variables is generated (Fig. 3).

2.2 Significant instants and events

Let us recall that the set T of the distinguished time-points of the system's behavior, which is generated by the simulation, is defined by $\bigcup_{i=1,m} T_i$, where T_i are the sets of the distinguished time-points associated with the variables which characterize the dynamical system. Obviously, some elements $t_j \in T$ represent instants where events which are significant for one variable but not for the others occur. Therefore, in order to avoid redundances in the causal explanations of a specific predicted behavior, the algorithm builds a set of significant time-points $T^* \subset T$. As we are aiming at giving explanations of the significant events of the state variables, which are defined by a change in their time derivative, we consider as significant time-points the instants $T^* := \{t_i \in T \mid \exists j : sign(\dot{x}_j(t_i)) \neq sign(\dot{x}_j(t_{i-1}, t_i))\}$, i.e. the end-points of time-intervals where the direction of change of a single state variable is constant. We call *significant events* the qualitative states in t_i and (t_i, t_{i+1}), where $t_i, t_{i+1} \in T^*$.

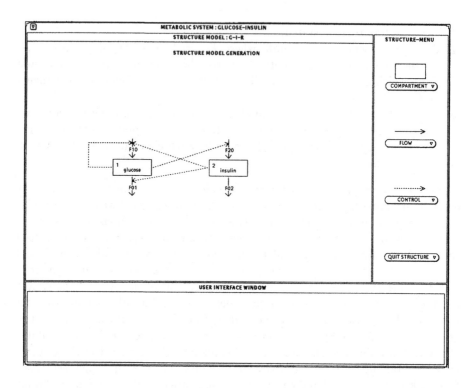

Fig. 2. A structure model of the glucose-insulin regulatory system

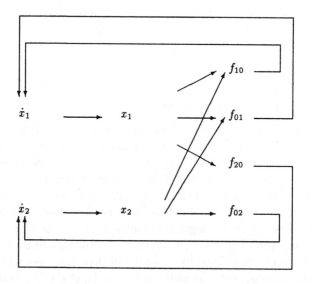

Fig. 3. Causal order graph associated with the structure model in Fig. 2

As we are interested in explaining the significant events, we only consider those transitions that concern a change in the evolution of the system. The set of possible transition rules, which are grounded on the hypothesis that all considered functions are continuously differentiable, are divided into two types, namely:

- P-transition - represents a transition from a state in a distinguished time-point to a state in an interval;
- I-transition - represents a transition from a state in an interval to a state in a distinguished time-point.

In order to determine the set T^* the algorithm considers the sequence of states $\{s_0, ..., s_n\}$ which defines the specific behavior to be explained, and singles out those states which have been determined by the occurrence of a significant transitions. More precisely:

- if the successor, s_{i+1}, of the state s_i is determined by a significant I-transition (denoted by I2, I5, I8, I9 in [9]) then s_{i+1} is a significant event;
- if the successor, s_{i+1}, of the state s_i is determined by a significant P-transition (denoted by P2, P3 in [9]) then s_i is a significant event.

As example, the elements of the set T^* associated with the behavior shown in Fig. 4, which has been predicted by a QSIM simulation of the glucose-insulin regulatory system in response to an increased value of glucose, are t_0, t_2, t_4.

2.3 Determination of causes

Applied to a behavior chosen by the user, the algorithm determines the causes of the changes of the state variables in the instants in T^*. The causes of the occurrence of significant events to be explained are determined by exploiting knowledge of both compartmental models and transition rules.
At a *first level* of explanation:

1. a significant I-transition of a state variable, i.e. a state variable x_i reaches a steady value ($\dot{x}_i = 0$), is caused by a perfect balancing of the entering and leaving flows associated with the i−th compartment, that is of the elements in $\mathcal{F}^i$;
2. a significant P-transition of a state variable, i.e. a state variable x_i leaves a steady state ($\dot{x}_i \neq 0$), is caused by a loss of balance between the entering and leaving flows associated with the i−th compartment.

The cause of such a change is explained by changes in the qualitative values of some elements in the set F. At a *more detailed level*:

3. a change in the state of the elements of $\mathcal{F}_i$ is caused by a change in the qualitative value of x_i, which may belong to a monotonicity interval different from the previous one;
4. a change in the state of x_i is caused by a change in $\dot{x}_i$.

Through the causal order graph and the functional relationships which link a state variable to flow variables in the behavior model, the effects of the change in the system are drawn out. Such effects turn out to be the causes of the next significant event. Thus, starting from the perturbation on the system, which is the external cause of changes in the system and does not need to be explained, the algorithm determines its effects, and then builds a chain of causes which explain the successive significant changes.

3 Presentation of explanation

Two different sub-modules of the EXPLAIN module, which can be selected through a sub-menu, deal separately with the presentation of causal explanation of a specific behavior and with the description about the structure and behavior model stored in the library.

3.1 Explanation of a specific behavior

The "specific behavior" sub-module has been designed to answer the user explanation requests about a specific predicted behavior, chosen within the set of all possible predicted behaviors of the pathophysiological system under study.
A user-friendly interface allows the user to ask for explanation by clicking the mouse on texts ad hoc presented in order to facilitate the question composition: the user is guided through the question formulation by menus which are generated so that the possible queries deal only with significant events. Two different kinds of questions are allowed: (1) *what* happened either to the system or to a variable at a given significant time-point or time-interval, and (2) *why* an event of a state variable occurs.

By default, a first explanation of the behavior consists in a verbal description of the QSIM plots which describe the evolution over time of all of the state variables. Such an explanation aims at pointing out the most significant state changes (change of direction, steady state values, etc.) and turns out to be of great utility and interest for users who are not familiar with the QSIM algorithm. Then, the user is allowed to ask for a causal explanation of the behavior of each state variable over the whole interval.

The "what happened" explanation provides information about the significant event of either a single state variable or all of the state variables which characterize the system behavior. More precisely, information about both the direction of change and the qualitative value of variables is given. The interpretation of what happened to the system at the initial time-point is of particular interest: the effects caused by the perturbation acting on the system are pointed out by highlighting the state changes of each variable which is directly influenced by the perturbed ones (Fig. 4).

The "why" explanation deals with the causes which determined the occurrence of a significant event of a variable x_i. At a first level of detail, the state

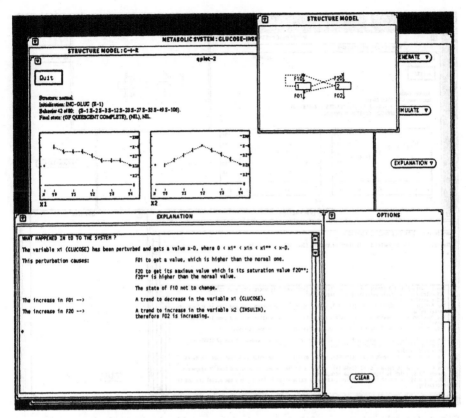

Fig. 4. Interpretation of the effects caused by the perturbation at the initial time-point.

of x_i is directly explained by the sign of its time derivative (see sections 2.1 and 2.3). In turn, the sign of $\dot{x}_i$, which represents the sign of the netflow in the compartment i, is determined by the changes in the elements of the subset $\mathcal{F}^i$. On request by the user, a causal explanation of the change in $\dot{x}_i$ is provided: all the influencing flow variables are analyzed. For each of them, both the functional dependencies on state variables and changes in their qualitative states are highlighted (Fig. 5).

3.2 Description of a system model

This facility aims at providing the user with a complete and clear description of the relevant features of the behavior model so that she can more easily understand the possible effects of the changes of the values of some variable . The user is allowed to ask for a description of a pathophysiological model built through the generation phase of QCMF and stored in the model library. The structure model is shown to the user, and information about the state variables is provided. More precisely, such information deals with:

1. the physiological meaning of the state variables;

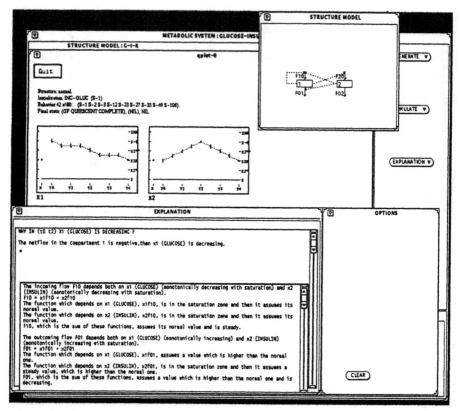

Fig. 5. Causal explanation of the behavior of the variable glucose in the interval $(t_0,\ t_2)$ at a second level of detail

2. the mass balance equation which characterizes the evolution over time of each state variables.

Moreover, for each functional relationship between flow and state variables, it is given (Fig. 6):

- a verbal description of their monotonicity properties;
- a plot of the flow variable versus the state variable, with their respective landmark values;
- the possible definition, through single valued auxiliary functions, of the dependence of a flow on multiple state variables.

4   Conclusion

The interpretation of the results obtained by simulating a pathophysiological model requires knowledge of the assumptions made during model formulation and simulation. As such information may not be explicit in the results to be interpreted, the interpretation is a difficult task [4, 10, 11]. We presented a

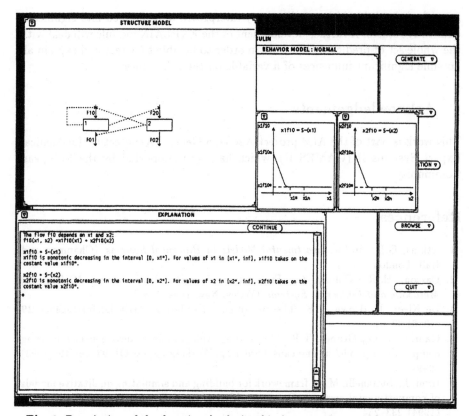

Fig. 6. Description of the functional relationship between flow and state variables

method for the generation of causal explanations of the simulation outcome, which reproduces the effects on a metabolic system caused by single or multiple pathophysiological mechanisms.

Such a method provides an operational procedure for inferring causal relations from an acausal description of the system behavior by QDEs. Our approach is clearly grounded on the compartmental structure model which allows us to state a physical definition of causality. This in turn can be applied to provide simple rules for detecting different types of dynamic dependency between the variables which characterize the behavior model. Therefore, feedback loops can be handled as they can be straigthforward determined from the compartmental structure. In order to avoid the generation of redundant explanations, the algorithm only considers significant transitions that deal with a change in the evolution of the system behavior.

Some work concerning with the user-interface has been described: the user may ask for both presentation of what happened in the whole system or to a specific variable during a significant instant or time interval and why a state has been achieved. Requests of explanation at different level of detail are allowed.

Although the method has been designed to explain qualitative simulation behaviors, it could be also used to explain numerical simulation results: the numerical data should be qualitatively interpreted and mapped into significant qualitative states. A further development of the EXPLAIN module will deal with the whole set of predicted behaviors in order to be able to detect and explain all possible significant influences of a variable on the other ones.

5 Acknowledgements

This work is part of the AIM project A2034, a General Architecture for Medical Expert Systems II (GAMES II), which has been supported by the European Community.

References

1. Atkins, G.L.: *Multicompartmental Models in Biological Systems.* (Chapman and Hall, London, 1974)
2. Carson, E. R., Cobelli, C., Finkenstein, L.: *The Mathematical Modeling of Metabolic and Endocrine Systems.* (Wiley, New York, 1983)
3. De Kleer, J., Brown, J. S.: Theories of causal ordering. Artificial Intelligence **29** (1986) 33–61
4. Gautier, P. O., Gruber, T.R.: Generating explanations of device behavior using compositional modeling and causal ordering. Working Papers QR 93, Seattle (1993) 89–97
5. Ironi, L., Stefanelli, M.: A framework for building and simulating qualitative models of compartmental systems. Computer Methods and Programs in Biomedicine **42** (1994) 233–254
6. Iwasaki, Y., Simon, H.A.: Causality in device behavior. Artificial Intelligence **29** (1986) 3–32
7. Iwaski, Y., Simon, H.A.: Theories of causal ordering: reply to De Kleer and Brown. Artificial Intelligence **29** (1986) 63–72
8. Iwaski, Y., Simon, H. A.: Causality and model abstraction. Artificial Intelligence **67** (1994) 143–194
9. Kuipers, B. J.: Qualitative simulation. Artificial Intelligence **29** (1986) 280–338
10. Lee, M., Compton, P.: Context-dependent causal explanations. Working Papers QR 94, Nara (1994) 176–186
11. Rickel, J., Porter, B.: Automated modeling for answering prediction questions: selecting the time scale and system boundary. Proc. Twelfth National Conference on Artificial Intelligence (AAAI-94), (AAAI/MIT Press, 1994) 1191–1198

Probabilistic Models

An Information-Based Bayesian Approach to History Taking

Giuseppe Carenini*   Stefano Monti*   Gordon Banks**

University of Pittsburgh, Pittsburgh, PA 15260

Abstract. Effective history-taking systems need to dynamically reduce the number of questions to ask. This can be done either categorically or probabilistically, by exploiting previous patient's answers. In this paper, we propose a probabilistic information-based history-taking strategy that combines synergistically two information-content measures for reducing the number of questions asked. We have applied this strategy to an existing history-taking system and some preliminary results seem to confirm our initial intuitions.

1 Introduction

History-taking systems collect information from patients about their condition, past-current treatments, habits, and family history [9].

To ask the whole predetermined set of questions available to the history taker, without regard to their relevance to the specific case, can be extremely slow and tedious. Patients become bored when they are asked too many questions, and when a patient is bored and feels the conversation is unnatural, the quality of the information provided is likely to decrease. For these reasons, two techniques have been developed in the past to reduce the number of questions asked by history-taking systems.

The first technique uses *branching strategies* to eliminate set of questions from the current interview. Categorical conditions are applied to either knowledge already collected about the patient or to knowledge about the medical domain and, depending on what condition is satisfied, different branches in a tree of possible questions are followed. Augmented Transition Networks (ATN) have been recently proposed as the formalism to implement this technique [6].

The second technique to reduce the number of questions asked and to sequence them more naturally uses control methods based on *probabilistic knowledge* [12]. Probabilities for each possible diagnosis are estimated after each question (or set of related questions) and the next set of questions asked is the one that has the highest probability of being relevant to the diagnostic process.

Our work proposes a general-purpose probabilistic information-based history-taking strategy that addresses two serious weaknesses of the probabilistic-based technique proposed in [12]:

---

* Intelligent Systems Program
** Department of Neurology

- The simple Bayesian hypothesis, which has been proved to be unrealistic in many clinical domains [2]. This hypothesis assumes that all the findings are conditionally independent given any disease instance.
- The method of selecting the next most relevant question, and eventually reducing the number of questions, merely consists of heuristically sensible probabilistic computations, which are not justified in the normative probabilistic and decision-theoretic framework.

In order to remove the requirements of the simple Bayesian hypothesis, we propose to represent diagnostic knowledge using the belief network formalism [5]. In belief networks, dependencies between findings can be represented in an effective and probabilistically sound way.

To address the second shortcoming, we tried to use the decision theoretic measurement of the *value of information* as proposed in the Pathfinder system [4][3]. Unfortunately, we found that in the case of our belief network the value of information was not effective enough in driving the question selection process from start to end. Therefore, we introduced to the selection process a second information-theory-based measure, the *expected disease entropy*. We integrated the value of information and the expected disease entropy measures into what we believe is an effective strategy for question selection and eventually for a reduction in the number of questions asked.

Finally, we performed an exploratory evaluation of this approach with a history-taking module previously developed as part of a system for patient education in the domain of chronic non-organic headaches [1].

2   Belief Network Construction

The first step in testing our proposal was to build the belief network for the domain of chronic non-organic headaches (see Figure 1). As knowledge sources for the knowledge acquisition phase, we used both current medical literature [8, 10], and the neurologist in our group. The mutually exclusive and exhaustive set of diseases (hypotheses) we considered are: Migraine, Cluster, and Tension headaches. The assumption that these diseases are mutually exclusive is a reasonable approximation in our case, since the only significant overlap is a 10% between migraine and tension. The assumption of exhaustiveness is also reasonable because Migraine, Cluster, and Tension headaches cover about 90% of all non-organic headaches. In assessing the conditional probability matrices for the network, we did not use any sophisticated knowledge acquisition techniques. Our expert expressed his subjective probabilities on a scale of six values ranging from impossible to certain[4]. We believe that all these approximations are acceptable

---

[3] In Pathfinder the value of information was used for selecting the next diagnostic-test, not for history-taking.

[4] 0 (never), 0.05 (rare), 0.2 (significant minority) 0.5 (about half), 0.8 (most patient), 1 (always)

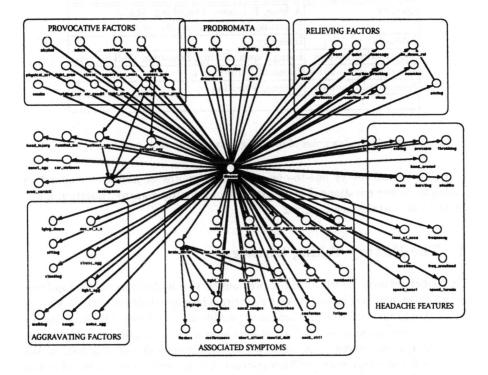

Fig. 1. The belief network for the clinical domain of chronic non-organic headaches.

for our purposes, because our main goal is to inform history-taking strategies, not to automate the diagnostic process.

Examining the network in Figure 1, the links diverging from the disease node correspond to findings, aggravating and relieving factors, and other patient features, whereas the links converging to the disease node are typical headache provocative factors. Although many links in the network can be given a causal interpretation, this is not true for all of them. For some nodes the connection simply states a correlation, and the choice of the link direction is solely determined by the availability of the corresponding conditional probabilities. Notice that in the belief network we can represent dependencies between findings that cannot be accounted for in the simple Bayesian model. The interaction between provocative factors nodes and the disease node can be reasonably approximated by a *noisy-OR* gate, as any trigger is likely to cause a headache and this likelihood does not diminish when several of the triggers prevail simultaneously. Since the disease node is not a binary variable, we have used a variation on the noisy-OR that permits the disease node to have any number of values [7, 11].

By design, all the nodes in the network, except for the disease node, correspond to questions in the history-taking system. As shown in Table 1, the

Questions	Nodes
Quality	Aching, Pressure, Throbbing, Sharp, Bursting, Viselike, Band_Around
Intensity	Intensity
Aggravating Factors	Lying_Down, Sitting, Standing, Walking, etc.
Speed of Termination	Speed_of_Termination
Associated Stomach	Nausea, Vomiting
...	...

Table 1. Examples of mapping between questions and nodes.

mapping is not always one-to-one, and some groups of nodes correspond to multiple-answer questions. For an example of a multiple-answer question asked by the history-taker see Figure 2.

3 The History-Taking Strategy

The belief network described in the previous section is the reasoning component on which we base our history-taking strategy. The general idea is that the history taker should first ask the questions that provide the most useful information. In terms of decision analysis, this corresponds to a history-taking strategy where the next question asked is the one with the maximum value of information. We tried such a strategy with our history-taker, but unfortunately we found it was not effective enough to drive the question selection process to a graceful completion. To address this problem, we introduced to the selection process a second information-theory-based measure, which we call expected disease entropy.

In this section, we first discuss the concept of value of information and how it can be applied to history-taking. Then, we briefly examine possible reasons for the failure of the history-taking strategy based solely on the maximization of the value of information. Finally, we describe how we integrated the value of information and the expected disease entropy in what we believe is an effective strategy for question selection eventually leading to a reduction in the number of questions asked.

Value of information for history-taking

With respect to the diagnostic problem, the value of a given piece of information is defined to be the difference in expected utility between best diagnoses before and after the piece of information is obtained. Following the approach proposed in [4], let q_{new} be a question we want to ask, and a_{new} a possible answer to that question. Let Φ be the set of question-answer pairs collected so far (and corresponding to clamped nodes in the network), $D = \{d_1, \ldots, d_n\}$ the possible diseases (migraine, cluster, and tension in our case), and let $\mu(d_i, d_j)$

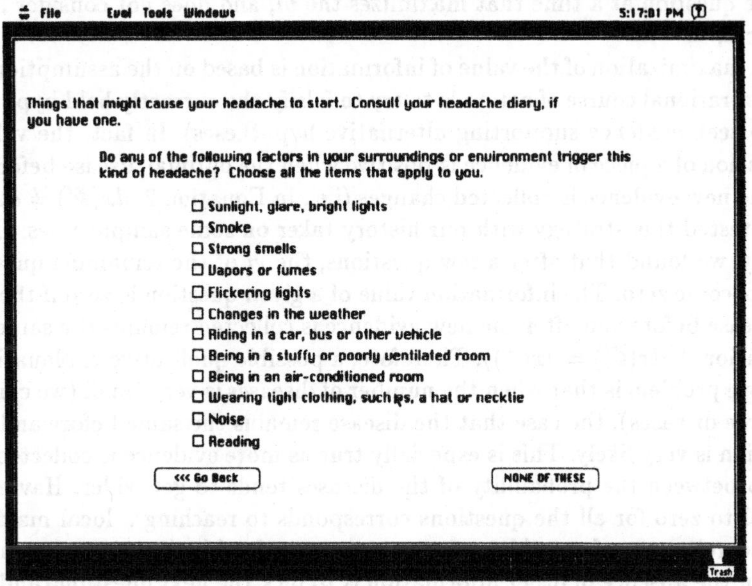

Fig. 2. An example of a multiple-answer question asked by the history-taking module.

---

denote the utility of diagnosing d_i while having disease d_j. With these notational conventions, the expected utility $e\mu$ of the disease d_k given Φ is:

$$e\mu(d_k|\Phi) = \sum_{d_j} p(d_j|\Phi)\mu(d_k, d_j) . \tag{1}$$

Given the expected utility of each disease, we can compute $dx(\Phi)$, the diagnosis with maximum expected utility given evidence Φ. Let Φ' be the set of question-answer pairs given by the set Φ and a new question-answer pair (q_{new}, a_{new}), i.e., $\Phi' = \Phi \cup \{(q_{new}, a_{new})\}$. The value of information vi of a non-answered question q_{new} is:

$$vi(q_{new}|\Phi) = \sum_{a_{new}} p(\Phi'|\Phi)[e\mu(dx(\Phi')|\Phi') - e\mu(dx(\Phi)|\Phi')] , \tag{2}$$

where the summation is over all the possible answers a_{new} to the question q_{new}. In other words, the value of information of q_{new} is the average over all the possible answers to q_{new} of the differences between expected utilities of best diagnoses before and after q_{new} is answered.

As we said, the vi-based history-taking strategy consists of iteratively selecting the question with maximum vi, in order to maximally increase the difference in expected utility between best diagnoses before and after the piece of information is obtained. Notice that this is a heuristic strategy that implements a greedy search in the space of possible sequences of questions[5]. It locally selects

---

[5] In our case, with around 80 evidences to be collected, and assuming binary values for such evidences, the size of the search space is about 80! .

just one question at a time that maximizes the vi, and does not consider longer sequences.

The maximization of the value of information is based on the assumption that the most rational course of action is to try to falsify the currently held hypothesis (i.e., to seek evidence supporting alternative hypotheses). In fact, the value of information of a piece of evidence is positive when the optimal disease before and after the new evidence is collected changes (i.e., in Equation 2, $dx(\Phi') \neq dx(\Phi)$).

We tested this strategy with our history taker on some sample cases. Unfortunately, we found that after a few questions, the vi of the remaining questions would become zero. The information value of a given question is zero if the optimal disease before and after the new evidence is collected remains the same (i.e., in Equation 2, $dx(\Phi') = dx(\Phi)$). Therefore, a possible qualitative explanation to the above problem is that when the number of diseases is very small (we consider just three diseases), the case that the disease remains the same before and after a question is very likely. This is especially true as more evidence is collected, and the gap between the probability of the diseases tends to get wider. Having the vi equal to zero for all the questions corresponds to reaching a local maximum in the search space of possible question sequences. In this situation, the only solution in the absence of other information is to pick the next question randomly, or according to a default ordering.

The expected entropy of the disease

To circumvent the problems in the use of vi, we introduce the concept of expected entropy of the disease. The entropy of a stochastic variable is a measure of its uncertainty. A question that maximizes the expected entropy of a disease variable is a question whose possible answers are more likely to reduce the probability of the current most probable disease. In other words, by maximizing the expected entropy of the diagnosis, we force the history taker to reconsider currently less likely alternative hypotheses. In terms of search space, this is an attempt to move the search away from the local maximum. Notice that this approach agrees with the rationale underlying the vi strategy. That is, it aims at finding counter-evidence to the currently most likely disease.

A formal definition of the entropy e of the disease variable D with respect to evidence Φ is:

$$e(D|\Phi) = \sum_{d_k \in D} -p(d_k|\Phi)log_2[p(d_k|\Phi)] , \tag{3}$$

and the expected entropy $\mathcal{E}_D$ of the disease variable for question q_{new} can be calculated as:

$$\mathcal{E}_D(q_{new}|\Phi)) = \sum_{a_{new}} p(a_{new}|\Phi)e(D|\Phi'), \qquad with \ \Phi' = \Phi \cup \{(q_{new}, a_{new})\} . \tag{4}$$

The $\mathcal{E}_D$-based history-taking strategy consists of selecting iteratively the question maximizing $\mathcal{E}_D$.

History-taking Strategy
    $\Phi \leftarrow \emptyset$,
   update_network($\emptyset$),
   $\mathcal{Q} \leftarrow$ initial set of questions
  while $\mathcal{Q} \neq \emptyset$ AND
            $p(dx(\Phi)) <$ *disease_threshold* AND
            $|\Phi| <$ *question_threshold*   /* optional */
     do
         select question q_{new} with max $vi(q_{new}|\Phi)$
         if $vi(q_{new}|\Phi) = 0$
            then select question q_{new} with $max(\mathcal{E}_D(q_{new}|\Phi))$
         answers $\leftarrow$ *make_question(q_{new})*
         $\Phi \leftarrow \Phi \cup$ *answers*
         $\mathcal{Q} \leftarrow \mathcal{Q} -$ *answers*
         update_network(Φ)
     end /* while */
 end /* History-Taking Strategy */

Table 2. Schematic description of the algorithm implementing the History Taking Strategy

Integration of expected entropy of the disease and value of information

The history-taking strategy we adopt combines the two selection strategies described above: vi-based and $\mathcal{E}_D$-based. It always starts by looking for the question with the highest vi. When only questions with null vi are left, it selects the question for which $\mathcal{E}_D$ is maximum, but it switches back to the vi-based selection whenever the value of information of some questions becomes greater than zero. In terms of search space, the vi-based strategy looks for a global maximum. When it reaches a local maximum, the $\mathcal{E}_D$-based strategy tries to move the search away from the local maximum.

In Table 2, we give a schematic description of the history-taking strategy. The algorithm first executes start-up operations: the belief network is initialized (*update_network($\emptyset$)*), that is, the probability of each node prior to the collection of any evidence is computed; the set Φ of asked questions is set to the empty set; and the set $\mathcal{Q}$ of remaining questions contains the whole set of questions available to the history taker. Once the initialization is complete, the algorithm starts selecting the questions to ask the patient. The selection of new questions terminates when one of the *while* conditions is violated. With the first condition, we test whether any questions remain. With the second condition, we ensure that the question selection process terminates when the probability of the most likely disease is above a given threshold (*disease_threshold*). We can also envision a third optional condition that terminates the question selection process when the number of questions asked is above a given threshold (*question_threshold*). The rationale behind this last *while* condition lies with the assumption that, in some application domains, it is possible to empirically determine a threshold on the number of questions needed, before being able to assert that the current most likely diagnosis will not change even if all the remaining questions are asked.

As previously mentioned in Section 2, some of the questions in our history-taking system are multiple-answer questions (see Table 1), i.e., questions with many items to be answered, and that correspond to sets of nodes in the network. For illustrative purposes, the strategy as described in Table 2 does not consider that the history-taking strategy has to treat multiple-answer questions differently.

To compute the vi (or the $\mathcal{E}$) of these questions, the average of the vi (or of the $\mathcal{E}$) of their items is taken. In order to prevent items with zero-vi from excessively lowering the cumulative vi of the question, such items are not considered in the computation, but they are still included in the question if it is selected.

4   Preliminary evaluation

We performed a preliminary evaluation of our history-taking strategy on five patient cases. Three patient cases were typical cases of migraine, cluster and tension headache respectively. The other two were typical ambiguous cases: migraine-tension and migraine-cluster. We did not consider the tension-cluster ambiguous case because of its unlikelihood.

The preliminary results of this evaluation confirm our intuition that the integration of the vi-based and $\mathcal{E}_D$-based selection strategies produces a history-taking strategy that behaves effectively both in typical and ambiguous cases. In each of the three typical cases the strategy avoided asking on average two thirds of the available questions, and they were avoided for the right reason. In fact, although the selection of questions based on expected entropy was activated, when the algorithm stopped because the probability of the most likely diagnosis was above the value of *disease-threshold* we set in the termination condition, the most likely diagnosis was the correct one for that case.

In the ambiguous cases our history-taking strategy kept on switching between selections based on vi and selections based on $\mathcal{E}_D$. It ended up asking all the available questions because the probability of the most likely disease did not converge to any stable value above the threshold *disease_threshold*.

5   Conclusion and Future work

In this paper we presented an information-based Bayesian strategy for history-taking that relies on two information-content measures, namely the value of information and the expected entropy of the disease. We applied this strategy to an history-taking system for headache patients previously developed as part of a system for patient education.

Although the results of our preliminary evaluation confirmed our intuition, we are aware that a more extensive evaluation is needed. Conclusive results can only come from an extensive testing of our strategy on a large number of cases, and from a better understanding of the relation between the diagnostic accuracy of the belief network, and the effectiveness of the history-taking strategy.

We also plan to investigate whether our strategy can be improved by adopting a less myopic search approach. As we said, our strategy is a greedy search in the space of possible sequences of questions. Its accuracy could be improved by allowing the strategy to look not just for the best next question, but for the best next sequence of n questions. Clearly, the tractability of the algorithm should be preserved.

Finally, as noted above it was probably the small number of diseases we considered that seriously limited the use of a strategy based exclusively on the vi. It would be interesting to explore the influence of the number of diseases considered on how the vi-based and $\mathcal{E}_D$-based strategies interact in the global history-taking strategy. In order to perform such an analysis, a more refined knowledge acquisition effort would be necessary. As previously mentioned, migraine, tension and cluster, account for about 90% of all chronic non-organic headaches. Other chronic non-organic headaches are quite rare, and their conditional probabilities would tend to be rather small and more difficult to assess.

Implementation notes

For the construction of the belief network we used the software package Ergo$^{©}$ for Macintosh from Noetic Systems. The package is provided with a robust and user-friendly interface that gave us great flexibility in the knowledge acquisition and belief network construction phase. However, we could not use the package for implementing our history-taking strategy because Ergo, at least in the version we have, does not provide any interface to programming languages. The solution to this problem was to develop our own program for carrying out belief network update, together with the procedure for translating the network developed with Ergo in a format that could be used by our program.

We implemented a network update algorithm based on recursive decomposition as described in [3]. Belief Network update by recursive decomposition is a *divide and conquer* method that carries out belief network update by recursively partitioning the network in simpler subnetworks.

The program runs on a SUN-SPARC20 workstation and is written in C++.

Acknowledgments

We would like to thank prof. Bruce Buchanam, director of the Migraine project, for allowing us to use the history taking module. We also thank prof. Greg Cooper for useful advice on the development of the belief network, Gil Ronen and the anonymous reviewers for providing useful comments on this manuscript.

References

1. B. G. Buchanan, J. Moore, D. Forsythe, G. Carenini, G. Banks, and S. Ohlsson. Using medical informatics for explanation in the clinical setting. Technical Re-

port CS-93-16, University of Pittsburgh, 1993. (To appear in the AI in Medicine journal).

2. G.F. Cooper. Current Research Direction in the Development of Expert Systems based on Belief Networks. *Applied Stochastic Models and Data Analysis*, 5:39–52, 1989.

3. G.F. Cooper. Bayesian belief-network inference using recursive decomposition. Technical Report KSL-90-05, Section of Medical Informatics, Stanford University, 1990.

4. D.E. Heckerman, E.J. Horwitz, and B.N. Nathwani. Toward Normative Expert Systems: Part I The Pathfinder Project. *Methods of Information in Medicine*, 31(2):90–105, 1992.

5. J. Pearl. *Probabilistic Reasoning in Intelligent Systems*. Morgan Kaufman Publishers, Inc., 1988.

6. A.D. Poon, K.B. Johnson, and L.M Fagan. Augmented Transition Networks as a Representation for Knowledge-Based History-Taking Systems. In *Proceedings of the 16th Symposium of Computer Applications in Medical Care*, pages 762–766, 1992.

7. M. Pradhan et al. Knowledge Engineering for Large Belief Networks. In *Proceedings of the 10th Conference on Uncertainty in Artificial Intelligence*, pages 484–490, San Francisco, California, 1994. Morgan Kaufmann Publishers.

8. J.R. Saper et al. *Handbook of Headache Management*. Williams and Wilkins, 1993.

9. W.V. Slack. A history of Computerized Medical Interviews. *M.D. Computing*, 1(5):53–59, 1984.

10. S. Solomon and S. Fraccaro. *The Headache Book*. Consumers Union of United States, 1991.

11. S. Srinivas. A Generalization of the Noisy-OR Model. In *Proceedings of the 9th Conference on Uncertainty in Artificial Intelligence*, pages 208–215, Washington D.C., 1993. Morgan Kaufmann Publishers.

12. H.R. Warner, B. Rutheford, and B. Houtchens. A Sequential Bayesian Approach to History Taking and Diagnosis. *Computers and Biomedical Research*, 5:256–262, 1972.

Medical Decision Making using Ignorant Influence Diagrams

Marco Ramoni[1] Alberto Riva[2] Mario Stefanelli[2] Vimla Patel[1]

[1] Cognitive Studies in Medicine
McGill Cognitive Science Centre
McGill University, Montreal, Canada
[2] Laboratorio di Informatica Medica
Dipartimento di Informatica e Sistemistica
Università di Pavia, Pavia, Italy

Abstract. *Bayesian Belief Networks* (BBNs) play a relevant role in the field of Artificial Intelligence in Medicine and they have been successfully applied to a wide variety of medical domains. An appealing character of BBNs is that they easily extend into a complete decision-theoretic formalism known as *Influence Diagrams* (IDs). Unfortunately, BBNs and IDs require a large amount of information that is not always easy to obtain either from human experts or from the statistical analysis of databases. In order to overcome this limitation, we developed a class of IDs, called *Ignorant Influence Diagrams* (IIDs), able to reason on the basis of incomplete information and to to improve the accuracy of the decisions as a monotonically increasing function of the available information. The aim of this paper is show how IIDs can be useful to model medical decision making with incomplete information.

1 Introduction

Over the past 20 years, decision theory and analysis played a dominant role in the effort of modeling medical decision making, and they were successfully applied to a wide variety of difficult clinical problems [10]. It is therefore not surprising that decision analysis techniques have been deeply exploited by the attempts of mechanizing medical decision making: some of the pioneering work in the field of Artificial Intelligence (AI) in Medicine used Bayesian and decision-theoretic methods to develop medical diagnostic expert systems. However, the interest for decision-theoretic methods decreased during the next few years, as fast as was growing the perception that these methods were both intractable, from the computational point of view, and too poor, from the representational point of view. This disenchantment motivated the development and the application of alternative (non-Bayesian) techniques to represent and handle uncertainty, such as *certainty factors* or the Dempster-Shafer theory [19]. However, during the past few years, the development of new knowledge representation and reasoning methods based on probability and decision theory, has gained back to decision analysis techniques the interest of AI researchers. Most of this work has focused on the use of direct acyclic graphs to represent conditional dependencies among stochastic variables; the results of this efforts are the knowledge representation and reasoning formalisms known as *Bayesian Belief Networks* (BBNs).

The theory of BBNs was summarized by Pearl [11], but they have been independently developed by several researchers during the past few years. A BBN is a direct acyclic graph in which nodes represent stochastic variables and arcs represent conditional dependencies among the variables. From a probabilistic point of view, they provide a straightforward way to represent dependency and independence assumptions among variables, thus making easier the acquisition of knowledge and reducing the amount of information needed to specify a BBN. BBNs are particularly appealing since they are based on a sound probabilistic semantics and they easily extend into a complete decision-theoretic formalism, called *Influence Diagrams* (IDs). IDs [5] provide compact representations of decision problems and they are an appealing complement to more traditional methods, such as table of joint probability distributions or decision trees, because they exploit the ability of BBNs to express conditional independence assumptions in graphical terms. IDs and BBNs have been successfully applied to a variety of medical problems [2].

Despite their considerable success, IDs bear some limitations that can make infeasible their application to real-world clinical problems. The most dangerous limitation lies in the fact that they still require a large amount of information: the number of conditional probabilities needed to specify a conditional dependency grows exponentially with the number of its parent variables. Current propagation algorithms require that all the conditionals probabilities defining a conditional dependency have to be known, as well as all the prior probabilities of the states of the root variables, before any reasoning process can start. To overcome this limitation, and maintain the appealing features of probabilistic soundness and graphical nature of BBNs, we have developed a class of BBNs able to reason on the basis of incomplete information. These BBNs are called *Ignorant Belief Networks* (IBNs) [15] and they have been used to develop a system to forecast blood glucose concentration in insuling dependent diabetic patiens, using the probabilistic information direclty extracted from a clinical database [17].

IBNs have been recently extended into a complete decision-theoretic formalism called *Ignorant Influence Diagrams* (IIDs) [13]. IIDs implement an inference policy, largely wished for in the literature about probabilistic reasoning systems, called *incremental refinement* policy [4], able to improve the accuracy of the solutions as a monotonically increasing function of the allocated resources and the available information. The aim of this paper is show how IIDs can be useful to model medical decision making when the large amount of information required by IDs is not available. This paper will briefly outline the theory and the properties of the IBNs, it will show how IBNscan be extended into IIDsand, finally, it will describe the decision procedures to be used when the available information is not sufficient to specify point-valued probability distributions. The properties of IIDs will be illustrated using a clinical decision problem: the assessment of the optimal dosage of Cyclosporine A to prevent the development of Graft Versus Host Disease (GVHD) after allogenic bone marrow transplantation.

2 Ignorant Belief Networks

The representation and use of incomplete information is a long standing challenge for AI researchers. During the past decade, they have developed a class of reasoning systems, called *Truth Maintenance Systems* (TMSs) [8], which are able to reason on the basis of incomplete information. TMSs are reasoning systems which incrementally record justifications for beliefs and propagate binary truth values along chains of justifications. Since beliefs usually incorporate measures of certainty, several attempts have been made to include probabilities in TMSs. TMSs that are able to reason on the basis of probabilistic rather than binary truth values are called *Belief Maintenance Systems* (BMSs) [3]. We have introduced a kind of BMS based on probabilistic logic, and therefore called *Logic-based BMS* (LBMS) [14]. Using the LBMS, we developed a new class of BBNs, called IBNs, able to reason with partially specified conditional dependencies (i.e. lacking some conditional probabilities) and interval probability values.

2.1 Belief Maintenance

TMSs are independent reasoning modules which incrementally maintain the beliefs for a general problem solver and enable it to reason with temporary assumptions with the growth of incomplete information. The use of a TMS endows the problem solver with the ability of assuming and retracting beliefs, detecting contradictions, and identifying the assumptions responsible for its conclusions. Three main classes of TMSs have emerged: *Justification-based* TMSs *Assumption-based* TMSs and *Logic-based* TMSs (LTMSs). A *Logic-based* BMS is an LTMS extended to probability, that is able to reason on the basis of continuous rather than binary truth values in the interval [0 1]. The LTMS uses a standard propositional language $\mathcal{L}$ defined by a set of *atomic propositions* $S = \{a_1, a_2, \ldots, a_n\}$ and by the standard Boolean operators $\neg$, $\vee$, $\wedge$, $\supset$, and $\equiv$. A *literal* l is an atomic proposition a_i or its negation $\neg a_i$. An atomic proposition a_i is a *positive* literal and the negation of an atomic proposition $\neg a_i$ is a *negative* literal. A *clause* C is a finite disjunction of literals $\bigvee_{i=1}^{n} l_i$. A *Conjunctive Normal Form* (CNF) *formula* f is a finite conjunction of clauses $\bigwedge_{i=1}^{n} C_i$. Any legal formula in a propositional language can be converted into a CNF formula.

The LTMS uses a forward-chained unit-resolution procedure called *Boolean Constraint Propagation* (BCP) to propagate the truth-values of propositions. When a formula is supplied to the LTMS, it is converted into CNF, thus becoming a set of clauses. Each clause acts as a constraint on the truth-values of the literals occurring in it. In order to extend the BCP from Boolean to probabilistic truth-values, we need a probabilistic interpretation of disjunction able to define which constraints are imposed by a clause over the (probabilistic) truth-values of the literals occurring in it. Probabilistic logic [9] provides a semantic framework, based on the concept of *probabilistic entailment*, extending the Boolean definition of satisfaction to a probabilistic one. However, since probabilistic entailment is unable to constrain the probability of formulas to point values, we preliminary need to introduce a new evaluation function over the formulas of our language $\mathcal{L}$. We call this function *labeling function* and we define it as a function $P(f) = [p_* \ p^*]$ such that for any formula f of $\mathcal{L}$, $p_* \in [0 \ 1]$, $p^* \in [0 \ 1]$, and $p_* \leq P_0(f) \leq p^*$. The function $P(f)$ assigns to f a *probability interval*, that is a convex set of probability distributions: we will call this assignment the *label* of f and we will denote with $P_*(f) = p_*$ and $P^*(f) = p^*$ the lower and the upper bounds of the interval $P(f) = [p_* \ p^*]$, respectively. A label $P(f) = [p_* \ p^*]$ will be satisfied for any subinterval of $[p_* \ p^*]$.

There are two constraints that are imposed by a clause over the labels of the literals occurring in it. The first constraint says that the label of a literal l_i in clause $\bigvee_{i=1}^n l_i$ is bounded by:

$$P_*(l_i) \geq P_*(\textstyle\bigvee_{i=1}^n l_i) - \sum_{j \neq i} P^*(l_j) + \sum_{\lambda_1=0}^1 \cdots \sum_{\lambda_s=0}^1 P_*(\bigwedge_{i=1}^s l_i^{\lambda_i}) \cdot \Delta_C(\bigwedge_{i=1}^s l_i^{\lambda_i}) \quad (1)$$

where the function Δ_c is defined as

$$\Delta_C(\bigwedge_{i=1}^s l_i^{\lambda_i}) = \max\{0, (\sum_{i=1}^s \lambda_i) - 1\}$$

The second constraint states that the label of the literal l_1 is bounded by:

$$\begin{aligned} P_*(l_1) &\geq \sum_{i=1}^{2^{s-1}} (1 - P^*(\neg l_1 \vee C_i)) \\ P^*(l_1) &\leq 2^{s-1} - \sum_{i=1}^{2^{s-1}} P_*(\neg l_1 \vee C_i) \end{aligned} \quad (2)$$

where $\{l_1, \ldots, l_s\}$ are a set of literals, and $\{C_1, \ldots, C_{2^{s-1}}\}$ are the clauses built from all the possible combinations of the negated and unnegated literals in the set $\{l_2, \ldots, l_s\}$.

In the LBMS, each proposition is labeled with a set of possible values, and the constraints (in our case, the application of the above defined constraints to the clauses) are used to restrict this set. The LBMS can exhibit this behavior because if a clause is satisfied for a given truth-value of a proposition $P(a_i) = [p_* \ p^*]$, it will be satisfied for any subset of $[p_* \ p^*]$. This property, which is implicit in the form of the inequalities in Formulas 1 and 2, implies a monotonic narrowing of the truth-values, thus ensuring the incrementality of the LBMS.

The most important feature of the LBMS is the ability to reason from any subset of the set of clauses representing a joint probability distribution, by bounding the probability of the propositions within probability intervals, and incrementally narrowing these intervals as more information becomes available. The LBMS is also endowed with the ability of retracting the label of an assumed proposition and, using a perfect constraint relaxation algorithm, propagate the effects of the retraction. Furthermore, the LBMS is able to detect inconsistencies in the network of clauses and to trace back the assumption responsible for a contradiction.

2.2 Representation

IBNs are *belief-maintained* BBNs based on the LBMS. The IBN acts as a knowledge representation formalism expressing the assumptions of conditional independence in the domain of application, and communicates the available conditional probabilities to the LBMS, lying at the lower level. These conditional probabilities are transformed into clauses relating the propositions of the LBMS which represent states of the stochastic variables of the IBN. Additional clauses are added to ensure the mutual exclusivity and exhaustivity of the states belonging to the same variable.

In a BBN, all the states of a stochastic variable are mutually exclusive and exhaustive: the probability values assigned to all the states in a variable have to sum to unit. In an IBN, when a variable is defined, each state is communicated to the LBMS as a proposition. Moreover, a set of clauses is installed to ensure that the states of the variable are mutually exclusive and exhaustive. For all propositions $a_1, \ldots, a_n$ in the LBMS representing the states of the variable, the disjunction $a_1 \vee \cdots \vee a_n$ and all the conjunctions $\neg(a_i \wedge a_j)$ (with $i \neq j$) are asserted as true in the LBMS. When a probability value is assigned to a proposition a_i representing a state of the variable, the LBMS receives the clause $P^*(a_{i+1} \vee \ldots \vee a_n) = \sigma$, where $\{a_{i+1}, \ldots, a_n\}$ is the set of proposition representing those states in the variable that are still *unknown*, and $\sigma = 1 - \sum_{k=1}^{i} P_*(a_k)$, i.e. the sum of the minimum probabilities of all *known* states in the variable.

Conditional dependencies among variables are defined by the conditional probabilities $P(a_i|a_{i+1}, \ldots, a_k) = [p_*\ p^*]$, among all the states of each variable. When the probability values of all states represented by the propositions $a_1 \ldots a_k$ is assigned, the two different clauses resulting from the application of the De Morgan's laws to $(a_i \wedge a_1 \wedge \ldots \wedge a_k)$ and $(\neg a_i \wedge a_1 \wedge \ldots \wedge a_k)$ are communicated to the LBMS. $P(a_i \wedge a_1 \wedge \ldots \wedge a_k)$ and $P(\neg a_i \wedge a_1 \wedge \ldots \wedge a_k)$ are calculated by a version of the Chain Rule extended to intervals:

$$P(a_i \wedge a_{i+1} \wedge \ldots \wedge a_k) = \prod_{j=1}^{k} P(a_j) \cdot P(a_i|a_{i+1}, \ldots, a_k) \qquad (3)$$

The direction of a conditional dependency can be reversed by using the *Inversion Rule* and applying the Formula 3 to the resulting conditionals [16].

2.3 Propagation

From the theory of TMSs, the LBMS inherits the concept of *consumer* [3]. A consumer is a forward-chained procedure attached to a proposition, that is fired when the truth-value of the proposition is changed. When a variable is defined in the IBN, for each proposition representing its states in the LBMS, two different consumers are defined. The first consumer is used to communicate to the LBMS the clause $P^*(a_{i+1} \vee \ldots \vee a_n) = \sigma$ above defined, in order to enforce the exhaustivity and exclusivity among states in a variable. A second consumer is used to encode the *conditional probabilities* among states, and it is defined when a conditional dependency is installed in the network. For each conditional probability $P(a_i|a_{i+1}, \ldots, a_k) = [p_* \ p^*]$ in a conditional dependency, a consumer is attached to each proposition $a_{i+1} \ldots a_k$. When it is fired, it applies the Chain Rule to the defined conditional and communicates the appropriate clauses to the LBMS. A prior probability assignment to a state in a variable is communicated to the LBMS by *assuming* the corresponding proposition with the assigned probability. When the proposition is assumed, the attached consumers are fired, thus starting the propagation process.

Using consumers, IBNs do not perform any computation themselves, but rather act as a high-level knowledge representation language, while the propagation of probabilities is performed by the LBMS. It is worth nothing that the LBMS both performs and drives the propagation, since consumers are attached to the propositions of the LBMS and are fired according to the changes occurring in their labels. Therefore, the computational cost of a propagation grows linearly in space and time with respect to the number of conditional probabilities, even if the number of conditional probabilities needed to specify a conditional dependency grows exponentially with the number of parent variables in the dependency. However, the incremental character of inference policy implemented by the IBNs will allow the decision maker to trade execution time with precision of solutions. There are two main properties of the IBNs that will be crucial in the development of the IID. IBNs converge toward point valued probabilities, and, when provided with all the conditionals defining a joint probability distribution, they behave as standard BBNs, returning point-valued probabilities, since the LBMS is complete for clauses representing joint probability distributions. Furthermore, the use of the LBMS enables the retraction of the assumed probabilities: in an IBN it is possible to change the probability of a state without re-compiling the entire network, and to propagate the new probability just over those variables actually affected by the change.

3   Ignorant Influence Diagrams

IDs [5] are a natural extension of BBNs. They allow the formulation of a decision problem into the sound and compact formalism of BBNs. In this section, we will illustrate how IBNs can be easily extended to a complete decision formalism, thus creating a new class of IDs called IIDs. IIDs inherit from the IBNs the ability to reason on the basis of incomplete information and to incrementally refine the accuracy of their conclusions as more information becomes available.

3.1   Representation

IDs are BBNs containing three different kinds of stochastic variables: *chance* variables (also called *state* variables), *decision* variables, and *value* variables (also called *preference* variables). These variables are related by the standard conditional dependencies. Fig. 1 shows an IID to assess the optimal dosage of Cyclosporine A in order to avoid the development of GVHD. The network is a simplified version of an ID to assess the optimal prophylactic intervention to prevent the development of GVHD after allogenic bone marrow transplantation [12].

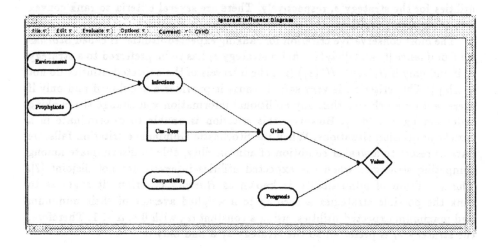

Fig. 1. An ID to assess the optimal GVHD prophylaxis, after Quaglini *et al.*

A chance variable represents a state of the world. It is basically a standard stochastic variable. BBNs are usually defined as "influence diagrams containing just chance variables" [5]. In Fig. 1, chance variables are depicted as oval nodes in the graph and they represent the sterility conditions of the **environment** in the operating room, whether a **prophylaxis** has been performed or not, the level of **compatibility** between the donor and the recipient of the organ, the risk of the **prognosis**, and the development of **gvhd**. A decision variable identifies a set of possible alternative actions available to the decision maker. In the IID of Fig. 1, the dosage of Cyclosporine A (**csa-dose**) is represented by a decision variable and is depicted as a square node. The decision problem here consists of the choice of a particular value for **csa-dose**. A set of actions that represents a possible solution for the decision problem is called a *strategy* or a *policy*. Different strategies lead to different outcomes. Value variables represent preferences or utilities of the decision maker for alternative outcomes. A decision problem may be represented into an ID as the problem of finding the strategy that leads to maximize the preferences expressed by the decision maker over the possible outcomes of the problem.

3.2 Decision Criteria

A standard ID is solved by choosing the strategy with the maximal *expected utility* of its outcomes. In the example depicted in Fig. 1, the expected utility of each dosage of **csa-dose** is the value assigned by the network to the *value* variable. The optimal strategy s will be the strategy with the highest expected utility. As IIDs propagate convex sets of probability distributions, they will also lead to convex set of expected utilities $U(s)$ for each strategy. We will denote with $U_*(s)$ and $U^*(s)$ the lower and upper bounds of the convex set of expected utilities for the strategy s, respectively. There are several criteria to rank convex sets of expected utilities.

The most conservative criterion for ranking expected utilities is called *Stochastic Dominance* [6] and dictates that a strategy s_i has to be preferred to a strategy s_j if and only if $U_*(s_i) > U^*(s_j)$ (i.e. the intervals of the expected utilities do not overlap). This criterion is very safe, because it provides a decision if and only if there will be no chance that any additional information will change the ranking order among s_i and s_j. However, this criterion is unable to discriminate in a variety of decision situations. When the Stochastic Dominance criterion fails, we have to resort to a weaker condition of admissibility, able to discriminate among competing strategies when the expected utilities intervals are not disjoint [7]. This condition of admissibility is known as *Hurwicz* criterion. It suggests to rank the possible strategies according to a weighed average of their minimum and maximum expected utilities, using a constant α, with $0 \leq \alpha \leq 1$. Therefore, the strategy s_i is preferred to the strategy s_j if and only if

$$U_*(s_i)(1-\alpha) + U^*(s_i)\alpha > U_*(s_j)(1-\alpha) + U^*(s_j)\alpha$$

The constant α, called *Hurwicz* value, can be thought as a *boldness* index representing the daring attitude of the decision maker. When $\alpha = 0$, the Hurwicz criterion reduces to well known *maximin* criterion. The *maximin* criterion prescribes to select the strategy having the highest minimum expected utility. Hence, a strategy s_i is preferred to a strategy s_j if and only if $U_*(s_i) > U_*(s_j)$. This criterion reflects the behavior of a cautious decision maker, who wants to be sure that, even if an unfavorable state of the world occurs, there is a known minimum payoff below which he cannot fall. When $\alpha = 1$, the Hurwicz criterion reduces to the *maximax* criterion. The *maximax* criterion adopts the opposite point of view than the maximin criterion. It consider only the maximum expected utilities of strategies and select the strategy with the highest. Then a strategy s_i is preferred to a strategy s_j if and only if $U^*(s_i) > U^*(s_j)$. This criterion reflects the standpoint of a daring gabler who cares just about the maximum payoffs of his strategies and can afford to stand possible losses.

3.3 Propagation

There are two traditional methods to evaluate an ID in order to assess the optimal strategy to be taken. The first consists in converting the influence diagram into a decision tree and solving the problem using the standard solution techniques for decision trees. The second method consists in eliminating variables from the ID through a sequence of value-preserving transformations [18].

IIDs adopt a different evaluation method [1]: for each possible strategy available to the decision maker, they propagate the probability values over the network and rank the strategies according to the expected utility intervals obtained

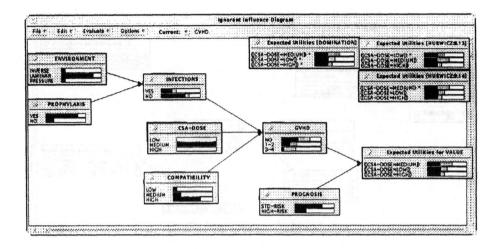

Fig. 2. The IID after the propagation of the strategy (csa-dose=medium).

during their evaluation. Then, the evaluation of an IID is performed by instantiating and retracting the possible strategies. This choice capitalizes on the ability of the IBN, based on the retraction capabilities of the LBMS, to change the probability value of a state without re-compiling the entire network, and to propagate the new probability just over those variables that are actually affected by the change. Furthermore, once the strategies have been evaluated, different criteria can be applied to rank them without re-evaluating the IID. Finally, this choice ensures us that IIDs will inherit all the properties of the IBNs, that they will monotonically converge toward point valued expected utilities, as more information become available, and that when all the information needed to specify a standard ID will be available, the IID will collapse on a standard ID, returning the same point-valued expected utilities.

4 Example

As an example, we will describe the evaluation of the IID depicted in Fig. 1. In their paper, the authors of the ID for GVHD prophylaxis write:

> The first assessment of model probabilities may suffer from a lack of medical knowledge, given that the immune system is not completely understood. [...] For example, infection development is reported as poorly predictable both in the literature and in medical experience.

The use of IIDs spares medical experts the effort of specifying a point valued probability distribution. The network of our example does not include 30% of the conditional probabilities relating **environment** and **prophylaxis** to **infection**, and 15% of the conditional probabilities relating **csa-dose, infection,** and **compatibility** to **gvhd**.

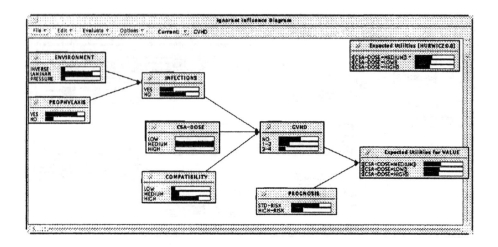

Fig. 3. The IID of the example with complete probabilistic information.

Fig. 2 shows the status of the IID after the propagation of the strategy (csa-dose=high). The pop-up windows over the variables graphically describe the probability interval (subset of [0 1]) associated to each one of their states. In each bar, the area between 0 and $P_*(a_i)$ is black, the area between $P^*(a_i)$ and 1 is white, and the area between $P_*(a_i)$ and $P^*(a_i)$ is gray. Thus, the width of the gray area is proportional to the ignorance about the probability. The value variable value reports the expected utility intervals for all the possible strategies of the example. The decision variable csa-dose reports its value in the strategy. The chances variables reports the probability values of each of their states. Fig. 2 shows also, in the upper right corner, the ranking of the possible strategies according to the different criteria described above. An asterisk indicates the strategies selected as optimal. The decision problem is solved when just one strategy is selected as optimal and marked with an asterisk. The pop-up windows show that the Stochastic Dominance criterion is unable to select only one strategy, and remains undecided among all of them. The weaker Hurwicz criterion is able to pick up a unique decision. However, the solution can change according to the Hurwicz value. We can analytically identify a decision threshold for the Hurwicz value to assess the sensitivity of the decision between the strategy s_i and s_j to the boldness attitude of the decision maker, according to the following formula:

$$\alpha = \frac{U_*(s_i) - U_*(s_j)}{U^*(s_j) - U_*(s_j) + U_*(s_i) - U^*(s_i)}$$

In our case, the expected utility intervals of the two competing strategies are $U(\text{csa-dose=low}) = [0.352\ 0.513]$ and $U(\text{csa-dose=medium}) = [0.333\ 0.633]$, and the decision threshold is $\alpha = 0.1367$. The pop-up windows in the upper-right corner show the changes in the strategies ranking around this value of

α. When all the probabilities required by a standard ID are specified, the IID behaves as a standard ID in returning point-valued probabilities, as shown in Fig. 3. Since both the Stochastic Dominance and the Hurwicz criteria reduce to the maximum expected utility criterion when they have to rank point-valued expected utilities, also the solution provided by the IID is identical to the one suggested by a standard ID. The ranking of the strategies is displayed in the upper right corner of Fig. 3.

It is worth noting that the region of the network describing the influence of environment and prophylaxis on the probability of developing an infection has not been re-evaluated during the process of propagation and retraction of the different strategies. This feature can be useful for those IDs containing large regions of chance variables (such as the description of a complex physiological process) that are not affected by the change in the strategy, and that, once evaluated, do not need to be reconsidered anymore. In this way, the modular nature of the IDs emerges automatically in the IID without any pre-compilation.

5 Conclusions

IDs are a powerful formalism to mechanize medical decision making, and they have been successfully applied to a wide range of medical areas. However, they require a large amount of information that is not always available to the decision maker. The acquisition of this large amount of information, either from human experts or from the statistical analysis of databases, usually represents one of the major challenges in the process of developing decision support systems based on IDs. This paper introduced a new class of IDs, called *Ignorant Influence Diagrams*, able to reason on the basis of incomplete information, and to incrementally refine the accuracy of their decisions as more information becomes available. However, when they are provided with complete probabilistic information, IIDs behave as standard IDs. Therefore, the use of IIDs represents a net gain with respect to the traditional IDs, since it allows to explicitly represent the actual lack of information in a particular decision problem, without loosing any capability of the traditional IDs when the needed information is available. Furthermore, the evaluation algorithm of IIDs, which deeply exploits the retraction capabilities of the LBMS, does not require the re-evaluation of the whole network for each possible strategy, but involves only those variables that are actually affected by the change in the strategy. Finally, the monotonic, incremental character of the refinement process in IIDs provides a way to trade the amount of computational time with the accuracy of the decisions. This feature makes IIDs a suitable formalism for real-time, resource-bounded decision tasks.

Acknowledgments

This research was supported in part by the AIM Programme of the Commission of the European Communities (A2034) and by a grant of the SSHRC of Canada

(410 92 1535) to Vimla Patel. Authors thank Riccardo Bellazzi and Silvana Quaglini for making available the influence diagram used as an example.

References

1. G.F. Cooper. A method for using belief networks as influence diagrams. In *Proceedings of the Conference on Uncertainty in Artificial Intelligence*, pages 55–63, 1988.
2. G.F. Cooper. Current research directions in the development of expert systems based on belief networks. *Applied Stochastic Models and Data Analysis*, 5:39–52, 1989.
3. B. Falkenhainer. Towards a general purpose belief maintenance system. In *Proceeedings of the 2nd Workshop on Uncertainty in AI*, pages 71–76, 1986.
4. E.J. Horvitz. Reasoning under varying and uncertain resource constraints. In *Proceedings of the International Joint Conference on Artificial Intelligence*, pages 1121–1127. Morgan Kauffman, San Mateo, CA, 1989.
5. E.J. Horvitz, J.S. Breese, and M. Henrion. Decision theory in expert systems and artificial intelligence. *International Journal of Approximate Reasoning*, 2:247–302, 1988.
6. H.E. Kyburg. Rational belief. *Behavioral and Brain Sciences*, 6:231–273, 1983.
7. I. Levi. *The Enterprise of Knowledge. An Essay on Knowledge, Credal Probability and Chance*. MIT Press, Cambridge, MA, 1980.
8. D. McAllester. Truth maintenance. In *Proceedings of the National Conference on Artificial Intelligence*, pages 1109–1115, 1990.
9. N.J. Nilsson. Probabilistic logic. *Artificial Intelligence*, 28:71–87, 1986.
10. S.G. Pauker and J.P. Kassirer. Decision analysis. *New England Journal of Medicine*, 316(5):250–258, 1987.
11. J. Pearl. *Probabilistic Reasoning in Intelligent Systems: Networks of plausible inference*. Morgan Kaufmann, San Mateo, CA, 1988.
12. S. Quaglini, R. Bellazzi, F. Locatelli, M. Stefanelli, and C. Salvaneschi. An influence diagram for assessing gvhd prophylaxis after bone marrow transplantation in children. *Medical Decision Making*, 14:223–235, 1994.
13. M. Ramoni. Ignorant influence diagrams. In *Proceedings of the International Joint Conference on Artificial Intelligence*, 1995. Forthcoming.
14. M. Ramoni and A. Riva. Belief maintenance with probabilistic logic. In *Proceedings of the AAAI Fall Symposium on Automated Deduction in Non Standard Logics*, Raleigh, NC, 1993. AAAI.
15. M. Ramoni and A. Riva. Belief maintenance in bayesian networks. In *Proceedings of Tenth Conference on Uncertainty in Artificial Intelligence*, pages 204–212, 1994.
16. M. Ramoni and A. Riva. Abduction with incomplete information. In *Proceedings of Eleventh Conference on Uncertainty in Artificial Intelligence*, 1995. Submitted.
17. M. Ramoni, A. Riva, M. Stefanelli, and V. Patel. An ignorant belief network to forecast glucose concentration from clinical databases. *Artificial Intelligence in Medicine Journal*, 1995. Forthcoming.
18. R.D. Shachter. Evaluating influence diagrams. *Operation Research*, 34:871–882, 1986.
19. G. Shafer. *A Mathematical Theory of Evidence*. Princeton University Press, Princeton, 1976.

Dynamic Propagation
in Causal Probabilistic Networks
with Instantiated Variables

O.K. Hejlesen, S. Andreassen and S.K.Andersen

Department of Medical Informatics and Image Analysis, Institute of Electronic Systems,
Aalborg University, Fredrik Bajersvej 7, DK-9220 Aalborg, Denmark

Abstract
An extension to the Hugin tool for implementing dynamic propagation in causal probabilistic networks with some instantiated variables is presented. The extension makes it possible to combine a dynamically defined network, implemented without the use of Hugin, with a static network implemented in Hugin, thereby reducing the calculation time by several orders of magnitude compared to using Hugin alone. The application of the dynamic propagation in a Diabetes Advisory System, DIAS, for giving advise on insulin dose adjustment in insulin dependent diabetes mellitus, is described as an example.

1 Introduction

Causal probabilistic networks (CPNs), also called Bayesian networks, have been used to build models of the physiology and pathophysiology of biological systems. These models can be used for diagnosis or for planning of therapy. For example, the MUNIN system diagnoses neuromuscular disorders, performing "at the level of an experienced clinical neurophysiologist" (Andreassen et al. 1989). As another example, the study of temporal aspects in medical reasoning led to the construction of DIAS, a system used for generating advice on insulin dosage in insulin dependent diabetes mellitus (Andreassen et al. 1994).

In CPNs all variables are stochastic variables, linked by conditional probabilities, which, in DIAS, can be seen as a translation of the differential equations describing the physiological relations. Therefore, inference in the CPN is not made by solving differential equations, but rather by making a Bayesian updating of the probability distributions associated with the stochastic variables. If we assume that all variables are discrete stochastic variables, then the updating is in principle very simple. In an algebraic notation (Jensen et al. 1990) the initialisation and the stochastic inference is determined by two equations, that specify apparently simple matrix operations. Unfortunately the number of elements in the matrices, the state space, is equal to the product of the number of states of all nodes in the CPN, which makes direct application of these equations impractical in CPNs with more than a few tens of nodes.

The Hugin tool (Andersen et al. 1989) has been developed to make Bayesian inference in CPNs computationally practical. This tool is based on the development of methods for local belief updating in CPNs (Pearl 1986, Lauritzen and Spiegelhalter 1988). Roughly speaking, Hugin divides the set of variables in the CPN into a number of subsets, called cliques. The equations for initialisation and updating can then conveniently be applied to the cliques, supplemented by procedures that propagate evidence between the cliques. The size of the cliques depends on the topology of the CPN, and in some cases the cliques generated by Hugin are so large that the computations remain impractical.

In our experience, this is the case when CPNs representing full scale medical problems are constructed. Two ideas will be proposed for reducing the computational effort. Both ideas will be illustrated, using one of the above mentioned medical applications, DIAS, as an example. Together, the two ideas reduced the requirements for storage space for the cliques' state space in the DIAS CPN from an estimated size of 10^{55} elements to 10^7 elements, which is within reach of current computer technology.

The first idea exploits the fact that when a stochastic variable in the network is instantiated, meaning that its current value is known, for example because it has been measured, then the CPN effectively changes topology: The d-separation properties of the CPN are modified (Pearl 1988). Under some circumstances, the CPN with some instantiated variables will generate smaller cliques.

The other idea is to allow a dynamic construction of CPNs. Instead of constructing one CPN that has to be able to deal with all possible relevant clinical situations, the CPN is divided into a static and a dynamic part. In the DIAS application the static part represents a model of the patient's glucose metabolism and the dynamic part represents the patient's insulin injections, which may not only be different in timing and type of insulin for different patients, but may also differ from day to day for the same patient. It turns out, that when a suitable set of variables has been instantiated, it becomes particularly simple to construct a part of the CPN dynamically.

In the following we shall first describe the computational problem, using the CPN model of glucose metabolism as an example. Then we will divide the model into a dynamic and a static part, and modify it using the knowledge that some variables have been instantiated. After a short general description of the calculation of probability distributions given evidence, i.e. the propagation procedure, we shall see how Hugin, applied to both the static and the dynamic part, would have propagated evidence. Finally we shall formulate a method for dividing the propagation procedure into a static part, performed by Hugin, and a dynamic part, performed without using Hugin. The proof of correctness of the method will be made by demonstrating that the proposed method performs calculations that are identical to those Hugin would have performed. However, it should be noted that the proposed method can be combined with other tools for calculation of probability distributions, i.e. that the method is not restricted to CPNs implemented in Hugin.

2 The Model

The central element in the diabetes advisory system is a CPN model of the glucose metabolism. Both the functionality of the system and the details of the model have been described previously (Andreassen et al. 1991, Hovorka et al. 1992, Andreassen et al. 1994). Consequently, the following description of the model is simplified and only contains the elements required to formulate the problem.

The model of glucose metabolism has two compartments. One compartment has the state variable CHO, which represents the reservoir of carbohydrates in the digestive system. In the model the variable appears in each "time slice". Fig. 1 shows three time slices, each separated by one hour. In the model arrows go from "parent" variables to "child" variables. The arrows indicate that for example the probability distribution for the carbohydrate content at time 1 (CHO-1) can be determined, conditional on the variable's parents: the carbohydrate content one hour earlier (CHO-0) and on the carbohydrate content of the meal eaten at time 1 (MEAL-1).

The other state variable, BG, represents the concentration of glucose in the blood. The arrows impinging on for example BG-1 state that the probability distribution for the blood glucose concentration at time 1 (BG-1) can be determined, conditional on four variables, CHO-0, BG-0, INS-SENS, and INS-ABS-0. INS-SENS represents the patients sensitivity to insulin, and INS-ABS-0 represents the average rate of insulin absorption during the first time slice.

The insulin is absorbed from the subcutaneous reservoirs created by the insulin injections, INS-INJ. Each injection creates a reservoir, and the arrows in the model state that the insulin absorption during a time slice depends on the absorption from the reservoirs created by all previous injections.

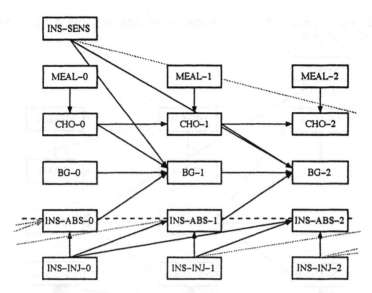

Fig. 1. Three time slices in the DIAS CPN. The dashed line divides the CPN into the upper "static" part that describes the metabolism of glucose, and the lower "dynamic" part that describes the injection and absorption of insulin.

3 The Problem

The DIAS CPN has about 500 nodes with an average of about 10 states each. The state space of DIAS thus has a size of about 10^{500}, a number that dwarfs the number of elentary particles in the universe. If Hugin is used to compile this CPN, i.e. to forming and initialising the cliques, this gives cliques with a total state space of about 10^{55} elements. Although this is a remarkable reduction, compared to the original 10^{500}, the problem is that it is still beyond current computer technology to store a state space of this size. This is in contrast to the 10^7 elements required to store the cliques from the part of the CPN above the dashed line, in the following referred to as the static part of the CPN. Adding the insulin injection variables, in the following referred to as the dynamic part of the CPN, creates many "cycles" (Jensen et al. 1990) which greatly increases the size of the cliques.

We want to develop a "dynamic" inference method that allows us to use Hugin to compile and propagate evidence in the static part of the CPN, while constructing and propagating in the dynamic part of the CPN using a method that can exploit the simplifications introduced by instantiating some of the variables in the dynamic part.

4 The Model Revisited

As mentioned, instantiation of variables may simplify a CPN, and as a particular simple example the insulin injection variables in Fig. 1 in the dynamic part are instantiated. This is reasonable in the sense that if we for example wish to use the CPN for simulating blood glucose, then it can be assumed that meals and insulin injections are known. Instantiation of the INS-INJ variables modifies the d-separations in the CPN (Pearl 1988). All d-connections going through the INS-INJ variables are eliminated. The CPN in Fig. 2 is constructed from the CPN in Fig. 1 by making additional copies of the INS-INJ variables, one extra copy for each INS-ABS variable that receives an arrow. Inspection of the two CPNs reveals that they have the same d-separation properties, and if all copies of a given

INS-INJ variable are instantiated to the same value, then the CPNs in Fig. 1 and Fig. 2 have identical properties.

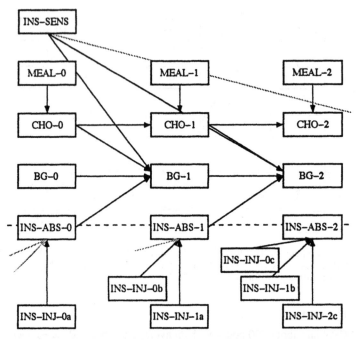

Fig. 2. Instantiation of the INS-INJ variables makes it possible to restructure the DIAS CPN, achieving a reduction of total state space to about 10^{50}. Each INS-INJ variable has been duplicated, once for each of its children.

At this point it is reasonable to view the ideas, described in this paper, as part of a three phase process reducing the number of elements required to store the tables: Phase one can be seen as the reduction from the 10^{500} elements, corresponding to the state space for the whole network in Fig. 1, to the state space of 10^{55} elements, corresponding to the estimated size of the tables holding the cliques after compilation of the network by Hugin. Phase two is the reduction of the state space when going from the network in Fig. 1 to the network in Fig. 2, the state space of the latter being estimated to about 10^{50} elements. Phase three is the reduction in size, to about 10^7 elements, obtained by implementing the "dynamic" inference method, to be described in the following.

5 The Problem Revisited

As the next step we restate the problem to be solved. We focus our attention on one of the INS-ABS variables. The INS-ABS variables lie on the border between the dynamic and the static part of the CPN. Fig. 3 shows a generalised version of the situation. In Fig. 3A the node C lies on the border between a set $D = \{D_1, D_2, ..., D_N\}$ of parent nodes belonging to the dynamic part of the CPN and a set of nodes $S = \{S_1, S_2, ..., S_M\}$ belonging to the static part of the CPN.

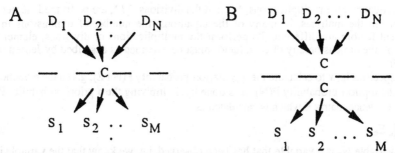

Fig. 3. A: The variable C lies on the border between the dynamic part of the CPN (with variables D_n) and the static part (with variables S_m). B: The CPN from A divided into a dynamic and a static CPN. The variable C on the border between the static and the dynamic part in A is duplicated and appears in both the dynamic and the static CPN.

To devise the dynamic propagation method, we will first detail the operations performed by Hugin during a propagation in the CPN in Fig. 3A. Subsequently, we shall devise the dynamic propagation method, in the CPN in Fig. 3B, such that the results from the two propagation methods are identical.

Before turning to the propagation in the networks in Fig. 3, the next section gives a short summary of the general propagation procedure. It should be noted that, in this summary, only the essential operations relating to the following sections are shown, and that a more comprehensive exposition of the algorithms and a full proof of correctness is given by Jensen et al. (1990).

6 Propagation in General

As mentioned in the introduction, propagation in a CPN is in principle simple. In a slightly modified form of the algebraic notation proposed by Jensen et al. (1990), the initialisation of a CPN with a set of variables $N = \{N_1, N_2, ..., N_N\}$ is specified by:

$$P(N) = \Phi \cdot P(T) \qquad \text{eq. 1}$$

and propagation of evidence E is specified by:

$$P(N \mid E) = P(N) \cdot E / P(E) \qquad \text{eq. 2}$$

In eq. 1 P(N) should be interpreted as a matrix, holding the joint probability distribution for the complete set of variables. P(N) is written in outline as a reminder that it is a matrix with a number of dimensions equal to the number of variables in N. The set of variables T $= \{T_1, T_2, ..., T_T\}$ is the set of top-variables in the CPN, i.e. variables that do not have parents. Thus, instead of having an associated table of conditional probabilities, each of them have an associated a priori probability $P(T_t)$. It is one of the properties of CPNs, that before instantiation of variables, all top-variables are independent. Therefore the joint a priori probability P(T) can be calculated as the product of the marginal a priori probabilities $P(T_t)$:

$$P(T) = \prod_t P(T_t) \qquad \text{eq. 3}$$

Φ represents the product of the conditional probability tables for the rest of the nodes R = $\{R_1, R_2, ..., R_R\}$, where R = N\T, and pa(R_r) represent the parents of the variable R_r:

$$\Phi = \prod_r P(R_r \mid pa(R_r)) \qquad \text{eq. 4}$$

In the equations the multiplications, "·", and the divisions, " / ", are performed element by element in the matrices. In many of the equations, the number of dimensions in the different factors are different. To perform the multiplications or divisions, element by element, the dimensionality of each factor must be extended as described by Jensen et al. (1990).

Eq. 2 specifies how to calculate the a posteriori probability $P(N \mid E)$, given the evidence E, from the a priori probability $P(N)$. It is done by multiplying the a priori probability $P(N)$ by the evidence matrix E which is calculated as:

$$E = \prod_e E_e \qquad\qquad\qquad\qquad \text{eq. 5}$$

If the variable N_e is a variable that has been observed, i.e. we know that the variable is in one of its p states with a probability of 1, then the likelihood vector, E_e, has p elements, where all elements are zero, except the element corresponding to the observed state. This element has the value 1. Extending the dimensionality of E and performing the multiplication, gives a matrix that is proportional to the a posteriori joint probability distribution for N. To make the elements in the matrix sum up to 1, it is normalised. The normalisation constant, i.e. the number that all elements should be divided by to make the elements sum to 1, is equal to the joint probability $P(E)$ of the evidence.

While these operations are conceptually very simple, it is obvious why they are not computationally convenient: As mentioned, the DIAS CPN has about 500 nodes with an average of about 10 states each, and the matrix $P(N)$ therefore has about 10^{500} elements.

The next two sections will describe the propagation in the networks in Fig. 3. First we will show how Hugin would have propagated in the network in Fig. 3A. Then we will devise an alternative method dividing the propagation, in Fig. 3B, into a propagation in the static part performed by Hugin and a propagation in the dynamic part performed without using Hugin.

7 Propagation by Hugin

In this section it will be stated how Hugin will perform the propagation in the CPN in Fig. 3A (Jensen et al. 1990). During the compilation of the CPN, the cliques are formed. Inside the cliques eq.s 1 and 2 are still used to initialise and update the matrices, but a mechanism is required for propagation of evidence between the cliques. This mechanism is provided by two procedures, called collect evidence and distribute evidence.

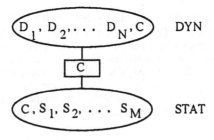

Fig. 4. Cliques and separation sets formed by compilation of the CPN in Fig. 3A.

Compilation

The first operation performed by Hugin is a compilation of the CPN into a junction tree. A junction tree consists of all the cliques, joined in a tree structure. A CPN can be compiled into many different junction trees, but we will assume that the junction tree has the structure given in Fig. 4. The junction tree in Fig. 4 has two cliques, DYN, with the nodes D and C

and STAT, with the nodes C and S. All cliques in the junction tree are separated by a separation set, which in this case contains the node C. The structure of the cliques is determined by the so-called elimination sequence, and a junction tree with the above mentioned cliques can be generated by eliminating the nodes in **D** first. A description of methods for compilation of CPNs is given by Kjærulff (1993).

Initialisation
Each clique has an associated data structure, called the belief table **B**. Hugin applies eq. 1 to the clique DYN during the initialisation and the result of the calculation is stored in the clique's belief table:

$$B(D_1, D_2, ..., D_N, C) = P(C \mid D_1, D_2, ..., D_N) \cdot \prod_i P(D_i) \qquad \text{eq. 6}$$

The symbol **B** has been chosen to distinguish it from the joint probability **P**. After initialisation and after a complete propagation the data structure **B** contains the joint probability for the clique, but this is not true during the different steps in the propagation. However, at this time we can note that $B(D_1, D_2, ..., D_N, C)$ was calculated according to eq. 1, and that **B** therefore at this time holds the joint probability of the DYN clique:

$$P(D_1, D_2, ..., D_N, C) = B(D_1, D_2, ..., D_N, C) \qquad \text{eq. 7}$$

Hugin similarly initialises the belief table for the clique STAT:

$$B(STAT) = \Phi_{STAT} \cdot P(T) = \Phi^*_{STAT} \cdot P(C) \qquad \text{eq. 8}$$

In eq. 8 the a priori probabilities from all top-variables, except C, have been lumped with the conditional probabilities, i.e. $\Phi^*_{STAT} = \Phi_{STAT} \cdot P(T \backslash C)$. Since $B(STAT)$ is calculated according to eq. 1 it holds the joint probability:

$$P(STAT) = B(STAT) \qquad \text{eq. 9}$$

The separation set also has an associated table, a separation table **S**. Hugin initialises the separation table by marginalisation of **B**:

$$S(C) = \sum_{D_1, D_2, ..., D_N} B(D_1, D_2, ..., D_N, C) \qquad \text{eq. 10}$$

We can again note that at this time the separation set holds the probability of C:

$$P(C) = S(C) \qquad \text{eq. 11}$$

Insertion of evidence
The next step in the propagation is the insertion of evidence in the cliques. This is done by carrying out multiplication of the first two factors in eq. 2, $P(N) \cdot E$. To do this we divide the evidence into evidence relevant for each of the cliques, such that $E = E_{DYN} \cdot E_{STAT}$. The result of the multiplication is stored in the cliques belief tables, and we shall denote the belief tables with evidence inserted by **B***:

$$B^*(D_1, D_2, ..., D_N, C) = B(D_1, D_2, ..., D_N, C) \cdot E_{DYN} \qquad \text{eq. 12}$$

$$B^*(C, S_1, S_2, ..., S_M) = B(C, S_1, S_2, ..., S_M) \cdot E_{STAT} \qquad \text{eq. 13}$$

Collection of evidence and normalisation
The insertion of evidence is followed by collection of evidence. We assume that evidence is collected from the DYN clique to the STAT clique. In that case the collect proceeds as follows. A new table for the separation set is calculated from the belief table for the clique DYN:

$$S^*(C) = \sum_{D_1, D_2, ..., D_N} B^*(D_1, D_2, ..., D_N, C) \qquad \text{eq. 14}$$

At this time the table for the separation set does not contain a probability, but we shall note without proof that the following relation holds:

$$P(C \mid E_{DYN}) = S^*(C) / P(E_{DYN}) \qquad \text{eq. 15}$$

Evidence is now propagated from the DYN to the STAT cliques by multiplying the belief table for STAT with evidence inserted by the ratio between the old and the new separation table:

$$B^{**}(C, S_1, S_2, ..., S_M) = B^*(C, S_1, S_2, ..., S_M) \cdot S^*(C) / S(C) \qquad \text{eq. 16}$$

As the last part of the collect procedure Hugin normalises the belief table for STAT by dividing it by a number, called the normalisation constant. We note, again without proof, that the value of the normalisation constant is equal to the joint probability of the inserted evidence.

$$B^{***}(C, S_1, S_2, ..., S_M) = B^{**}(C, S_1, S_2, ..., S_M) / P(E) \qquad \text{eq. 17}$$

After normalisation we can note that B^{***} contains the joint probability, such that:

$$P(C, S_1, S_2, ..., S_M \mid E) = B^{***}(C, S_1, S_2, ..., S_M) \qquad \text{eq. 18}$$

Distribution of evidence
To finish the propagation the distribute evidence procedure is executed. This procedure first calculates a new separation table by marginalisation of the belief table for the STAT clique:

$$S^{**}(C) = \sum_{S_1, S_2, ..., S_N} B^{***}(C, S_1, S_2, ..., S_M) \qquad \text{eq. 19}$$

Finally, the evidence from the STAT clique is propagated back to the DYN clique. This is done by multiplying the belief table for the upper clique by the ratio between the two separation tables $S^{**}(C) / S^*(C)$:

$$B^{**}(D_1, D_2, ..., D_N, C) = B^*(D_1, D_2, ..., D_N, C) \cdot S^{**}(C) / S^*(C) \qquad \text{eq. 20}$$

This completes the propagation of evidence in the junction tree in Fig. 4.

8 Dynamic Propagation by Instantiation

We now divide the CPN in two parts, the dynamic and the static part. We let the variable C appear in both parts (Fig. 3B). Compilation of the dynamic part yields a clique identical to the clique DYN in Fig. 4 and compilation of the static part can yield a clique identical to the clique STAT in Fig. 4. It is now clear how we in principle can perform a complete propagation where Hugin is only used to propagate in the static part of the CPN. We simply perform all the calculations concerning the DYN clique without using Hugin, and let all calculations concerning the STAT clique be performed by Hugin. Stated in more detail, we do the propagation in 3 steps:

Dynamic propagation, version 1:

Step 1) Outside Hugin:

Initialise:	$B(DYN) = P(C \mid D) \cdot P(D)$	(eq. 6)
	$S(C) = \sum_D B(DYN)$	(eq. 10)
Insert evidence:	$B^*(DYN) = B(DYN) \cdot E_{DYN}$	(eq. 12)
Calc. separator:	$S^*(C) = \sum_D B^*(DYN)$	(eq. 14)

Step 2) Hugin:

Initialise	$B(STAT) = \Phi^*_{STAT} \cdot S(C)$	(eq. 8)
Insert evidence:	$B^*(STAT) = B(STAT) \cdot E_{STAT}$	(eq. 13)
Transfer evidence:	$B^{**}(STAT) = B^*(STAT) \cdot S^*(C) / S(C)$	(eq. 16)
Normalise:	$B^{***}(STAT) = B^{**}(STAT) / P(E)$	(eq. 17)
Calc. separator:	$S^{**}(C) = \sum_S B^{***}(STAT)$	(eq. 19)

Step 3) Outside Hugin:

Calculate:	$B^{**}(DYN) = B^*(DYN) \cdot S^{**}(C) / S^*(C)$	(eq. 20)

This three step procedure leaves us with two problems: The first is that performing the calculations outside Hugin in itself does not make them easier. In the case of DIAS it remains impractical to calculate $S(C)$. This leads to the second problem, since step 2 calls for initialisation of the static part with the a priori probability of C equal to $S(C)$, which apparently is not available.

To solve these two problems we first note that it may be easier to calculate $P(C \mid E_{DYN})$ than $S(C)$. $S(C)$ is the probability distribution for C with the evidence inserted in the dynamic part. If that is true, then we can propose the following version of the dynamic propagation method:

Dynamic propagation, version 2:

Step 1) Outside Hugin:
Calculate: $P(C \mid E_{DYN})$

Step 2) Hugin:

Initialise:	$B(STAT) = \Phi^*_{STAT} \cdot S'(C)$,	(eq. 8)
	where $S'(C)$ is a randomly chosen positive probability distribution.	
Insert evidence:	$B^*(STAT) = B(STAT) \cdot E_{STAT}$	(eq. 13)
Transfer evidence:	$B^{**}(STAT) = B^*(STAT) \cdot P(C \mid E_{DYN}) / S'(C)$	(eq. 16)
Normalise:	$B^{***}(STAT) = B^{**}(STAT) / N$,	(eq. 17)
	where N is the normalisation constant required to make the elements of B^{***} sum up to 1.	
Calc. separator:	$S^{**}(C) = \sum_S B^{***}(STAT)$	(eq. 19)

Step 3) Outside Hugin:

Calculate:	$B^{**}(DYN) = B^*(DYN) \cdot S^{**}(C) / S^*(C)$	(eq. 20)

To show that the two versions of the propagation give the same result, we can insert eq.s 16, 13 and 8 into eq. 17. For version 1 and 2 this gives, respectively:

Version 1:

$$B^{***}(STAT) = \Phi^*_{STAT} \cdot S(C) \cdot E_{STAT} \cdot S^*(C) / (S(C) \cdot P(E))$$
$$= \Phi^*_{STAT} \cdot E_{STAT} \cdot S^*(C) / P(E) \qquad \text{eq. 21}$$

Version 2:

$$B^{***}(STAT) = \Phi^*_{STAT} \cdot S'(C) \cdot E_{STAT} \cdot P(C \mid E_{DYN}) / (S'(C) \cdot N)$$
$$= \Phi^*_{STAT} \cdot E_{STAT} \cdot P(C \mid E_{DYN}) / N \qquad \text{eq. 22}$$

Insertion of eq. 15 into eq. 22 gives:

$$B^{***}(STAT) = \Phi^*_{STAT} \cdot E_{STAT} \cdot S^*(C) / (P(E_{DYN}) \cdot N) \qquad \text{eq. 23}$$

From eq. 21 and 23 we can see that the expressions for B^{***} are identical, except for the normalisation factors, $P(E)$ and $P(E_{DYN}) \cdot N$, respectively in version 1 and 2. Since we also know that B^{***} was normalised in both versions, the two expressions must be identical, and accordingly:

$$P(E) = P(E_{DYN}) \cdot N \qquad \text{eq. 24}$$

Since:

$$P(E) = P(E_{DYN}) \cdot P(E_{STAT} \mid E_{DYN}) \qquad \text{eq. 25}$$

it follows that:

$$N = P(E_{STAT} \mid E_{DYN}) \qquad \text{eq. 26}$$

We have now completed the task of devising an alternative method for propagation in the CPN in Fig. 3A, and we shall now return to the DIAS example to apply the method.

9 Dynamic Propagation by Instantiation in DIAS

In Fig. 2 the dynamic part of the DIAS CPN has been restructured and divided into a number of small dynamic CPNs, each consisting of an insulin absorption variable and its parents, i.e. the insulin injection variables. The proposed dynamic propagation algorithm can be used to propagate in each of the small dynamic CPNs.

In DIAS the insertion of evidence $E_{DYNAMIC}$ in the dynamic part will correspond to instantiation of all the insulin injection variables. The specification of the insulin absorption, given the insulin injections is such that P(INS_ABS | INS_INJ) can be calculated by adding a few normally distributed variables (Andreassen 1992a). The assumption in version 2 of the dynamic propagation method, that $P(C \mid E_{DYNAMIC})$ is easy to calculate is thus valid, making step 1 simple.

In this connection it can be noted that in applications where the insertion of the evidence $E_{DYNAMIC}$ does not make the calculation of $P(C \mid E_{DYNAMIC})$ trivial, Hugin can be used to calculate $P(C \mid E_{DYNAMIC})$. This can be done by first restructuring the dynamic part to exploit the modified d-separation properties of the CPN, and then using Hugin to compile and propagate in the restructured dynamic part of the CPN.

It can also be noted that in the DIAS example step 3 can be omitted. In the dynamic part all variables, except INS_ABS are instantiated. The probability distribution for INS_ABS is already known from the propagation in the static part of the CPN, so nothing new can be learned from the execution of step 3. This only leaves step 2, which is a normal propagation in the static part of the CPN. As previously mentioned, a propagation in the static part of the CPN only requires storing of a state space of 10^7, which is unproblematic.

So far we have only discussed the simplifications that can be obtained by exploiting that some variables in the dynamic part of the CPN have been instantiated. The idea of dividing the CPN into two communicating parts, a *dynamic* and a *static* part, has a value in itself. By choosing the names dynamic and static for the two parts we have tried to express that in some situations there may be a part of the CPN that can be used in all the situations where the CPN is applied, and another part of the CPN that must be modified to apply to the situation at hand. In the case of DIAS the *static* part represents the model of glucose metabolism. This model is applied to all patients.

The *dynamic* part represents the model for absorption of insulin from the subcutaneous injections. In the model in Fig. 1, and as a consequence also in the model in Fig. 2, it is implicitly assumed that injection only take place on the hour. For practical use this is too crude a time resolution, and it would be necessary to expand the CPN with insulin injection variables that represent possible insulin injections with a time resolution of for example 15 min. With the dynamic propagation method outlined above it would be possible to handle such a CPN, but it would be both conceptually clumsy and tedious to construct the CPN. Instead we chose to construct the dynamic part of the CPN such that we first conceptually build a CPN with the desired time resolution. Subsequently we omit all insulin injection variables that represent points in time where injections did not take place for the patient currently being considered. This means that the *dynamic* part needs to be recompiled for each patient, but the separation of the dynamic and the static part also means that the *static* part does not need to be recompiled.

In DIAS, Hugin was not used to do the calculations in the dynamic part, and the simplifications due to the omission of insulin injection variables were instead used to simplify the calculation of P(INS_ABS I INS_INJ) outside Hugin.

10 Discussion

In many applications, the algorithms offered by Hugin make it possible to do propagation of evidence in large CPNs. In the DIAS application the total state space is about 10^{500}. Hugin makes it possible to propagate evidence in DIAS, by dividing the DIAS CPN into cliques that together have a total state space of about 10^{55}. While this an astronomical reduction in the memory required to store the state space, it is still impractical.

A dynamic propagation algorithm was proposed that reduced the size of the state space to be stored to about 10^7, which is within reach of current computer technology. This reduction was achieved in two stages. In the first stage the CPN was divided in a static and a dynamic part, and the knowledge, that some of the variables in the dynamic part were instantiated, was used to restructure the CPN. This reduced the size of the state space by a factor of 10^5 to about 10^{50}. This is still too large for practical computations, and a further reduction to a state space of 10^7 was achieved by simplifying the calculation of the probability distribution of the variables on the border between the static and the dynamic part. This calculation could also be simplified as a consequence of the knowledge that some variables in the dynamic part are instantiated. At the same time this makes it easy to build the dynamic part to suit the individual patient, without needing to recompile the static part of the CPN.

A state space of 10^7 is still a bit bulky, but can be managed. In DIAS the assumption that the variables, representing insulin injections, are instantiated is usefull. The assumption is valid when the CPN is used to simulate the time course of blood glucose. It is also true when the CPN is used to adjust the doses of the injected insulin. In that situation the insulin injections are considered to be decision variables, i.e. variables whose value can be determined by a decision. In the traditional influence diagram approach to therapy planning, the decision variables are not considered to be instantiated (Howard and Matheson 1984), but in an alternative approach to planning (Andreassen 1992b) it is possible to assume that the decision variables are instantiated. This makes it possible to use the DIAS CPN also for planning of insulin therapy (Andreassen et al. 1994).

The savings in memory and computation time achieved by the application of the proposed dynamic propagation algorithm depends on the actual application, in this case DIAS. It is likely that substantial simplifications can also be obtained in many other applications by dividing the CPN into a static and a dynamic part and exploiting the knowledge about which variables that are going to be instantiated.

11 References

1) Andersen, S.K., Olesen, K.G., Jensen, F.V., and Jensen, F., 1991. HUGIN - A shell for building Bayesian belief universes for exper systems. In: *Proceedings of IJCAI 89 2*, pp. 1080-1085.

2) Andreassen, S., Jensen, F.V., Andersen, S.K., Falck, B., Kjærulff, U., Woldbye, M., Sørensen, A.R., Rosenfalck, A., and Jensen F. 1989. MUNIN - An expert EMG assistant. *Computer-Aided Electromyography and Expert Systems*, (ed. Desmedt, J. E.) Elsevier, pp 255-277.

3) Andreassen, S., Hovorka, R., Benn, J., Olesen, K.G., and Carson, E., 1991. A model-based approach to insulin adjustment. In: *Lecture Notes in Medical Informatics*, (eds. M. Stefanelli, A. Hasman, M. Fieschi, and J. Talmon), vol. 44, pp 239-249, Proc. of AIME '91, Springer Verlag.

4) Andreassen, S., 1992a. Knowledge representation by extended linear models. In: *Deep Models for Medical Knowledge Engineering* (ed. E. Keravnou), Elsevier, pp 129-145.

5) Andreassen, S., 1992b. Planning of therapy and tests in causal probabilistic networks. *Artificial Intelligence in Medicine*, 4, 227-241.

6) Andreassen, S., Benn, J.J., Hovorka, R., Olesen, K.G., and Carson, E.R., 1994. A probabilistic approach to glucose prediction and insulin dose adjustment - Description of a metabolic model and pilot evaluation study. *Computer Methods and Programs in Biomedicine*, 41, 153-165.

7) Jensen, F.V., Lauritzen, S.L., and Olesen, K.G., 1990. Bayesian updating in causal probabilistic networks by local computations. *Computational Statistics Quarterly*, 4, 269-282.

8) Howard, R.A.,□and Matheson, J.E., 1984. Influence diagrams. In: *The principles and application of decision analysis* (eds. Howard, R.A.,□and Matheson, J.E.), vol. II, Ch. 37, Srategic decision group, pp. 719-762.

9) Hovorka, R., Andreassen, S., Benn, J.J., Olesen, K.G., and Carson, E.R., 1992. Causal probabilistic network modelling. An illustration of its role in the management of chronic diseases. *IBM Systems Journal*, 31(4), 635-648. Reprinted in: *1993 Yearbook of Medical Informatics* (eds. J.V. Bemmel and A.T. McCray), Stuttgart: Schattauer-IMIA, 1993, pp. 328-340.

10) Kjærulff, U., 1993. *Aspects of efficiency improvements in Bayesian networks*, Ph.D. Thesis, Aalborg University.

11) Lauritzen, S.L., and Spiegelhalter, D.J., 1988. Local computations with probabilities on graphical structures and their application to expert systems. *J. Royal Statist. Soc.*, B50, 157-224.

12) Pearl, J., 1986. Fusion, propagation and structuring in belief networks. *Artificial Intelligence*, 29, 241-288.

13) Pearl, J., 1988. *Probabilistic reasoning in intelligent systems; Networks of plausible inference*, Morgan Kaufmann.

Patient Management and Therapy Planning

Alerts as Starting Point
for Hospital Infection Surveillance and Control

Edith Safran[1], Didier Pittet[2], François Borst[1], Gérald Thurler[1], Marcel Berthoud[1], Pascale Schulthess[1], Pascale Copin[2], Valérie Sauvan[2], Anna Alexiou[2], Ludovic Rebouillat[1], Mathieu Lagana[1], Jean-Philippe Berney[1], Peter Rohner[3], Raymond Auckenthaler[3] and Jean-Raoul Scherrer[1]

[1]Centre d'Informatique Hospitalière; [2]Groupe de Prévention et Contrôle de l'Infection; [3]Laboratoire Central de Bactériologie; Hôpital Cantonal Universitaire, 24 rue Micheli-du-Crest, CH-1211 Geneva 14

Abstract

Expert systems methodology investigated within the GAMES-II European Union research project was applied in the Hôpital Cantonal Universitaire of Geneva to infection control, and in particular to the control of methicillin-resistant *Staphylococcus aureus* (MRSA) outbreak.

For detecting MRSA patients, alerts were implemented by periodic queries to the databases of the hospital information system (HIS). Two mechanisms of alerts were established. The 'lab alert' is based on recent laboratory results and detects all patients with a positive result for MRSA found the day before. The 'readmission alert' identifies patients previously known to have been colonized by or infected with MRSA if and when they are readmitted to the hospital. When useful, alert outputs include guidelines about infection control measures to be taken. Receipt of the alert is the initial step for additional patient data collection by the infection control nurses. These data are being recorded in a patient chart structured to be linked to HIS data.

The 'lab alert' detected, in the period from March till end of October 1994, 1415 lab results positive for MRSA corresponding to 307 infected or colonized patients in the hospital, other medical institutions or ambulatory care. The main advantage of this alert lies in improved work organization and efficacy for the infection control nurses.

The 'readmission alert' detected 150 admissions of patients previously known to have been colonized by or infected with MRSA over a period from June 9 till October 31, 1994. The 150 admissions corresponded to 107 different patients. Main task of this alert was found in the field of prevention.

Computer alerts using data routinely stored in the HIS were found very useful for several medical tasks, in particular for prevention and work organisation, and as starting point to create an extended computerized patient record. Alerts represent an application of telematic computer technology to improve quality of care and reduce health care costs at low additional computer efforts, and their use should be promoted as starting point for a medical action.

1. Introduction

Nosocomial infections represent an important public health problem. They are a major cause of increased morbidity, mortality and costs in hospitals [1-4]. A report published in 1978 even suggests that there were four times as many nosocomial infections in US per year as admissions for acute myocardial infarction [5], and it was suggested that nosocomial infections could be placed among the 10 leading causes of death in the US [6].

The need for computer systems as an aid for nosocomial surveillance is well recognized [7-10]. Within the GAMES-II European Union research project [11], the role of Geneva was precisely to investigate this medical domain in relation to expert system methodology. This was performed by testing methods and tools developed within the project. In particular, practical medical work was matched to a model of reasoning, the Select and Test model [11]. Computer alerts [12-14] were found as corresponding to abduction reasoning and were applied to the running computer environment of the Hôpital Cantonal Universitaire of Geneva [15].

An alert system was implemented and was applied to fight MRSA outbreak, an important nosocomial problem in many hospitals and ambulatory settings of medical institutions.

Staphylococcus aureus was identified as causing hospital acquired infection as long ago as 1883-4 [16]. Short after introduction of methicillin, in the early 1960s, methicillin-resistant *Staphylococcus aureus* (MRSA) appeared and since then number of patients affected are escalating in hospitals. Transmission of *Staphylococcus aureus* is mainly due to direct contact mediated by transient carriage on the hands of nurses and medical staff [17-20].

Early and appropriate therapy and isolation measures for MRSA cases are important to avoid dissemination, with the consequences of increased mortality, morbidity or costs. In addition, since MRSA tends to become resistant to multiple antibiotics, it is important to control the use of antibiotics still active against this organism, in order to prevent the appearance of resistance.

Control policies should then aim at limiting the number of cases, and this task can be particularly well assisted by computer technology.

2. Method

Several populations of MRSA patients were considered: hospitalized patients with new lab results positive for MRSA, patients once detected by the hospital lab and who are readmitted to the hospital and patients who were never detected by the hospital lab. Data useful for detecting the first two populations of patients are microbiological data and patient admission-movement data which can be found in the HIS databases.

Alerts established were based on periodic queries to the clinical microbiology laboratory database or the admission/discharge patient database of the Hospital Information System (HIS) of the Hôpital Cantonal Universitaire of Geneva, DIOGENE [15,21,22]. This was possible within its distributed architecture with laboratories relational database systems.

All alerts are printed in the hospital infection control team offices. In case of emergency readmission to the hospital, the alert is also sent to the emergency hospital admission center.

2.1. MRSA Alert on the Clinical Microbiology Laboratory Database

Among hospitalized patients, new MRSA cases are detected by the clinical microbiology laboratory, if microbiological exams have been performed. Lab results are then routinely sent to the care unit where usually the personnel is not specially trained to take in charge MRSA cases.

The role of the infection control team is to give in the care unit specialized advice for each case detected, and to check compliance with control measures. For that, the team needs to be systematically advised of new cases. For this purpose, a daily alert was established on the DIOGENE bacteriological database. Everyday at 1:15 pm, an alert is printed in the infection control team office. It contains a list of patients with positive result for MRSA the previous day and the same morning; this list includes new cases and results of screening samples. With this list, the infection control nurses go to the ward to check the newly detected patients and decide actions to be taken, in collaboration with the ward personnel there, such as decontaminate the patient, request additional laboratory exams and take isolation measures.

The same mechanism of alert is used to produce a list of screening sample performed with negative results for MRSA.

These two alerts, the first with positive results for MRSA, and the second with negative screening results, allow to prospectively monitor the efficiency of the treatment procedures during the entire hospital stay, allowing iterative adjustments in treatment, which are sometimes even made by direct phone calls between the infection control nurse and the nurse on the ward caring for the MRSA patient.

The hospital laboratory giving services also to hospital ambulatory services, other medical institutions and some private physicians, the 'lab alert' detects not only hospitalized patients, but also patients in other institutions or ambulatory care.

2.2. 'Possible MRSA' Alert on the Patient Admission/movement Database: Detection of Previously Known Cases upon Readmission to the Hospital

Patients detected by the 'lab alert' may remain MRSA carriers, and in case of readmission to the hospital, control measures, such as isolation precautions and lab exams, should be taken immediately, in order to prevent transmission of MRSA and possible subsequent infection of other patients. Persistant MRSA carriage may also indicate specific changes in antibiotic perioperative prophylaxis. Persistance of the carrier state may be caused by lack of decontamination or by spontaneous recolonisation after decontamination [18]. Recent data demonstrated that MRSA carriers could remain colonized as long as two years.

In order to adequately reuse information obtained through laboratory database, a computer alert was established to detect patients known to have been previously colonized or infected with MRSA, if and when they are readmitted to the hospital. This alert was implemented taking into account information both from the patient admission/discharge database and from a 'MRSA file' containing the identification numbers of previously detected cases. When an identification number of an admitted or discharged patient matches with a number present in the 'MRSA file', an alert is generated and produces the printing of an advice in the infection control team offices.

The alerts are also directly printed in the emergency center for emergency cases. In this case, the alert outputs also include guidelines concerning isolation measures to take, microbiological samples to be obtained and infection control team telephone numbers for further information.

Update of the 'MRSA file' with new cases detected is automatically performed daily using the 'lab alert' output.

3. Results

The 'lab alert' runs since March 1994 and detected up to end of October a total of 307 infected or colonized patients, among whom 209 were either hospitalized or treated in ambulatory care in the hospital. A total of 1415 laboratory results positive for MRSA have been found, representing an average of 4.6 per patient. The 'lab alert' has important consequences for the care given to the diseased patient, for protection of non-diseased patients, for reporting of infection, and for teaching to the personnel. Its main consequence is for rationalization of the infection control nurses work since every day the nurses wait upon the alert to take care of all MRSA patients detected by the 'lab alert' and they also use information received through alerts for monitoring the patients.

The 'readmission alert' detected between June 9 and October 31, 1994, 150 admissions of patients previously known to have been colonized by or infected with MRSA. These admissions corresponded to 107 different patients. The alert has an important role for protecting other patients from acquiring MRSA infection, and its main task is therefore in the field of prevention.

4. Discussion

The alerts described in this paper are extremely useful to improve work organization of the infection control team, as well as of the emergency department and ward teams.

Before the implementation of the readmission alert, average time for recognition of potential MRSA carriers in the hospital was 45 days (median 25, range 0-225) [D. Pittet, manuscript in preparation], whereas now immediate alerting upon patient readmission allows infection control strategies to be set up immediately.

Alerts are the starting point for monitoring and treating the patients from an infection control perpective. Additional data on patients detected by the alerts are collected by the infection control nurses in the wards and recorded in a computerized patient record which is structured to be linked to HIS data. The resulting extended patient record, based on data at our institution, will help more objectively define the natural history of MRSA infection and will allow further definition of at risk patient profile constructed with local risk factors for this infection.

For example, some specific factors are already known to predispose to MRSA infection. These factors include older age, preexisting comorbidities, previous hospitalization, poor functional status, heavy patient census while critically ill patients are colonized with MRSA, and treatment with multiple antibiotics (in a study reported in 1993, acquisition of MRSA nasal carriage occurred in patients given a mean of 5.1 antibiotics [23]).

Knowledge on MRSA patient profile will be the basis for suggesting guidelines to treating physicians and might be used to build rules for an expert system.

5. Conclusion

Computer alerts using data from the hospital information system (HIS) were applied to MRSA epidemics and found very useful for several medical tasks, such as work organization, care given to the diseased patient, protection of non-diseased patients, reporting of infection, and teaching to the personnel. Their main utility was found in the field of prevention and work organization.

The alerts are used to design a system centered around patients detected, as recently proposed by Charles Safran [24]. The alerts are starting points for medical actions - prevention, treatment and monitoring - , as well as for establishing an extended computerized patient record which might be linked to an expert system.

Alerts systems represent an application of telematic computer technology reusing information to improve quality of care and reduce health care costs at low additional computer efforts. They are a valuable tool for surveillance in clinical epidemiology [25,26]. Their use should be promoted as starting point for a medical action.

6. Acknowledgements

This work is part of the AIM (Advanced Informatics in Medicine) European Union project A-2034 entitled "GAMES-II: A General Architecture for Medical Expert Systems", and was supported in part by the Swiss government OFES (Office fédéral de l'éducation et de la science), contract AI010.

7. References

[1] Haley R.W. : The nationwide nosocomial infection rate. A new need for vital statistics. Am J Epidemiol, 1985, 121, 159,167.

[2] Martone W.J., Jarvis W.R., Culver D.H., Haley R.W. : Incidence and nature of endemic and epidemic nosocomial infections. In: Hospital Infections (Ed. Bennet J.V., Brachman P.S.), 3rd ed., Little, Brown & Company, Boston, 1992, pp. 577-96.

[3] Pittet D. : Nosocomial bloodstream infections. In : Prevention and Control of Nosocomial Infections (Ed. Wenzel R.P.), 2nd ed. Williams & Wilkins. Baltimore, 1993, pp. 512-55.

[4] Pittet D., Tarara D., Wenzel R.P. : Nosocomial bloodstream infection in critically ill patients: excess length of stay, extra costs, and attributable mortality. JAMA., 1994, 271, 1598-601.

[5] United States National Center for Health Statistics. Utilization of short-stay hospitals: annual summary for the United States, 1978, Hyattsville, MD: National Center for Health Statistics, 1978 (Data from the National Hospital Discharge Survey, Series 13, no. 46). (DHEW publication no. (PSH)80-1797), quoted by [1].

[6] Centers for Disease Control. Ten leading causes of death in the United States: 1977. Atlanta, GA: Centers for Disease Control, 1980. (CDC publication no. (CDC)99-725), quoted by [1].

[7] Evans R.S., Larsen R.A., Burke J.P., Gardner R.M., Meier F.A., Jacobson J.A., Conti M.T., Jacobson J.T., Hulse R. : Computer surveillance of hospital-acquired infections and antibiotic use. JAMA, 1986, 256, 1007-11.

[8] Wenzel R.P., Pfaller M.A. : Infection control: the premier quality assessment program in United States hospitals. Am J Med, 1991, 91 (suppl.3B), 27-31.

[9] Perl T.M. : Surveillance, reporting, and the use of computers. In: Prevention and Control of Nosocomial Infections, (Ed. Wenzel R.P.), 2nd ed, Williams & Wilkins, Baltimore, 1993, pp. 139-76.

[10] Reagan D.R. : Computer use in infection control. In: Prevention and Control of Nosocomial Infections. (Ed. Wenzel R.P.), 2nd ed, Williams & Wilkins. Baltimore, 1993, pp. 981-92.

[11] Stefanelli M., Bellazzi R., Berzuini C., Bugliesi M., Calzadilla J., Hunter J., Kolary P., Lanzola G., Leaning M., Mantas J., Moustakis V., Qualini S., Ramoni M., Ruggieri C., Zambon F. : GAMES: A general architecture for medical expert systems. In: Advances in Medical Informatics (Ed. Noothoven van Goor J., Christensen J.P.), IOS Press, 1992, pp. 133-140.

[12] Hripcsak G., Clayton P.D., Pryor T.A., Haug P., Wigertz O.B., Van der Lei J. : The Arden syntax for Medical Logic Modules. In : Proc. of Fourteenth Annual Symposium on Computer Applications in Medical Care, (Ed. Miller R.A.), IEEE Computer Society Press, New York, 1990, pp. 200-4.

[13] Rind D.M., Safran C., Phillips R.S., Slack W.V., Calkins D.R., Delblanco T.L., Bleich H.L: The effect of computer-based reminders on the management of hospitalized patients with worsening renal function. In : Proc. of Fifteen Annual Symposium on Computer Applications in Medical Care. (Ed. Clayton P.D.), McGraw Hill, Inc., New York, 1992, pp. 28-32.

[14] Rind D.M., Safran C., Phillips R.S., Wang Q., Calkins D.R., Delblanco T.L., Bleich H.L., Slack W.V. : Effects of computer-based alerts on the treatment and outcomes of hospitalized patients. Arch Intern Med, 1994, 154, 1511-7.

[15] Safran E., Borst F., Thurler G., Pittet D., Lovis Ch., P. Rohner P., Auckenthaler R., Scherrer J.-R. : An evolutive alert system for diagnosis of nosocomial infections based on archiving distributed databases. In: Artificial Intelligence in Medicine, (Ed. Andreassen S., Engelbrecht R., Wyatt J.), IOS Press, Amsterdam, 1993, pp. 180-4.

[16] Rosenbach J. Microorganismen bei den Wundinfektionskrankheiten des Menschen. Wiesbaden 1884, quoted by [19].

[17] Wenzel R.P., Nettleman M.D., Jones R.N., Pfaller M.A. : Methicillin-resistant Staphylococcus aureus: implications for the 1990s and effective control measures. Am J Med 1991, 91, 3B-221S-238S.

[18] Panlilio A.L., Culver D.H., Gaynes R.P., Banerjee S., Henderson T.S., Tolson J.S., Martone W.J. : Methicillin-resistant *Staphylococcus aureus* in U.S. hospital, 1975-1991. Infection Control and Hospital Epidemiology, 1992, 13, 582-6.

[19] Gordon J. : Clinical significance of methicillin-sensitive and methicillin-resistant *Staphylococcus aureus* in UK hospitals and the relevance of povidone-iodine in their control. Postgrad Med J 1993, 69 (Suppl. 3), S106-S116.

[20] Mulligan M.E., Murray-Leisure K.A., Ribner B.S., Standiford H.C., John J.F., Korvic J.A., Kauffman C.A., Yu V.L. : Methicillin-resistant Staphylococcus aureus: a consensus review of the microbiology, pathogenesis, and epidemiology with implications for prevention and management. Am J Med 1993, 94, 313-328.

[21] Scherrer J.-R., Baud R., Hochstrasser D., Ratib O. : The DIOGENE Hospital Information System. MD Computing, 1990, 7, 81-9, 1990.

[22] Safran E., Pittet D., Borst F., Thurler G., Schulthess P., Rebouillat L., Lagana M., Berney J.-P., Berthoud M., Copin P., Rohner P., Lew D., Auckenthaler R., Scherrer J.-R. : Alertes informatiques et qualité des soins : application à la surveillance des infections hospitalières. Revue Médicale de la Suisse Romande, 1994, 114, 1035-43.

[23] Boyce J.M., Landry M., Deetz T.R., Dupont H.L.: Epidemiologic studies of an outbreak of nosocomial methicillin-resistant *Staphylococcus aureus* infections. Infect. Control. 1981, 2 : 110-116.

[24] Safran C: Editorial. MD Computing 1994, 11: 133-4.

[25] Sackett D.L., Haynes R.B., Guyatt G.H., Tugwell P. : Clinical epidemiology. A basic science for clinical medicine. 2nd ed. Little, Brown and Company, Boston, 1991.

[26] Garnerin Ph., Saidi Y., Valleron A.-J. : The French Communicable Diseases computer network: a seven year experiment. In : Extended clinical consulting by hospital, (Ed. Parsons D.F., Fleisher C.M., Greenes R.A.), Annals of the New York Academy of Sciences, 1992, 670, 29-42.

Cooperative Software Agents for Patient Management

Giordano Lanzola, Sabina Falasconi and Mario Stefanelli
University of Pavia, Dept. of Informatics and Systems Science
Medical Informatics Laboratory
Via Abbiategrasso 209 - 27100 Pavia - Italy
{giordano,sabina,mario@ipvstefa.unipv.it}

Abstract. Managing patients in a shared-care context is a knowledge intensive activity. To support cooperative work in medical care, computer technology should either augment the capabilities of individual specialists and enhance their ability of interacting with each other and with computational resources. Thus, a major shift is needed from centralized first generation Hospital Information Systems to distributed environments composed of several interconnected agents, cooperating in maintaining a full track of the patient clinical history and supporting health care providers in all the phases of the patient management process. In this paper we describe a methodology for implementing a network of cooperating software agents aimed at improving the health care delivery process. Moreover, a preliminary computational prototype exploiting the proposed methodology is also illustrated.

1 Introduction

The management of a patient in a shared-care context is a knowledge intensive activity. Health care providers must be able to exchange information and share a common understanding of the patient's clinical evolution. Medical care delivery, however, has several characteristics that make such interaction difficult to support, since diverse and complex forms of information are involved. In fact, medical knowledge can be viewed as distributed on different hierarchical levels of sciences: from the lowest level, where atoms and molecules behavior is explained, to the highest level, where the time course of a disease is described [1]. Moreover, providing health care services requires the expertise of highly qualified medical, technical and administrative personnel, often appointed to use specialized computational resources and very expensive equipment. Therefore new organizational infrastructures should be pursued for exploiting an advanced integration of physical and knowledgeable resources in an effort of reducing the overall cost-efficiency ratio of the health care delivery process.

Such distinctive features of medicine have led to the development of several sites, each one trained on a particular medical specialty. The patient lifelong medical record is often highly fragmented because of the many clinical settings he/she has been visiting in his/her past, and the information needed to assess his/her clinical status turns out to be often incomplete, incorrect or unavailable. Thus, a major shift is needed from first generation Hospital Information Systems (HISs), mainly intended as simple centralized information repositories, to a distributed environment composed of several interconnected agents which actively cooperate in maintaining a full track of the patient clinical history and supporting care providers in all the phases of patient management.

In this paper we discuss a methodology for implementing a network of cooperating software agents aimed at improving the health care delivery process, and we also

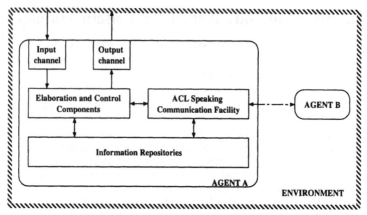

Figure 1. A general software agent model.

illustrate a set of tools we developed based on the proposed methodology. The underlying idea entails the incremental construction and maintenance of a consistent *patient model* shared by each cooperating agent. That model allows either storage and retrieval of patient information as well as generating a suitable interpretation based on that information and formulating the proper advice for helping professionals in accomplishing their daily tasks (e.g. medical, legal, insurance or financial assessments, population studies, etc...).

2 Collaboration as Agent Interaction

The Multi-Agent paradigm fits the depicted scenario. Computer-based systems and human operators can be seen as agents in that, while sharing physical resources, they encapsulate specialistic information (data/knowledge repositories), information processing capabilities and procedures determining their external behavior.

Activities and services explicated by agents take place in a common environment, which in principle can be as large as the area covered by worldwide telecommunication networks. A realistic vision involves a number of relatively small communities of agents (including humans) connected with each other through a network, to cover non internally supported needs. Starting from the seminal Minski's ideas [11], these communities of agents which highly interoperate with each other can be called *agencies*, in that an external observer may consider the community as a sort of "structured" agent offering multiple services.

An emerging notion of computer-based agents is that of *software agents*. From a software engineering point of view, a software agent can be defined as an at least semi-autonomous process able to interoperate with other processes (running on the same or on a separate machine) through a suitable Agent Communication Language (ACL) [4]. That is, we assume an intuitive general software agent model, shown in Fig. 1, in which each "isolated" entity possesses one or more conveniently represented information repositories. Using that information the agent may perform processing activities by means of opportune elaboration and control components, without any coordination with other agents. The information flows along the agent I/O channels facing the environment (e.g. sensors and actuators; software interfaces) adopting a representation which is specific to the agent itself. Cooperation with other agents is achieved by extending the representation mechanisms of this isolated entity with ACL communication primitives. An essential feature of such primitives is that they must allow agent integration at the knowledge level, that is, in a manner independent of implementation related aspects. For the sake of simplicity, the inter-agent communication facility is shown as a self-contained module in the figure.

A general trend is emerging aiming at appropriately specifying the meaning of technical terms and the definitions of the entities involved in information transactions among agents. In order to cope with the babel of existing knowledge and data representation schemes, standardized formats and languages along with appropriate conversion mechanisms have been developed to store large domain terminologies, libraries of reusable ontologies and also comprehensive bodies of declarative knowledge, to be shared among cooperating systems. Within the Knowledge Sharing Effort (KSE) project sponsored by Darpa [12] a first-order-logic flavored "interlingua" called Knowledge Interchange Format (KIF) has been developed for expressing knowledge-level statements. Moreover, a language for the specification of ontologies named Ontolingua [6] has been defined based on KIF. We used this language to implement a preliminary library of medical ontologies, described in [2,17], in order to state the ontological commitments for our envisioned medical agency: that is, a set of knowledge-level common definitions on which all the agents agree and base their interactions. The KSE group is also developing an agent communication protocol, called Knowledge Query and Manipulation Language (KQML) [3], consisting of a set of notification (*tell*, *untell*) and request/reply (*ask-if*, *reply*) primitives. This approach allows to linguistically decouple the communication modes (expressed in KQML) from the information conveyed (expressed for example in KIF). From the KQML effort we took inspiration while developing our experimental ACL.

Various agent types can be discerned according to the main role they play within the agents community. *Library/Server* agents provide other agents with information elements from their own repositories, structured as bodies of sharable and reusable components, that can be employed as building blocks for agent construction or expansion. Instances of this category can be terminology servers, ontology servers, problem solvers servers, data servers and so on, each server being named after the kind of information chunks it's able to offer. *Application agents* are endowed with specialistic competence (e.g. Knowledge Based Systems); thus they can reply to queries pertaining to their expertise field. *Personal agents* (also called personal assistants, softbots, knowbots, etc.) are tools, usually combined with adaptive user interfaces, tailored to meet single user's wishes and requirements. For example, they retrieve information (e.g. Usenet Netnews) along the net, and filter, select and display it in behalf of the user and according to his/her favorite styles.

3 The Construction of Health Care Software Agents

The need of software agencies supporting health care delivery emerges from the increasing availability and demand of the second generation HISs in which continuously updated patient records stored in databases on different computers (e. g.

belonging to different clinical divisions) are made available across the network to integrated KBSs. Diverse views and elaboration facilities of patient data should also be made available to end-users characterized by different interests, working conditions and access permissions.

Besides the several commercial products available that allow the configuration of graphical user interfaces and databases (e.g. relational or object-oriented ones), general knowledge engineering methodologies may be very useful in the design and implementation of the agencies. An example is given by the modelling approach of the KADS methodology [19], aiming at facilitating through an appropriate multi-level conceptual framework the knowledge acquisition and formalization phases of a system development process. The PROTEGE-II framework [5] also focuses on conceptual models for knowledge acquisition realized through mappings between domain and problem-solving method ontologies.

The GAMES-II project [18], which constitutes our reference framework, has led to the development of a meta-system named M-KAT that we used as an agency building tool. M-KAT allows the knowledge engineer to implement a medical KBS by configuring a set of heterogeneous medical problem solvers, constructed from knowledge representation formalisms and suitable methods (production rules, influence diagrams, qualitative/quantitative dynamic models etc.) and able to interact through a blackboard, as illustrated in [9]. Different problem solvers are viewed as black boxes by the central module invoking them; interactions are possible thanks to a common epistemological model of the medical tasks at hand, on the basis of which the central blackboard is organized. Such epistemological model is defined by an application ontology and an inference model, respectively organizing the application knowledge and illustrating how the application ontology may be exploited in the problem solving process. It represents the result of a preliminary epistemological analysis performed by the knowledge engineer who can be helped in this task by suitable software tools.

A set of tools have been constructed within GAMES-II which incorporate a library of medical ontologies written in the standard language Ontolingua, from which it is possible to configure application ontologies. Those tools also allow to specialize the general inference model according to the basic medical tasks. The main advantage of maintaining a separate server of specialized ontologies and inference models is that different medical agents, whether built with M-KAT or not, may commit to the same general ontological and inferential structures, so that their communication is notably facilitated. A medical terminological server, such as GALEN [16] or an opportune application built upon UMLS [10], could be very useful during the epistemological analysis phase, making it possible to compare, map or refer one's terminological choices to extensive, uniform and controlled vocabularies and classification schemes.

4 A Networked Architecture of Agents

The simplest way of classifying software agents accounts for the type of support they are able to provide to the whole community, as illustrated in Fig. 2.

Ontology repositories are mandatory in a distributed environment, and every other agent in the network must be linked at least to one of those. Agents supplying the

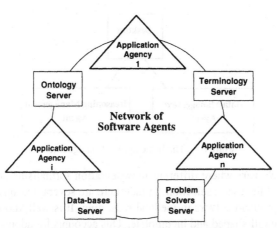

Figure 2. A diagram illustrating the Network of Software Agents.

ontology have the burden of defining a common semantics which once adopted by two or more agents can be used as the basis for successfully exchanging and interpreting either data or knowledge chunks. Of course, in a distributed environment several ontology servers may be present, defining different and possibly incompatible ontologies. However, if two or more agents intend to exchange some information among themselves it is required that either they share the same ontology or the adopted ontologies overlap so that they actually share the same semantics for that particular subset of their application domains encompassing the information to be exchanged.

The definition of an ontology is only possible if we agree upon a standard terminology for expressing the concepts of interest. Defining entities, features and attributes as well as the relationships occurring among those in a given domain may only be accomplished starting from a basic set of well understood terms. Moreover in medicine, just as in many other experimental sciences, a generally adopted standard is lacking, so it is quite usual that the same term is used by members of different communities with a slightly different meaning. This is a potential cause of misunderstanding which is quite easily overcome by human agents by accurately referring each term to the underlying context of the discourse. While such a behavior is suitable for humans, it is obviously impossible to be adopted by software agents, and a mean for clearly and uniquely expressing the adopted terminology is therefore mandatory. This task is accomplished by specialized agents acting as *terminology suppliers*, and their use is required in order to be able to construct a suitable ontology.

Both terminology and ontology servers may be considered as the fundamental agents on top of which all the remaining ones are constructed. *Data base servers*, for example, play the role of distributed data repositories exploiting either a relational or an object oriented model.

Within our architecture a problem solver may be defined as a specialized chunk of application knowledge represented through a specific formalism and combined with a method for exploiting it. A cooperating network of agents for implementing patient management should not commit itself to any specific problem solver, allowing instead to choose among a wide set of those. To this aim we foresee the existence of several

Figure 3. The basic application agency.

Problem Solver Servers where different representation formalisms and methods are made available for the user's selection. In fact there are several reasons supporting the use of different problem solvers in medical reasoning. It is well known that medical knowledge is often ill-shaped and incomplete. This accounts for adopting, even within the same medical task, the representation formalism which is best applicable given the context at hand and the goal to be solved. Moreover, distinct reasoning techniques may also have very different computational requirements and explanation capabilities, so that even when coping with a very specific application domain it is almost impossible to identify a single criterion for ranking them. Finally, each reasoning technique is strongly dependent on the particular ontology adopted so that in many cases some of them may turn out to be definitely inappropriate.

4.1 The Application Agency

All the agents discussed so far are able to provide a generic functionality to the whole network, which is not biased towards any specific class of users. Sometimes it is useful to distinguish between *server agents,* which provide an on-line service by answering incoming queries, and *library agents,* which provide instead an off-line service by making available a set of reusable components. The construction of an application will therefore need the contributions of at least one server/library agent for each of the four types described in the previous paragraph. Moreover, different classes of users in most of the cases will virtually use the same set of agents available on the net, so sharing data among themselves. However, despite the fact that different users share the same data, they definitely need distinct views on those to properly exploit the distributed architecture in their daily work. This aspect concerning the customization of a user access to the network is accomplished through the existence of another entity, that is the *application agency.* As shown in Fig. 3, an application agency is composed of several agents falling into three distinct classes: *desktop agents, data management agents* and *reasoning management agents.* By using this framework physicians, nurses, hospital administrators, researchers and many more operators working in the health care services will share the same information, although they will use it in different ways and for different purposes according to the nature of their professional duties.

As its name suggests, a desktop agent resides on the user workstation and it mainly consists of a set of graphical procedures. It acts as an operational interface between the user and the two other agents, thus shaping the view seen by its user while interacting with those. In fact, both the data management agent and the reasoning management agent of an agency must be connected with a suitable set of ontology providers. This

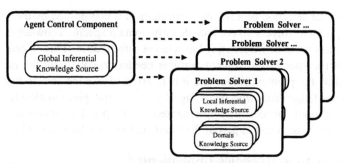

Figure 4. The internal architecture of a reasoning management agent.

has the purpose of defining a common domain model for the application itself. So when the desktop agent is subsequently connected to the data management agent, the database servers available on the network will no longer play the role of repositories of unstructured information, but the data available will be organized, interpreted and used according to that model. The data management agent, for example, may successfully exploit the ontology to make sure that any new data acquired fits into the context defined by the already existing information. In case of a discrepancy between the value supplied and the expected one, the data management agent may send out a message to the agent requesting the storage of the new datum, looking for a further confirmation.

In the same way a reasoning management agent must be configured by selecting and integrating into the application a suitable set of problem solvers exploiting the most appropriate reasoning techniques for interpreting the available data and assessing the current situation. As several researchers point out [14, 18, 19], problem solving is an intense reasoning activity involving each time either the dynamic selection of a goal to pursue and the subsequent choice of a particular method which is best applicable given the context. We therefore identify two different knowledge types involved with the construction of an agent implementing the problem solving process. One type complements the application ontology, and is aimed at defining all the relationships allowing to hypothesize the solutions of a problem given its features as well as at deriving the expected consequences from each hypothesis. This is usually referred to as the application knowledge. The other type concerns instead the procedures able to manage the available application knowledge in order to solve problems, and is referred to as inferential knowledge. The inner structuring of a reasoning management agent actually reflects this two-layered organization, as indicated in Fig. 4. The leftmost box representing the inferential knowledge is built as a separate module whose goal is to bring about the overall control regime of the agent itself. This entails selecting a suitable strategy to reach a solution for a given problem in sight of the capabilities available within the agent. Each time the agent is invoked, the control is passed to its inferential knowledge which is activated. As a consequence a sequence of tasks will be dynamically planned, each one involving the invocation of some specialized problem solvers, which are shown in the right part of the figure. The inferential knowledge also has the duty of stopping the process either when a satisfactory solution is achieved, or when it realizes that no problem solver is applicable nor able to provide new useful solution elements.

Although the selection of a reasoning strategy is accomplished by the inferential knowledge, problem solvers do not include only domain specific knowledge, but they also encompass a small inferential knowledge submodule. This is required for providing all the specialized control features needed to cope with the formalisms implemented by the problem solver and it is referred to as a method. In fact such an inferential knowledge deals with modeling of specific strategies too closely tied with the representational formalisms implemented by the problem solver itself which therefore cannot be included as part of the inferential knowledge of the whole agent.

5 The Application Agencies Implemented

As a first attempt to implement a community of cooperating software agents for improving patient care we developed a set of three prototype applications with the aim of covering only some of the most important tasks involved with hospital management. Each application agency is able to help its user, that is a physician, a nurse or an administrator, in accomplishing their daily duties. An agency consists of three separate cooperating agents, as described in Section 4.1, and each agent is actually implemented as a different process on a Unix workstation. The agents communicate among each other through the usual TCP/IP connection available with Unix systems by means of a protocol closely resembling KQML developed within the KSE project. That protocol allows exchanging messages in terms of *queries* and *replies*, and it is structured on several layers, each one responsible for coping with a different aspect of the transaction. While the lower levels are responsible for routing and dispatching the messages considered just as plain byte sequences, the higher level ones are responsible for interpreting a message contents and undertaking the proper action needed to achieve the replies. For implementing the reasoning management agent a specialized tool called M-KAT [7] has been used, which allows an easy acquisition and modeling of the inferential knowledge [8], and supports an easy integration of different problem solvers.

5.1 The Medical Problem

Our application domain was set in the area of Acute Myeloid Leukemia (AML). Leukemias are a heterogeneous group of diseases arising from a malfunctioning in the growth and maturation process of the bone marrow stem cells. Acute Leukemias usually represent some rapidly progressing forms of Leukemias, requiring a prompt planning and administering of a suitable therapy, as any delay may turn out to have a great influence on the patient life expectancy. They may be divided into Acute Lymphoblastic Leukemia (ALL) and AML according to the nature of the involved cells. We limit ourselves to the management of AML only. In both cases however a further classification into subtypes is accomplished according to the cells morphology, and one of the most widespread criteria to do that is commonly known as the FAB (French American British) classification. According to FAB, AML is classified into seven groups named FAB-M1 through FAB-M7.

Only two therapies are available for AML, which are Bone Marrow Transplantation (BMT) and Chemotherapy. The former has the greatest probability of success, but it is only applicable under certain circumstances including the possibility of finding a suitable donor. Chemotherapy instead has quite a general applicability, although it may have long term negative side effects as a consequence of the toxicity which is inherent to the treatment itself. The initial classification of the AML is very helpful for planning the therapy. More specifically, if the AML diagnosis falls in some FAB groups the selection of chemotherapy is the more appropriate, as the risks associated with BMT are too high. For some other FAB groups BMT is desirable instead, mainly as there is a high risk of a worse relapse of leukemia connected with the chemotherapy. Finally, a couple of FAB groups do not suggest on their own a specific treatment. Additional information must be gathered, including the response to a preliminary chemotherapy treatment.

5.2 The Physician Application Agency

The management of AML is a very complex task requiring a careful and intensive patient monitoring over a period spanning several months. Additionally the choice of a suitable therapy for AML requires different reasoning techniques. This is especially true if the initial FAB classification doesn't indicate a specific treatment on its own. In that case the use of some probabilistic reasoning technique might be useful, as there is still no clear understanding of the underlying mechanisms allowing the physician to foresee the future evolution of either BMT or chemotherapy, in order to select the therapy on a solid pathophysiological basis.

The agency grants the physician full access to the patient data base so that he/she may select patients, add or delete visits, and use the configured reasoning management

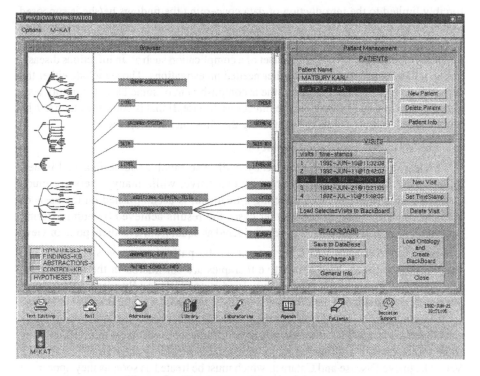

Figure 5. The desktop of the physician agency.

agent for having advice about the three main tasks involved in medical reasoning, that is *diagnosis*, *therapy planning* and *monitoring* [15]. The ontology will give him/her a thorough overview of the medical problem in order to support the use of all the different reasoning techniques required depending on the context and made available by the reasoning management agent. Moreover, the whole task of managing AML patients is inherently adaptive, and at any time the physician agency must be able to retract some of the previously established conclusions if new evidence is gathered suggesting to do so.

The desktop of the physician agency, which is illustrated in Fig. 5, shows the ontology definition in terms of a taxonomic graph on the left, while the interface with the data base is on the right. In the bottom part of the screen several icons are located allowing the user to access some additional services.

5.3 The Nurse Application Agency

No matter if BMT or chemotherapy is selected, the patient will undergo a set of side treatments which will cause him/her to fall into a severe immunodeficiency state. As he/she is more likely to develop infectious diseases, a careful monitoring activity must be planned in search for any possible alerting finding which, in a normal state, would have been flatly discarded. This activity is best carried out by the nurse who spends much more time than the physician at the bedside, and develops a more friendly interaction with the patient. The nurse only has a partial view of the problem which is focused each time on a specific different subtask. His/Her interaction with the database will be forcibly limited to the introduction of data concerning the findings he/she is supposed to collect on his/her daily duty, and the only problem solver used by the reasoning management agent will be based on the production rule formalism. If something is noticed which suggests a possible outset of a complication such as an infectious disease, he/she will be asked to perform further needed investigations. Those include either the acquisition of additional findings or the accomplishment of some basic laboratory tests aimed at confirming or ruling out the infectious disease. If that is confirmed then he/she will be asked to drop a message addressing the physician desktop agent.

5.4 The Manager Application Agency

Finally, treating patients affected by AML also involves some financial issues. Some of the involved therapies are very expensive, while many others are much cheaper. While in many cases selecting a treatment is accomplished on the basis of medical concerns only, in some other cases the results achieved by administering several alternative treatments may be considered as equal from a medical point of view.

Before performing BMT, for example, the patient must be submitted to a conditioning regimen. Two alternative therapies are considered in the literature for achieving that goal, which are Total Body Irradiation (TBI) and Busulfan. TBI requires a complex equipment and therefore turns out to be very costly, while Busulfan is instead a cheap drug. As far as their conditioning effect is of concern, they may be definitely considered as equivalent, although they have different side effects. While TBI has only some minor ones, Busulfan has major side effects including Interstitial Pneumonia, Veno Occlusive Disease and Cataract, which must be treated as soon as they appear. At first sight busulfan might be preferred due to its lower cost, but a careful analysis should also account for the costs of treating any possible side effect arising as a consequence of the treatment itself. According to the simple example provided, it transpires that sometimes the selection of a therapy may only take place within a broader context and must be therefore accomplished by the hospital administrator.

The ontology available within the manager agency will not allow him/her to access data concerning the clinical history of the patients, as he/she is only interested in financial and managerial aspects. All the entities such as laboratory tests, specialized exams and treatments will be considered by the ontology of his/her agency only in terms of their expenses and are supposed to be ranked according to a suitable cost vs. effectiveness analysis performed by the reasoning management agent. The agency will use the results for generating daily reports on the overall financial status of the clinical setting as well as on the management of each single division within the hospital.

6 Conclusions

The main breakthrough of this paper is to propose a unified methodology integrating into a single framework several already existing techniques. In fact the methodology we propose heavily borrows both from the technologies of databases and distributed computing and builds on already existing work on object oriented programming and artificial intelligence. The blackboard model for problem solving [13] is recognized as one of the most promising approaches to the implementation of multi-agent reasoning, and has been widely used for implementing several applications in AI. Our approach further extends the blackboard notion, considering it as a shared communication mean among all the agents available on the network. This gives rise to the D-CBA (Distributed Control Blackboard Architecture), which is described in [9].

Nevertheless this research effort is just at its earliest development stages, as there are many issues still left open. We didn't address so far the problem of data security and authentication as well as communication management and resource allocation, which may nevertheless become crucial in a real communication environment. One of the foremost limitations shown by the agencies implemented is represented by the adoption of a fixed schema for connecting their component agents. So far, each agent must know in advance which are the other ones cooperating with it and their exact location on the network, and the whole agency fails if any component becomes unavailable at a give time. However, a real implementation should also tackle the problem of negotiating services on the network, so that agencies are automatically and dynamically built using the resources currently available on the network. Finally, the possibility of representing ontologies in a computational way should be investigated to a greater extent. This also involves the possibility of implementing translators from different ontological representations which use different terminologies so that a wide set of agents may be used interchangeably for the same purposes even if they adopt different domain models.

Acknowledgments. This work is part of the AIM project A2034 entitled "GAMES-II: a General Architecture for Medical Expert Systems", and was supported by the Commission of the European Communities. It is also supported by a MURST grant.

References

[1] Blois, M. S. (1988). Medicine and the nature of vertical reasoning. *New England Journal of Medicine,* 13(38):817-851.

[2] Falasconi, S., Stefanelli, M. (1994). A library of medical ontologies. In Mars, N. J. I., editor, *Proceedings of the ECAI94 Workshop Comparison of Implemented Ontologies,* pages 81-91, Amsterdam.

[3] Finin, T., Weber, J., Wiederhold, G., Genesereth, M. R., Fritzson, R., McKay, D., McGuire, J., Pelavin, P., Shapiro, S. and Beck, C. (1993). Specification of the KQML Agent Communication Language.Technical Report EIT 92-04, Enterprise Integration Technologies, Palo Alto, CA.

[4] Genesereth, M. R., Ketchpel, S. P. (1994). Software agents. *Communications of the ACM,* 7(37):48-53.

[5] Gennari, J. H., Tu, S. W., Rothenfluh, T. E. and Musen, M. A. (1994). Mapping domains to methods in support of reuse. In *Proceedings of the Knowledge Acquisition Workshop KAW-94,* Banff, Canada.

[6] Gruber, T. R. (1993). A translation approach to portable ontology specifications. *Knowledge Acquisition,* 5:199-220.

[7] Lanzola, G. and Stefanelli, M. (1992). A Specialized Framework for Medical Diagnostic Knowledge-Based Systems. *Computers and Biomedical Research,* 25:351-365.

[8] Lanzola, G. and Stefanelli, M. (1993). Inferential knowledge acquisition. *Artificial Intelligence in Medicine,* 5:253-268.

[9] Lanzola, G. and Stefanelli, M. (1993). Computational Model 3.0. Technical Report GAMES-II Deliverable 25, Laboratory of Medical Informatics, University of Pavia, Italy.

[10] Lindberg, D., Humphreys, B. and McCray, A. (1993). The Unified Medical Language System. In van Bemmel, J., editor, *1993 Yearbook of Medical Informatics,* pages 41-53, International Medical Informatics Association, Amsterdam.

[11] Minski, M. (1985). *The society of mind,* Simon and Schuster, New York.

[12] Neches, R., Fikes, R. E., Finin, T., Gruber, T. R., Patil, R., Senator, T. and Swartout, W. (1991). Enabling technology for knowledge sharing. *AI Magazine,* 12:36-56.

[13] Nii, H. P. (1986). Blackboard Systems: The blackboard model of problem solving and the evolution of blackboard architectures (part i). *AI Magazine,* 38-53.

[14] Puerta, A. R., Tu, S. W. and Musen, M. A. (1992) Modeling Tasks with Mechanisms, *International Journal of Intelligent Systems.*

[15] Ramoni, M., Stefanelli, M., Magnani, L., and Barosi, G. (1992). An Epistemological Framework for Medical Knowledge-Based Systems. *IEEE Transactions on Systems, Man, and Cybernetics,* 6(22):1361-1375.

[16] Rector, A. L., Solomon, W. D., Nowlan, W. A. and Rush, T. W. (1994). A terminology server for medical language and medical information systems. Technical Report, Department of Computer Science, University of Manchester.

[17] van Heijst, G., Falasconi, S., Abu-Hanna, A., Schreiber, G. and Stefanelli, M. (1995). A case study in ontology library construction. *Artificial Intelligence in Medicine.*To appear.

[18] van Heijst, G., Lanzola, G., Schreiber, G. and Stefanelli, M. (1994). Foundations for a Methodology for Medical KBS Development. *Knowledge Acquisition,* 6:395-434.

[19] Wielinga, B. J., Schreiber, A., Th. and Breuker, J., A. (1992). KADS: a modelling approach to knowledge engineering. *Knowledge Acquisition,* 4:5-53.

High Level Control Strategies
for Diabetes Therapy

Alberto Riva and Riccardo Bellazzi

Laboratorio di Informatica Medica
Dipartimento di Informatica e Sistemistica
Università di Pavia, Pavia, Italy

Abstract. The project we describe here is aimed at assisting out-patients affected by Insulin Dependent Diabetes Mellitus. Our approach exploits the usual scheme of diabetic patients management, based on (i) a periodic evaluation of the patients' metabolic control performed by the physician, and (ii) patient-tailored tables for self-adjustments of insulin dosages. Following this scheme we have defined a system built on a two-levels architecture, that can be conveniently implemented in a telemedicine context. The *High Level Module* exploits both medical knowledge and clinical information in order to assess an insulin protocol, defined in terms of insulin timing, type, and total amount. The High Level Module exchanges information with the *Low Level Module* in order to define the control actions to be taken at the low level, as well as to periodically evaluate protocol adequacy on the basis of patient data. The goal of the Low Level Module, whose characteristics can be chosen by the High-Level Module, is to suggest the next insulin dosage, depending on the actual blood glucose measurement and a certain pre-defined insulin delivery protocol. In this paper we outline the overall organization of the system and we describe in detail the methodology and the strategies exploited by the high-level module.

1 Introduction

The conventional therapy of *Insulin-dependent diabetes mellitus* (IDDM) tries to control the blood glucose levels (BGL) of the patients through subcutaneous insulin injections several times a day. The injections are usually planned according to an open loop strategy, decided by the physicians according to their experience. When possible, an effective strategy is to add closed loop control, managed by the patients themselves. To this aim the patient is taught how to select the insulin dosage using decision tables tailored by the physician. Blood glucose control is therefore obtained as the interaction of (at least) two basic control tasks, the first one being the definition of an insulin administration protocol and the second one being the closed-loop adjustment of the insulin doses.

Many *decision support systems* (DSS) have been studied to assist patients and physicians in the design of a therapy. After some preliminary studies aimed at rationalizing the definition of the decision tables on the basis of mathematical models of glucose metabolism and of clinical experience [10, 9, 4], expert systems were developed in order to assist the management of IDDM [5, 3], as well as to provide an educational tool for patients and physicians [6]. These advisory systems may be broadly classified into two categories, that reflect the distinction between the two control tasks outlined above. The *day by day* advisory systems aim at assisting the patients during their every-day monitoring activity, by suggesting the next insulin dosage depending on the actual blood glucose levels; an example is the DIAS system proposed by Andreassen [1]. The *visit by visit* advisory systems aim at assisting the physicians in their periodic evaluation of the patients' glucose metabolism. One of the most complete realizations is represented by the AIDA system [7], that conjugates heuristic and model-based reasoning.

The system we propose integrates the above mentioned tasks in a comprehensive framework. It is meant to assist the patients in their routine self-monitoring activity as well as the physicians in assessing the basal insulin regimen and the diet plan. Note that since the goal of completely replacing the physicians' monitoring and control activity is unrealistic and probably undesirable, we prefer to limit the scope of our decision support system to the *routine* management of the patients. The contribution that we expect from our system is therefore the automatization of those decisions that are needed to handle normal situations, so that the physician may devote more resources to handling dangerous and exceptional episodes and the patients may achieve a better quality of life even without continuous assistance from an expert.

1.1 A Dual-Level Control Architecture

In order to precisely define the architecture of our system, it is essential to understand the structure of the decisions involved in IDDM management: *who* usually takes decisions, *what decisions* are to be taken, and *what control rules* are employed.

The two "natural" decision-makers in the IDDM management problem are the *physician* and the *patient*. The two control agents are hierarchically interrelated. The physician decides the insulin therapy, the diet and also what are the *decision rules* for insulin adjustment that the patient must follow during self-monitoring. Thus, the physician establishes a *policy* for the patient's self-management, that is summarized in a *protocol*. The tasks of the physician are very complex, and the knowledge that is used to accomplish them must be as deep as possible. On the contrary, the task of the patient usually reduces to consulting a set of predefined decision tables and the required medical knowledge could be limited to an understanding of the basic action-reaction processes related to diabetes control.

Although the definition of a policy is a complex reasoning process, the control strategies that the physician applies may be broadly classified into two categories. The *Feed-back/feed-forward control* strategy usually exploits injections of regular (fast-acting) insulin in order to compensate a meal ingestion seen as a known disturbance (feed-forward control), or to obtain a fast decrease of the BGL taking into account the actual error between the desired and the measured glycemia (feed-back control). The *Predictive control* strategy is usually applied to the therapeutic planning of intermediate and long-acting insulin, that are characterized by slow onset rates and delayed peaks of action. The presence of a *delay* is known to complicate the design of a control system, and is often associated with instability problems. In clinical practice the physician copes with these problems using a *prediction* of the blood glucose response to an insulin injection. The simplest way to perform a prediction is to observe the effect of a certain insulin injection and suppose that a future insulin injection of the same type, of the same amount and at the same time of the day will have the same effect (*cyclo-stationarity* assumption). The control action will therefore depend on the error between the prediction and the desired effect.

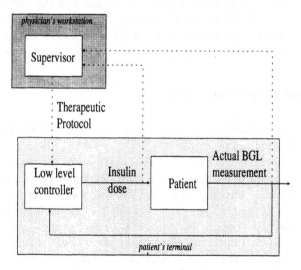

Fig. 1. The overall scheme for intelligent patient monitoring

The system we propose reflects the decision architecture outlined by the above analysis, and is therefore based on the cooperation between two distinct control agents, a *Low-level control module* (LLCM) and a *High-level control module* (HLCM). The two modules are arranged in a hierarchical networked architecture that is distributed between the patient's house and the clinic, and is primarily aimed at increasing the frequency and quality of the information interchange between the patient and the physician. In other words, the patient will view the system as a mean to obtain quicker and more accurate assistance by the physician, and the physician will consider the system as the key for controlling the metabolic behavior of a large number of patients. The architecture of the system is sketched in Figure 1.

While the HLCM, located in the physician's workstation, assists the physician in the definition of the basal insulin regimen and diet through a periodic evaluation of the patient's data, the LLCM allows automatic data collection and transmission from the patient's house to the clinic, and assists the patients in their self-monitoring activity, suggesting the insulin dosage adjustments. In other words, the low level module must decide the supplementary dose of regular insulin following the control tables contained in the protocol.

Since the two agents play different roles in the decision and control processes, they must handle and communicate different kinds of knowledge. It is therefore necessary to define in detail what is the knowledge possessed by each module, what are the formalisms used to represent it, and how the exchange of information between the modules takes place. The HLCM output is the result of a high-level reasoning process and defines a *protocol*, that is, a set of prescriptions that is communicated to the LLCM in order to bound the space of its admissible actions. The LLCM produces as its output an action that is directly performed on the patient, and communicates both the action and the current metabolic state of the patient to the HLCM. In this way the HLCM can follow the actual insulin therapy and the patient's response to it. This paper will focus mainly on the HLCM, while a possible implementation of the LLCM is described in [2].

2   Protocol Representation

In this section we will provide a set of parameters able to formally characterize an insulin administration protocol.

Number of injections. At the highest level of abstraction, a protocol is characterized by the number of daily insulin injections. This number must obviously be kept as low as possible, although in critical situations a high number of injections can yield a better control. Usual values for this parameter range from 2 to 4.

Insulin types. Several types of insulin are available, characterized by different onset and peak action times, and effective duration. Two or more insulin types can be administered at the same time, thus combining their effects to achieve the desired blood-glucose profile. For each injection, the protocol must therefore specify the relative amounts of the various kinds of insulin.

Injection times. Insulin injections take place at well-defined times of the day, that usually coincide with (but are not limited to) the main meals (e.g. breakfast, lunch, dinner). The protocol specifies the injection times as a set of "qualitative" time-points, while the correspondence between these and actual times depends on the patient's habits.

Total dose. All the insulin doses are expressed as fractions of the daily total dose, that depends primarily on the weight of the patient. The protocol must indicate the total dose and its maximum allowed variation (in units per kilogram).

Glucose set-point and ranges. The protocol specifies the desired blood glucose value in each time-point, as well as the highest and lowest admissible values.

Diet. The amount of caloric intake per day with an indication of the relative contribution of each meal.

Physical exercise. A qualitative indication of an extra physical activity.

Low-level controller and control law. The HLCM can select a low-level controller implementation among the available ones (e.g. fuzzy controller, minimum variance controller, optimal controller) and the appropriate control strategy.

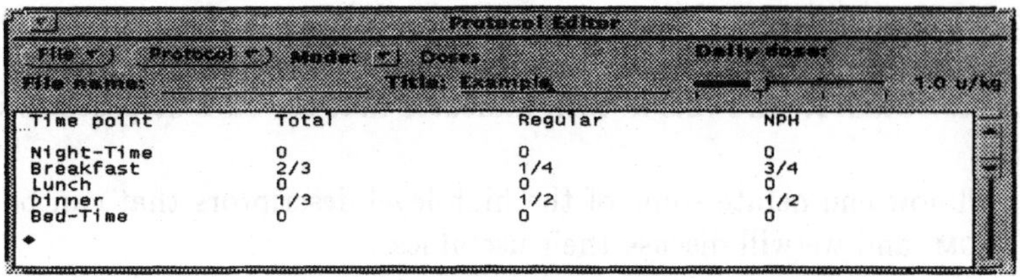

Time point	Total	Regular	NPH
Night-Time	0	0	0
Breakfast	2/3	1/4	3/4
Lunch	0	0	0
Dinner	1/3	1/2	1/2
Bed-Time	0	0	0

Fig. 2. An example insulin administration scheme.

Figure 2 shows a window of a Protocol Editor developed by the authors. In particular, an example insulin administration scheme is shown. The qualitative time-points considered are Night-time (NT), Before Breakfast (BB), Before Lunch (BL), Before Dinner (BD), Bed Time (BT). The scheme suggests a total amount of daily insulin of 1 Unit per Kg, to be taken in two injections a day. 2/3 of the total insulin are to be taken at BB, and 1/3 at BD. 1/4 of the BB dose is regular, and 3/4 are NPH; 1/2 of the BD dose is regular, and 1/2 is NPH.

3 The High-Level Control Module

The goal of the *high-level control module* is to determine the optimal insulin-administration protocol on the basis of the available data, to communicate it to the low-level controller, and to monitor the performance of the resulting control system. The HLCM is required to perform a kind of *batch processing* over the available data, and to arrange things so that the LLCM may run for up to several days without external intervention (except for unpredictable dangerous events). The inputs to the reasoning procedures of the HLCM will therefore be *high-level descriptors* that characterize the current situation without taking the single episodes into account.

The protocol parameters we have listed in the previous section define a multidimensional *protocol space* through which one can move in order to adjust a given protocol. Namely, a protocol adjustment can affect the injection time-points ("time axis"), the relative amounts of insulin to be administered on every injection ("dose axis"), or the daily total dose. When exploring the protocol space, several constraints must be satisfied. For example, the insulin injections cannot be too numerous, or occur at undesirable moments of the day.

3.1 Data Interpretation

The HLCM will base its reasoning process on two different sources of information: the data coming from the patient (physiological parameters and measurements) and the data coming from the LLCM, describing the control actions it performed during the period of interest. The reason for this is that the HLCM is required to monitor not only the patient's general response to the therapy, but also the performance of the current protocol as enforced by the LLCM: even if the patient's situation is apparently satisfactory, an analysis of the control actions performed by the LLCM could reveal that a high number of protocol adjustments were necessary to obtain it, and this in turn indicates that the current protocol is not optimal.

We will now enumerate some of the high-level descriptors that can be used by the HLCM, and we will discuss their usefulness.

The blood glucose *Mean Value* and *Variance* are the simplest high-level descriptors. The former can, in principle, indicate whether the basal insulin level is too high or too low. In practice, however, the blood glucose sampling rate is so low that this piece of information is of little value: since blood glucose measurements are usually carried out in very particular moments of the day, the average is likely to be biased. The latter can be used as a rough indication of the quality of the control action: while a certain amount of oscillation is inherent in the blood glucose dynamics, a high variance suggests that the physiological system is heading towards instability. For example, it could be caused by a rebound effect from a hypoglycemic episode or by a wrong timing of the injections.

The *Blood Glucose Modal day* (MBG) is a "characteristic daily blood glucose pattern" [5] that summarizes the patient's response to the on-going therapy. The MBG provides a high-level description of the expected behavior of the system, that is not based on a physiological model but is calculated on the basis of the available data. The goal of the MBG is to filter out the exceptional, non-representative episodes from the blood glucose profile, and to indicate when undesirable events are "normally" most likely to occur. Such information will be used by the HLCM to detect problems in the therapy, to identify their source and to adjust the relevant component of the protocol.

The same applies to the *Control Actions Modal day* (MCA) and to the *Glucose Error Modal day* (MGE). The MCA is a high-level description of the pattern of action of the LLCM. If a component of the protocol is not optimal, the LLCM will be forced to perform periodic adjustments in order to prevent possible dangerous events. The MGE pattern describes the difference between the desired and the actual BGL in each time-slice. This information is more useful than the mean value, because the choice of the set-point in each time-slice can take into account the physiological variations of the BGL during the day.

Under the cyclo-stationarity assumption, the analysis of the MCA and MGE patterns can reveal the "weak spots" of the protocol by pointing out where the control action was most often required and where the BGL does not follow the desired behavior. Failure to recognize such patterns, on the other hand, does not mean that there are no problems in the control strategy, but simply that these methods are not adequate to detect them. In such a case it is necessary to resort to other indicators, like the well-known M index [11].

As we shall see in the following, the reliability and the usefulness of the modal day profiles are highly dependent on the amount of data collected from the patient since the most recent protocol change. In general, the predictions based on the MBG and the MCA will be affected by a high degree of uncertainty, and *ad hoc* methodologies, able to deal with incomplete information, will be **required to handle them.**

3.2 Reasoning

The reasoning process of the HLCM is based on an innovative methodology called *Logic-based Belief Maintenance* [8]. A problem solver built on a logic-based belief maintenance system (LBMS) represents knowledge as a network of propositional sentences connected by logical clauses, and assigns probabilistic truth-values to the sentences and to the clauses. The main feature of the methodology is the ability to reason with underspecified probabilistic models; that is, it is not necessary that all the probabilities that describe the probabilistic model be known for the reasoning process to develop. The propagation algorithm, that enjoys linear-time complexity, will not in general be able to assign precise probabilistic values to the sentences, but will constrain them between bounds that will monotonically converge to the correct value as the supplied information increases. A problem solver based on this methodology is therefore able to deal with a variety of knowledge representation formalisms (production rules, belief networks, influence diagrams, etc.) endowing them with the power to explicitly represent an incomplete probabilistic specification of the problem.

We will now briefly describe how the LBMS can be useful in the reasoning module of the HLCM by showing how it can represent the MBG. If we subdivide the 24-hour period into a set of consecutive non-overlapping time-slices (such as the ones shown in the example protocol in section 2) and the possible blood glucose values into a set of qualitative ranges, the modal day can be represented as a matrix of logical propositions $P(t, s)$, each of which represents the sentence "the qualitative blood glucose value in time-slice t is s". What we are looking for is a set of truth-value assignments to the propositions that will enable us to determine the probability that the blood glucose value during a particular time-slice lies in a particular range. Initially all the propositions have the truth-value [0 1] meaning that, since we have no information at all, we are not able to put bounds on their probability value. In the course of the monitoring process, the data coming from the patient is used to update the bounds on the probabilities of the sentences. Since the width of the interval is a monotonically decreasing function of the available information, the bounds are guaranteed to converge: if enough data is collected, the range of uncertainty in the probability values can be arbitrarily reduced.

There are several motivations in favor of the use of this methodology. First of all, it is essential to be able to reason in the presence of missing data and incomplete information, since the volume and the quality of the available data is very little. Moreover, our approach integrates the *learning, forecasting* and *monitoring* tasks in a unified framework.

The learning technique we used is a slight modification of the one proposed by Lauritzen and Spiegelhalter [12] that provides bounds on the probability of an event on the basis of the direct count of its occurrences.

Given a set of propositions that represent the possible discrete values of a stochastic variable and a truth-value assignment to each of them, there are several ways to determine what is the value "forecasted" on the basis of the available data, taking into account the fact that the truth-values are represented by probability intervals. The simplest way is to look for the state that has the highest upper bound, weighted by the inverse of the amount of ignorance [8]. In this way, we score the variables by their maximum probability and the certainty about the probability (this is equivalent to the posterior mode for usual probability distributions). Another way is to calculate an *expected range* for the value of the variable. While the usual expected value is calculated as the sum of the products between a numerical value associated to each state and its probability, in our case this must be done twice, choosing the probability values inside the intervals in order to minimize and maximize the result.

The system provides a way to continuously monitor the network of propositions and to react to changes in their probabilities. For example, when the minimum probability of the proposition that indicates a VERY-LOW glucose level becomes too high, an alarm can be raised signaling the risk of hypo-glycemia in the corresponding time-slice.

3.3   Decision Taking

The decision process starts with an evaluation of the high-level descriptors calculated from the available data. The mean blood glucose value and the trend component of the blood glucose time series give a first indication of the adequacy of the total daily insulin dose. For example, if the mean BGL is found to be above the normal range, and a time-series analysis reveals that it is not decreasing, it will be necessary to increase the total dose. The variance and the M index, on the other hand, are related to irregularities in the BGL profile and may suggest an inadequacy of the insulin doses and timings prescribed by the protocol.

In the latter case, it is necessary to resort to the modal day profiles described above to determine what are the protocol components that need to be changed and what are the required adjustments. The modal day profiles are interpreted according to heuristic rules that reduce the set of the admissible protocol adjustments according to various kinds of "context" constraints (e.g. the total number of injections cannot be too high, some injection times are preferable over others).

The selection of the possible adjustments is based on two main decision tools. The first is the concept of *competent time-slice*: each time-slice in the insulin administration protocol will be competent for the time-slices in the MBG that it directly affects. In other words, given the dynamics of the different insulin types, an intake of regular insulin will be competent for the time-slices that cover the subsequent six hours, an intake of NPH insulin will be competent for the time-slices that cover the subsequent twelve hours, and so on. Therefore, when a problem is detected in a particular time-slice t of the MBG, the possible adjustments will be the ones affecting the insulin doses in the time-slices competent over t.

The second tool is a set of spaces of admissible predefined protocols, provided by medical expertise; each space consists of all protocols with the same daily number of injections. We define the notion of distance between the protocol spaces (as the difference of the number of injections) and inside the protocol spaces, as a function of the relative amounts of the insulin dosages in each injection. For example, consider the simple case of two injections per day and only two types of insulin (regular and NPH). If we denote with x and y the relative amounts of regular insulin in the first and second injection of the current protocol respectively, the distance between this and a new protocol described by x' and y' is simply the Euclidean distance $\sqrt{(x - x')^2 + (y - y')^2}$.

Since several different adjustments could be suggested for a particular time-slice, a first screening based on the context constraints is performed. Each of the remaining adjustments can be interpreted as a movement in the protocol space characterized by a direction and a distance from the current one. In order to minimize the differences between the current protocol and the new one, we will prefer the adjustments that lead to protocols closest to the current one, according to the above defined metric.

If the amount of information is sufficiently high, the technique here described can be replaced by purely quantitative ones that rely on the ability to correctly identify a predictive model of the glucose metabolism. In this case the selection of the adequate adjustment can be performed by directly simulating the effect of each one of the protocols under consideration and choosing the one that yields the best BGL profile (optimal control). In general, different decision tools will be available, ranging from the most qualitative to the most quantitative ones, and the one to be used will be selected according to the amount of ignorance to be dealt with.

4 An Example

We have applied the above described learning technique to one of the patients of the data set provided by the AAAI 1994 Spring Symposium organizing committee (namely, patient #30). A preliminary analysis of the 1175 available BGL measurements (over a period of six months) gives the following results: average BGL = 152.45, BGL standard deviation = 51.16, M index = 90289. These parameters indicate that the protocol is acceptable but not optimal (in particular, the M value is too high), and don't give any useful hint about the source of the problems. We will now show how more sophisticated techniques based on the modal day profiles can provide more detailed information. In order to extract the modal day representation of the blood glucose and the insulin doses we used seven time-slices (Breakfast, Mid-morning, Lunch, Mid-afternoon, Dinner, Bedtime, and Night-time), five qualitative levels for the blood glucose (<70, 70-120, 120-180, 180-300, and >300 mg/dl) and five qualitative levels for the insulin doses (<5, 5-10, 10-15, 15-20, >20 units). The results of the analysis are shown in Figure 3. In each box, the horizontal bars provide a graphical representation of the probability intervals associated with the corresponding proposition. The black region is the area below the lower bound, the white region is the area above the upper bound, and the gray region is the area where the actual probability value can lie. Therefore, the width of the gray area is proportional to the igno-

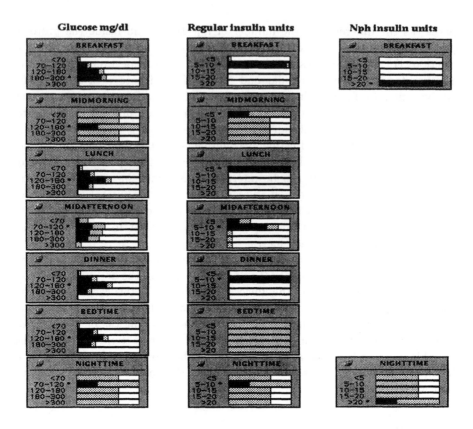

Fig. 3. The blood glucose and control actions modal days derived from the analysis of the patient data described in the example.

rance about the proposition. The asterisk marks the value predicted according to the mode technique.

It is important to note that the observations are mostly concentrated around particular time-points (i.e., the meals) and are scarce in the "intermediate" time slices (e.g., Mid-morning, Night-time). This is reflected in the width of the corresponding ignorance regions in the figure. Therefore, when trying to achieve a high temporal resolution it will be necessary to integrate the directly observed data with "indirect" information that covers a longer time span (such as glycosuria).

The insulin regimen is nearly constant over the observed period: regular insulin injections are scheduled in coincidence with the meals, some are present in the afternoon and occasionally during the morning and the night; NPH insulin is taken only at breakfast and occasionally in the night.

Concerning the glucose modal day, the main problem is represented by the high hyper-glycemia probability at breakfast time, while in all the other time-slices the mode belongs to a safe range. The source of this problem can be easily identified by looking at the competent time-slices: in particular, we can note that after the regular insulin dose at dinner, no other insulin intakes are prescribed by the protocol. Moreover, there is no evidence of hypo-glycemic states during the night, even with the little available data. The suggested adjustment therefore consists in the addition of an NPH dose to the injection at dinner: this choice minimizes the number of protocol changes since it does not change the number of injections but only the relative amounts of regular and NPH doses in the dinner time-slice (to determine the actual amount of the NPH dose it would be necessary to know at least the weight of the patient, that was not available in this particular case). The system could also suggest additional measurements during the night, once or twice a week, in order to evaluate the adequacy of the proposed adjustment.

5   Discussion

The methodology we propose is aimed at performing an intelligent patient monitoring based on a strong link between quantitative and qualitative techniques, and is here applied to the problem of IDDM patients management. The system is organized in two physically and logically distinct layers: a low-level control module that performs day-by-day patient management by adjusting the insulin dosages, and a high-level module that deals with the long-term management of the patient by choosing the appropriate treatment protocol. In this paper we outlined the organization and the operation of the high-level module whose fundamental tasks are to interpret the available data, to reason on the abstracted data, and to take decisions regarding the protocol to be enforced. The *data interpretation* task extracts high-level metabolic and statistical parameters from a collection of individual measurements communicated by the low-level controller; the *reasoning* task applies a logic-based belief-maintenance algorithm in order to evaluate the state of the patient and the performance of the low-level controller; the *decision* task exploits the results of the first two tasks to choose or to adjust a protocol using heuristic or model-based techniques.

The high-level controller realizes a form of *reactive planning* by periodically revising the control strategy to be applied at the low-level on the basis of the measured inputs and outputs of the system. Drawing an analogy with control systems, it can be seen as the supervisor of an adaptive control system, able to change both the low-level strategy and the patient representation used to plan the low-level strategy itself.

A key issue of the proposed architecture regards its possible implementation in a telemedicine context. While the high-level controller is designed run in a centralized medical workstation, the low-level controller is suitable to be implemented on a portable device able to communicate the glucose time series together with the corresponding insulin deliveries to the high level controller. Unlike previously proposed decision support systems for IDDM management, our methodology tries to achieve a higher flexibility by exploiting different tools on each layer, according to the quality and the amount of the patient data. Instead of searching for the overall optimal solution of the problem, the goal of our approach is therefore to provide a *service* to both the patient and the physician in order to improve the quality of the treatment.

Acknowledgments.

The authors thank sincerely Marco Ramoni for his methodological support and Mario Stefanelli for revising the draft of this paper. This work is partially supported by the European Community, through the AIM project 2034 GAMES II - A General Architecture for Medical Expert Systems.

References

1. S. Andreassen, J. Benn, R. Hovorka, K.G. Olesen, E.R. Carson, A probabilistic approach to glucose prediction and insulin dose adjustment: description of metabolic model and pilot evaluation study, Computer Methods and Programs in Biomedicine, 41 (1994) 153-165.
2. R. Bellazzi, C. Siviero, M. Stefanelli, G. De Nicolao Adaptive controllers for intelligent monitoring To appear in: Artificial Intelligence in Medicine Journal.
3. M. P. Berger, R. A. Gelfand, P.L. Miller, Combining statistical, rule-based and physiologic model-based methods to assist in the management of diabetes mellitus, Computers and Biomedical Research, 23 (1990) 346-357.
4. J. Beyer, J. Schrezenmeir, G. Schulz, T. Strack, E. Kustner, G. Shulz, The influence of different generations of computer algorithms on diabetes control, Computer Methods and Programs in Biomedicine, 32 (1990) 225-232.
5. T. Deutsch, E.D. Lehmann, E.R. Carson, A.V. Roudsari, K.D. Hopkins, P.H. Sönksen Time series analysis and control of blood glucose levels in diabetic patients Computer Methods and Programs in Biomedicine, 41 (1994) 167-182.
6. T. Hauser, L.V. Campbell, E.W. Kraegen and D.J. Chisholm, Glycaemic response to an insulin dose change: computer simulator predictions vs mean patient responses, Diabetes Nutrition and Metabolism, 7 (1994) 89-95.
7. E.D. Lehmann, T. Deutsch, E.R. Carson, P.H. Sönksen, AIDA: an interactive diabetes advisor, Computer Methods and Programs in Biomedicine, 41 (1994) 184-203.
8. M.Ramoni, A. Riva, M. Stefanelli, V. Patel, Forecasting glucose concentration in diabetic patients using ignorant belief networks, Proceedings of the AAAI Spring Symposium on Artificial Intelligence in Medicine, Stanford, CA, 1994.
9. E. Salzsieder, G. Albrecht, E. Jutzi, U. Fischer, Estimation of individual adapted control parameters for an artificial beta-cell, Biomed. Biochem. Acta, 43 (1984) 585-596.
10. A. Schiffrin, M. Mihic, B.S. Leibel, A.M. Albisser, Computer Assisted Insulin Dosage Adjustment, Diabetes Care, 8 (1985) 545-552.
11. J. Schlichtkrull, O. Munck, M. Jersild, The M-value, an index of blood sugar control in diabetics, Acta Med. Scand. 177, 95-102.
12. D. Spiegelhalter, A. Dawid, S. Lauritzen, R. Cowell, Bayesian Analysis in Expert Systems. Statistical Science, 8 (1993) 219-283.

Therapy Planning Using Qualitative Trend Descriptions

Silvia Miksch[1], Werner Horn [1,2], Christian Popow [3], Franz Paky[4]

[1] Austrian Research Institute for Artificial Intelligence (ÖFAI)
Schottengasse 3, A-1010 Vienna, Austria
E-mail: silvia@ai.univie.ac.at
[2] Department of Medical Cybernetics and Artificial Intelligence, University of Vienna
[3] NICU, Division of Neonatology, Department of Pediatrics, University of Vienna
[4] Department of Pediatrics, Hospital of Mödling

Abstract: This paper addresses a method of therapy planning applicable in the absence of an appropriate curve-fitting model. It incorporates knowledge about data points, data intervals, and expected qualitative trend description to arrive at unified qualitative descriptions of parameters (temporal data abstraction). Our approach benefits from derived qualitative values which can be used for recommending therapeutic actions as well as for assessing the effectiveness of these actions within a certain period. It results in an easily comprehensible and transparent concept of therapy planning. Furthermore, we improved the system model of data interpretation and therapy planning by using importance ranking of variables, priority lists of attainable goals, and pruning of contradictory therapy recommendations.

Our methods are applicable in domains where an appropriate curve-fitting model is not available in advance. We have applied them in the field of artificial ventilation of newborn infants. The utility of our approach is illustrated by VIE-VENT, an open-loop knowledge-based system for artificially ventilated newborn infants.

1 Introduction: The Needs For Special Therapy Planning Concepts

The care of critically ill patients in intensive care units (ICUs) is complex, involving interpretation of many variables, comparative evaluation of therapy options, and control of patient-management parameters. The technical improvement of the ICUs' equipment makes a huge amount of data available to the medical staff, and even skilled physicians frequently suffer from this information overload. Additionally, there are increased demands on the quality control and quality assurance ([8]: EURISIC-European User Requirements for Intensive Care).

Knowledge-based systems for intelligent alarming, diagnosis, monitoring and therapy planning have separately been developed. In contrast to diagnosis, which tries to find the best explanation for the actual situation of a patient, monitoring and therapy planning imply actions: *monitoring* indicates observing the course of a patient's condition under a given therapy, and assessing whether the selected therapeutic action is effective and the predicted improvement of the patient's condition occurs. *Therapy planning* involves selecting which therapeutic actions may improve the patient's condition, predicting the outcome, and adopting a therapeutic plan according to some explicitly defined preferences on the predicted condition of the patient [14]. Recent studies pointed out the challenge and the need to integrate all these activities within a unique framework, especially when dealing with dynamic domains ([1], [2]).

Control theory or statistical analysis is useful to allow for a straightforward mapping of monitoring data and appropriate therapeutic actions. In view of the lack of appropriate curve-fitting models for predicting the time course of clinical variables, as in the case of artificial ventilation, these methods are unsuitable. However, it is possible to express expected trends, like "the $P_{tc}O_2$ value should reach the normal region in approximately 10 to 20 minutes". We therefore tried to overcome the missing of an appropriate curve-fitting model by applying a dynamic temporal data abstraction mechanism based on spot and trend data analysis as well as expected qualitative trend descriptions to arrive at unified qualitative descriptions of variables. These qualitative values are used in the system model of data interpretation and therapy planning. An advantage of using qualitative values is their unified usability in the system model, no matter of which origin they are.

In the first part of this paper we introduce the system architecture of VIE-VENT. In the second part we focus on important components of the temporal data abstraction process to arrive at unified qualitative descriptions of data *points* and *trends*. In the third part we explain VIE-VENT's therapy planning module based on the interpretation of the patient's health condition, on pruning of therapeutic actions, and on verifying whether therapeutic actions are effective.

2   VIE-VENT's System Architecture

Developing VIE-VENT, an open-loop knowledge-based monitoring and therapy planning system for artificially ventilated newborn infants [10], we incorporated alarming, monitoring, and therapy planning tasks in one system to overcome some of the limitations of existing systems. The data-driven architecture of VIE-VENT consists of several modules: data selection, data validation, data abstraction, data interpretation and therapy planning. All these steps are involved in a single cycle of data collection from monitors. VIE-VENT is especially designed for practical use under real-time constraints at neonatal ICUs (NICUs) and the various components are built in analogy to the clinical reasoning process.

On the one hand VIE-VENT's whole input data set can be divided into continuous and discontinuous data. Continuous data (e.g., blood gas measurements, like $P_{tc}O_2$, S_aO_2, $P_{tc}CO_2$, and ventilator settings, like PIP, F_iO_2) are taken from the output of the data selection module every 10 seconds. Discontinuous data are entered into the system on request by the user depending on different conditions (e.g., critical ventilatory condition of the neonate, elapsed time intervals, missing monitoring data). The system output consists in primarily therapeutic recommendations for changing the ventilator setting. Additionally, VIE-VENT gives warnings in critical situations, as well as comments and explanations about the health condition of the neonate.

On the other hand VIE-VENT's variables can be divided into dependent (e.g., blood gas measurements, chest wall expansion) and independent variables (e.g., ventilator settings). The therapeutic actions are composed of independent variables.

VIE-VENT monitors the patient during the whole artificial ventilation process. We divide the whole period into four phases: initial phase, controlled ventilation, weaning, returning to spontaneous breathing. Transition from one phase to the next is handled by rules depending on the amount of artificial ventilation needed.

3 Preconditions for Temporal Data Abstraction and Therapy Planning

The problem of planning artificial ventilation of newborn infants - as in other medical fields, like pediatric growth [6] - lies in the lack of an appropriate curve-fitting model to predict the development of physiological variables from actual measurements. Therefore our first effort was to approximate the growth of continuously assessed measurements such as $P_{tc}O_2$, $P_{tc}CO_2$, and S_aO_2 using a simple linear regression model ($E(Y) = a + k\,X_i$) where $E(Y)$ is the expected value, X_i are the observed data points, a is a constant value (offset), and k is the growth rate). We assumed that observations are mutually independent and have the same variance.

Choosing this simple linear regression model was influenced by practical clinical reasons: the only important characteristics of variables used by physicians for a real therapeutic action are on the one hand increases, decreases, or zero changes of variables, and on the other hand too slow, too fast, or reasonable changes of variables. Therefore it would be superfluous to calculate a curve-fitting model of higher order with additional features for our purpose.

An appropriate curve-fitting function would be an exponential function with variables improving towards the normal range after a therapeutic action. On the basis of this concept we compare the actual trend with a stepwise linearized function representing the exponential curve. This decreases the complexity of a comparison of exponential functions and ensures responsiveness of the system. Additionally, the linear approximation is reasonable and applicable for small time intervals.

Based on physiological criteria, four kinds of trends of our 10 seconds data samples can be discerned:
  (1) *very short-term* trend: sample of data points based on the *last* minute,
  (2) *short-term* trend: sample of data points based on the last *10* minutes,
  (3) *medium-term* trend: sample of data points based on the *last 30* minutes,
  (4) *long-term* trend: sample of data points based on the *last 3* hours.
Comparing different kinds of trends is a useful method of assessing the result of previous therapeutic actions, of detecting if oscillation is too rapid, and of isolating the occurrence of artifacts (compare [9]).

4 Data Abstraction

The aim of the data abstraction process is to arrive at unified qualitative descriptions of data points and trend data. It transforms quantitative measurements into qualitative values, which can be used in the system model for data interpretation and therapy planning. An advantage of using qualitative values is their unified usability in the system model, no matter of which origin they are. Adaptation to specific situations can easily be done by using specific transformation tables without changing the model of data interpretation and therapy planning. Additionally, by using qualitative values an easily comprehensible and transparent system model can be developed.

VIE-VENT uses five different kinds of data abstraction: transformation of quantitative data points into qualitative values, transformation of trend data, dynamic calibration of values, context-sensitive adjustment of qualitative values, and smoothing for data oscillating in the neighborhood of thresholds. The data abstraction process dealing with spot data (only data points are involved in contrast to data abstraction based on data intervals as well as combinations of data points and data intervals) has been described in [9]. We will focus on the first two components.

4.1 Transformation of Data Points (Data-Point-Transformation Scheme)

The transformation of quantitative data points into qualitative values is usually performed by dividing the numerical range of a variable into regions of interest. Each region stands for a qualitative value. The region defines the only common property of the numerical and qualitative values. It is comparable to Shahar's et al. [13] "point temporal abstraction".

The basis of the transformation of the blood gas measurements are *data-point-transformation schemata* relating single values to seven qualitative categories of blood gas abnormalities (qualitative *data-point*-categories):

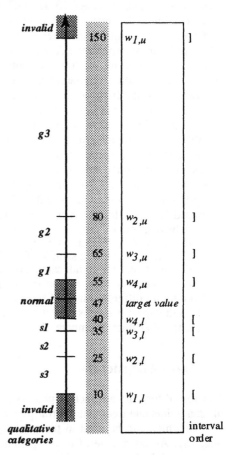

g3	...	extremely
g2	...	substantially
g1	...	slightly
normal	...	target range
s1	...	slightly
s2	...	substantially
s3	...	extremely

above the target range

below the target range

These data-point-transformation schemata are defined for all kinds of blood gas measurements depending on the blood gas sampling site (arterial, capillary, venous, transcutaneous) and the mode of ventilation (IPPV, IMV). The different modes of ventilation require specific predefined target values depending on different attainable goals. Fig. 1 shows the scheme of $P_{tc}O_2$ during IPPV. For example, the transformation of the transcutaneous $P_{tc}O_2$ value of 91 mmHg during IPPV results in a qualitative $P_{tc}O_2$ value of $g3$ ("extremely above target range"). The $w_{i,x}$ -values divide the qualitative regions. The transformation of trends is based on these qualitative data-point-categories, which are described in the following section.

Fig. 1. Data-point-transformation scheme of $P_{tc}O_2$ during IPPV

4.2 Transformation of Trend Data (Trend-Curve-Fitting Scheme)

The transformation of trend data into qualitative values is based on the combination of qualitative data-point-categories and the qualitative descriptions of the expected behavior of a variable (*expected qualitative trend descriptions*; e.g., "variable $P_{tc}O_2$ is moving one qualitative step towards the target range within 10 to 30 minutes"). These *trend-curve-fitting schemata* transform the quantitative trend values into ten qualitative categories guided by physiological criteria (Fig. 2).

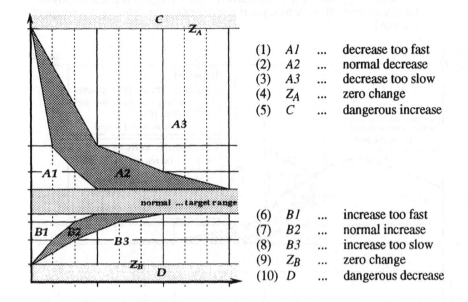

(1)	$A1$	...	decrease too fast
(2)	$A2$	...	normal decrease
(3)	$A3$	...	decrease too slow
(4)	Z_A	...	zero change
(5)	C	...	dangerous increase

(6)	$B1$	...	increase too fast
(7)	$B2$	...	normal increase
(8)	$B3$	...	increase too slow
(9)	Z_B	...	zero change
(10)	D	...	dangerous decrease

Fig. 2. Trend-curve-fitting scheme of $P_{tc}O_2$

A qualitative trend-category depends on the relative position of corresponding data points. For example, if a $P_{tc}O_2$ data point is classified as *g1*, *g2* or *g3* (" ... above target range") we would expect a therapeutic intervention to result in a decrease of type A2 as "normal" trend. If the data point lies in the target range ("*normal*") no therapeutic action is recommended.

As an example, Fig. 3 gives the trend-curve-fitting scheme of $P_{tc}O_2$ where we have reached a value of 91 mmHg after 43 minutes. The x-axis describes the discrete granularity of the representation in minutes. The y-axis shows the $P_{tc}O_2$ levels. It indicates the quantitative values of data points (at thresholds horizontal dotted lines are drawn) and their corresponding qualitative categories are listed on the right side. Based on the guiding principle depicted in Fig. 2, we compute the actual curve for selecting between the different qualitative categories. The stripped area A2 shows the expected normal development. The qualitative trend-categories are written in bold, capital letters and determine if an additional therapeutic action should be recommended (visualized with light-gray arrows in Fig. 3 and described in chapter 5).

An appropriate approach to classify trend data is to transform the curve (borders of the dark gray area) shown in Fig. 3 to an exponential function and to compare it with the actual growth rate. We used a dynamic comparison algorithm to classify the trend data, which performs a stepwise linearization of the expected exponential function to overcome complexity. It consists of two steps:

Step one: calculates the actual growth rate k_a using the linear regression model explained in chapter 3 and two thresholds of the growth rate k_1 and k_2 depending on the relative position of data point; k_1 and k_2 are used for discerning the qualitative trend-categories

Step two: classifies the qualitative trend-category depending on the actual growth k_a, the two thresholds k_1, k_2 and the qualitative region where the data point belongs. In addition to k_1 and k_2 we use an ε-range around zero to classify a trend as "Z_A" and "Z_B", respectively. The ε-range is created on physiological grounds in order to support a wider range for defining no change of a variable.

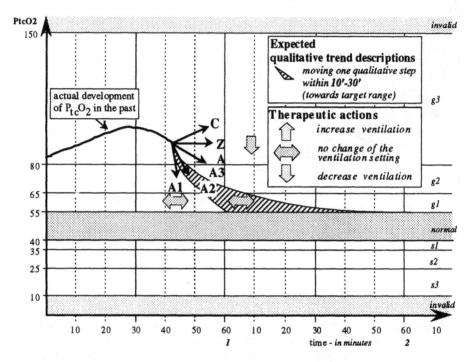

Fig. 3. Trend-curve-fitting scheme of $P_{tc}O_2$

Step one: The value space of a variable consists of an upper and a lower qualitative region divided by the target range. In the following we explain the algorithm used in the upper region. For the lower region the conditional elements ($\leq$, $>$) have to be changed to ($\geq$, $<$).

Let[1] w_1, w_2, w_3, and w_4 be the thresholds of the qualitative data-point-categories above target range (as shown in Fig. 1), where $\infty > w_1 > w_2 > w_3 > w_4 > 0$. Let a_t be the actual value at current time t, and $[t_{min}, t_{max}]$ the expected time interval for normal moving of one qualitative step towards the target range. The two thresholds of the growth rate k_1 and k_2 are then calculated in Fig. 4.

Step two:
    if the data point belongs to the region above the target value **then**
        if $(k_a > \varepsilon)$ **then** (qual_trend_category is C)
        else   **if** $(|k_a| \leq \varepsilon)$ **then** (qual_trend_category is ZA)
                else   **if** $(k_a < k_1)$ **then** (qual_trend_category is A1)
                        else   **if** $(k_a < k_2)$ **then** (qual_trend_category is A2)
                                else (qual_trend_category is A3)

---

[1] In contrast to figure 1, the second index has been eliminated to increase readability.

$$\text{if} \quad (a_t \le w_3) \text{ then} \qquad k_1 = \frac{(w_4 - w_3)}{t_{min}} \text{ and } k_2 = \frac{(w_4 - w_3)}{t_{max}}$$

$$\textbf{else} \quad \text{if } (a_t > w_2) \text{ then} \qquad k_1 = \frac{\left(\dfrac{(w_2 - w_3)}{(w_1 - w_2)}(a_t - w_2) + w_3 - a_t\right)}{t_{min}} \text{ and}$$

$$k_2 = \frac{\left(\dfrac{(w_2 - w_3)}{(w_1 - w_2)}(a_t - w_2) + w_3 - a_t\right)}{t_{max}}$$

$$\textbf{else} \qquad k_1 = \frac{\left(\dfrac{(w_3 - w_4)}{(w_2 - w_3)}(a_t - w_3) + w_4 - a_t\right)}{t_{min}} \text{ and}$$

$$k_2 = \frac{\left(\dfrac{(w_3 - w_4)}{(w_2 - w_3)}(a_t - w_3) + w_4 - a_t\right)}{t_{max}}$$

Fig. 4. Calculation of the two thresholds

The result of this process are instantiations of qualitative trend descriptions for each dependent variable (namely blood gas measurements) for each kind of trend, and for each mode of ventilation. For explanations the qualitative values as well as the corresponding numerical values are stored in a `qualitative_trend` template.[2] E.g.,

```
(qualitative_trend
     (BG_name PtcO2) (kind_of_trend very-short) (mode_of_vent IMV)
     (qualt_trend_category A3)
     (numerical_growth -0.03)(numerical_const 93.5)
     (k1 -0.39) (k2 -0.13))
```

5   Therapy Planning

The therapy planning module consists of formulation of therapeutic actions based on the interpretation of monitoring data, of pruning of therapeutic actions, and of verifying whether the therapeutic actions are effective. The data interpretation module has been described in [10].

5.1   Formulation of Therapy Recommendations

The qualitative abstraction of monitoring variables allows for creation of simple rules to activate therapeutic recommendations. Rule `R5-therapeutic-actions` (Fig. 5) gives an example of such a rule. The rule syntax is defined in Clips notation (v6.02, COSMIC/NASA), a forward chaining rule and/or object based development system.

The essential preconditions to trigger therapeutic actions depend on the qualitative trend-category (expressed as `qualt_trend_category` for a short-term trend) and the qualitative data-point-category (expressed as `qual_data_point_category`). If the qualitative data-point-category does not belong to the set {normal} and the qualitative

---

[2] A corresponding template specifies the expected qualitative trend description.

trend-category belongs to the set {A3, ZA, C}, then an action is recommended (e.g., to decrease ventilator settings). Amount and frequency of an action depend on the degree of abnormality of the blood gas measurement (e.g., *s3* is worse than *s2*, therefore a larger amount of change is recommended) and the strategy of ventilation (e.g., aggressive or conservative). These features have been described in more detail in [10]. The second fact ventilations_phase in the left-hand-side (LHS) of the rule refers to the mode of ventilation (i.e., IMV) and indicates that this rule belongs to the set of rules dealing with this phase of ventilation.[3] The right-hand-side (RHS) of rule R5-therapeutic-actions specifies the therapeutic actions. Each action-fact includes the kind of the recommended action and an explanation of the circumstances: the fact (reason oxygenation) refers to the depending process "oxygenation" of the system model of ventilation[4], (BG ?BG) refers to the depending variable, namely the blood gas measurement, and (kind ?x) determines which particular action has to take place (e.g., (kind dec-pip) means that a decrease of the peak inspiratory pressure (PIP) is recommended).

```
(defrule R5-therapeutic-actions
              "activate-therapeutic-action-PtcO2-oxygenation"
      (phase (kind therapy_recommendation))
      (ventilations_phase (mode imv))
      ?f1 <- (thp_recommendation oxygenation)
      (qualt_trend_category          (BG_name PtcO2)
                        (kind_of_trend short)
                        (qualt_trend_category A3|ZA|C))
      (qual_data_point_category (kind PtcO2) (value ~normal))
      =>
      (retract ?f1)
      (assert  (action (reason oxygenation) (BG PtcO2) (kind dec-fiO2))
               (action (reason oxygenation) (BG PtcO2) (kind dec-peep))
               (action (reason oxygenation) (BG PtcO2) (kind dec-pip))
               (action (reason oxygenation) (BG PtcO2) (kind dec-ti))))
```

Fig. 5. Example: Rule R5-therapeutic-actions

5.2 Assessment of Therapeutic Actions

The duty of the assessment of therapeutic actions is to assess their therapeutic efficiency. VIE-VENT defines a therapeutic action as "effective" or "ineffective" based on the ordering of the qualitative trend-categories and a delay-time.
   (1) *ordering of the qualitative trend-categories*:
        qualitative *upper* region: C ≺ ZA ≺ A3 ≺ A2 and A1 ≺ A2
        qualitative *lower* region: D ≺ ZB ≺ B3 ≺ B2 and B1 ≺ B2
        In the qualitative upper region the best attainable category is A2. There are two possible ways to reach the category A2: first, it is possible to get there from the most severe (worst) category C to ZA, then to A3 and finally A2. Second, from category A1.
        The same concept is used in the qualitative lower region for categories D, ZB, B3, B2, and B1 respectively.
   (2) *delay-time* for all observed dependent variables is 10 minutes.

---

[3] The set of rules is divided into rules which hold for all phases of ventilation and rules which hold only for a specific phase. The phase of ventilation is related to the mode of ventilation as mentioned in chapter 2.

[4] VIE-VENT's system model of ventilation is divided into two processes: oxygenation and ventilation. Different parameters are involved in these two processes (compare [10]).

If a therapeutic action has taken place, we would expect the qualitative trend-category to improve by one step in the ordering direction after the delay-time. If no qualitative improvement can be detected then the action is assessed as "ineffective" and the book-keeping procedure is activated. The book-keeping procedure collects and counts all previously ineffective therapeutic actions. If the book-keeping procedure verifies that a previous therapeutic action has *twice* failed to improve the patient's clinical condition, it forwards a signal to the "therapy evaluator" (namely the component dealing with the priority lists of attainable goals, compare chapter 5.3.2) to trigger an alternative action.

5.3 "Therapy Evaluator": Pruning of Therapy Recommendations

The process "formulation of therapy recommendations" renders a list of possible therapeutic actions according to the independently observed monitoring variables, e.g.,

```
Example 1:
    ((action (reason oxygenation) (BG PtcO2)    (kind dec-fio2))
     (action (reason oxygenation) (BG PtcO2)    (kind dec-peep))
     (action (reason oxygenation) (BG PtcO2)    (kind dec-pip))
     (action (reason oxygenation) (BG PtcO2)    (kind dec-ti))
     (action (reason ventilation) (BG PtcCO2)   (kind inc-pip))
     (action (reason ventilation) (BG PtcCO2)   (kind inc-f))).
```

As can be seen in example 1, it is possible that according to the blood gas measurements, contradictory (e.g., dec-pip and inc-pip) and/or too many therapeutic actions are recommended. Therefore the duties of the "therapy evaluator" are to rank and to prune therapeutic actions. VIE-VENT distinguishes three different kinds of therapy evaluation: importance ranking of variables, priority lists of attainable goals, and pruning of contradictory therapy recommendations. These methods are applied in the order as they are mentioned in the following.

Our approach differs from the existing mechanisms for assessing and comparing costs and benefits of therapeutic actions (e.g., the decision-theoretic approach in VentPlan ([5], [12]) by using heuristic knowledge to rank and to prune important or unimportant therapeutic actions.

5.3.1 Importance ranking of Variables

Importance ranking[5] of variables specifies which therapeutic actions should take place first according to the qualitative categories of the dependent variable. VIE-VENT uses the following rules:

(1) invasive blood gases are more reliable, thus more important, than transcutaneous blood gases;
(2) SaO_2 is more important than $P_{tc}O_2$;
(3) therapeutic actions depending on the qualitative data-point category " ... below target range" (s1, s2, s3) are more important than therapeutic actions depending on the qualitative data-point category " ... above target range" (g1, g2, g3);
(4) within the range of the qualitative data-point category " ... below target range" (s1, s2, s3) therapeutic actions aiming to improve PO_2 or S_aO_2 are more important than therapeutic actions depending on PCO_2;

---

[5] In the data validation component a similar method is used to validate the input data as reliable, called *reliability ranking*.

5.3.2 Priority Lists of Attainable Goals

The priority lists of attainable goals specify the order in which the variables should be changed. In particular the priority lists rank which variable should reach which value space next. The value range of the variables is divided into a valid and an invalid range. Additionally, the valid range is divided in subintervals and each subinterval indicates a specific attainable value space. The first variable in the priority list is chosen according to its value space. If a signal is received that the previous therapeutic actions had twice resulted in an insufficient improvement of the patient's condition (compare chapter 5.2), the alternative variable or the next variable is chosen. The goals are separately defined for dependent (e.g., blood gas measurement $P_{tc}CO_2$) and independent (e.g., ventilator setting PIP) variables.

The global goal for the dependent variables, namely blood gas measurements, is to arrive at the normal range as soon as possible according to physiological criteria. The normal ranges of the different blood gas measurements are determined in the data-point-transformation schemata. It is different depending on the sampling site and the mode of ventilation. This results in a different target range for each variable (e.g., (PO_2, (transcutaneous, IMV), 40, 55), (PCO_2, (arterial, IPPV), 49, 35)).

The global concept of priority lists of the independent variables consists of a *general order of variables* and a *particular priority list of attainable value spaces* which has to be reached first depending on the direction of change, the number of available variables, and the relation of available variables expressed in corresponding intervals of value space. In VIE-VENT the attainable goals for the independent variables, namely the ventilator settings, depend on the two processes of the system model: ventilation and oxygenation. The two ventilator settings peak inspiratory pressure (PIP) and frequency (f) as well as the optional information about the chest wall expansion are involved in the ventilation process.

The *general order* for increasing is try "f before PIP" and the *global order* for decreasing is try "PIP before f". If the value of chest wall expansion is available then this additional precondition is added (e.g., if (chest wall expansion is normal) then increase f before PIP, if (chest wall expansion is small) then increase PIP before f). The *particular priority list* of attainable value spaces for PIP and f is listed in Table 1.

In the case of decreasing, PIP has to be first decreased to 45 (45 is the plausible upper limit of PIP), then f has to be decreased to 150 (150 is the plausible upper limit of f), then PIP has to be decreased stepwise to 40 then f has to be decreased stepwise to 120, and so on. For example, let PIP be 28, f be 45, and chest wall expansion be normal, then PIP will stepwise be decreased to 15 before starting to decrease f to 40. The actual amount of change depends on the degree of blood gas abnormality, namely the qualitative data-point-category. The priority list of attainable goals just determines which variable has to be taken first depending on the attainable intervals

PIP (cm H$_2$O)	f (breaths/minute)
10	20
15	40
20	60
25	80
30	100
35	120
40	135
45	150

Table 1. The particular priority list of attainable value spaces for PIP and f.

In the case of oxygenation a similar *general order* for increasing and decreasing and a *particular priority list* of attainable value spaces for the ventilator settings F_iO_2, PIP, PEEP and T_i are defined to rank the therapy recommendations.

5.3.3 Pruning of Contradictory Therapy Recommendations

The pruning of contradictory therapy recommendations identifies the important recommendations from a set of conflicting therapy recommendations. Currently, we apply a straight forward strategy to deal with contradictory therapy recommendations because most of the possible occurrences are already handled by the other features of the "therapy evaluator" which are applied first.

```
HEURISTIC:
If      an increase as well as a decrease of the same variable is
        recommended
then we delete both therapeutic actions.
```

In the example 1 (see above), both actions (inc-pip) and (dec-pip) are deleted.

6   Related Work

In the past decade several strategies have been developed to support monitoring and therapy planning. The drawback of most approaches is that they were developed for low-frequency data (like in pediatric growth monitoring where new data are collected several months or a year apart (e.g., TrenDx [6]), in diabetes where new data arrive three or four times a day (e.g., [11]), or in artificial ventilation management using only invasively determined variables (e.g., GUARDIAN [7]). Monitoring and therapy planning of high-frequency data of ICU patients require different strategies for data validation, monitoring and therapy planning.

A comparable, but closed-loop approach is NeoGanesh/Ganesh [3, 4]. It is a rule-based system designed for weaning of artificially ventilated adults. Its temporal model consists of aggregation and forgetting to choose an appropriate treatment plan. Our temporal data abstraction covers and extends parts of RESUME's [13] temporal abstraction of time-stamped data. E.g., we extended the "point temporal abstraction" mechanism of RESUME with expected qualitative trend descriptions to arrive at unified qualitative values. The main purpose of RESUME is temporal data abstraction and trend detection. It lacks on activating therapeutic actions and assessing the benefits of therapeutic actions. The same drawback holds for DIAMON-1 [15], a two-stage monitoring system based on fuzzy sets.

7   Conclusion

We demonstrate a method to improve monitoring and therapy planning in the absence of an appropriate a priori curve-fitting model. Our approach benefits from derived qualitative values of data points, data intervals and expected qualitative trend descriptions (temporal data abstraction). These qualitative values can be used for recommending therapeutic actions as well as for verifying whether these actions are effective within a certain period. Additionally, we improved the system model of data interpretation and therapy planning by using importance ranking of variables, priority lists of attainable goals, and pruning of contradictory therapy recommendations. Our approach results in an easily comprehensible concept of therapy planning.

Acknowledgment

The current phase of the project is supported by the "Jubiläumsfonds der Oesterreichischen Nationalbank", Vienna, Austria, project number 4666. We greatly appreciate the support given to the Austrian Research Institute of Artificial Intelligence (ÖFAI) by the Austrian Federal Ministry of Science and Research, Vienna.

References

1. Barahona P., Christensen J.P.(eds.): *Knowledge and Decisions in Health Telematics, The Next Decade*, IOS, Amsterdam, 1994.

2. Console L., Molino G., Torasso P.: Some New Challenges for Artificial Intelligence in Medicine, in Barahona P., Christensen J.P.(eds.), *Knowledge and Decisions in Health Telematics*, IOS, Amsterdam, 1994.

3. Dojat M., Brochard L., Lemaire E., Harf A.: A Knowledge-Based System for Assisted Ventilation of Patients in Intensive Care Units, *International Journal of Clinical Monitoring and Computing*, 9, pp.239-50, 1992.

4. Dojat M., Sayettat C.: Aggregation and Forgetting: Two Key Mechanisms for Across-Time Reasoning in Patient Monitoring, in Kohane I.S., et al.(eds.), *AI in Medicine: Interpreting Clinical Data*, AAAI Press, Menlo Park, pp.33-36, 1994.

5. Farr B.R., Fagan L.M., Decision-theoretic Evaluation of Therapy Plans, in Kingsland L.C.(ed.), *Proceedings of the Thirteenth Annual Symposium on Computer Applications in Medical Care (SCAMC-89)*, IEEE Computer Society Press, Washington D.C., pp. 188-92, 1989.

6. Haimowitz I.J., Kohane I.S.: Automated Trend Detection with Alternate Temporal Hypotheses, in Bajcsy R.(ed.), *Proceedings of the 13th International Joint Conference on Artificial Intelligence (IJCAI-93)*, Morgan Kaufmann, San Mateo, CA, pp.146-151, 1993.

7. Hayes-Roth B., Washington R., Ash D., Hewett M., Collinot A., Vina A., Seiver A.: Guardian: A Prototype Intelligent Agent for Intensive-Care Monitoring, *Artificial Intelligence in Medicine*, 4(2), pp. 165-66, 1992.

8. Kari A.: Quality Control and Quality Assurance in Finland, in Metnitz P.G.H.(ed.), *Patientdaten Management System auf Intensivstationen*, Workshop Notes, Wiener Intensivmedizinische Tage (WIT 94), Vienna, 1994.

9. Miksch S., Horn W., Popow C., Paky F.: Context-Sensitive Data Validation and Data Abstraction for Knowledge-Based Monitoring, in Cohn A.G. (ed.), *Proceedings of the 11th European Conference on Artificial Intelligence (ECAI 94)*, Wiley, Chichester, UK, p. 48-52, 1994.

10. Miksch S., Horn W., Popow C., Paky F.: VIE-VENT: Knowledge-Based Monitoring and Therapy Planning of the Artificial Ventilation of Newborn Infants, in Andreassen S., et al. (eds.): *Artificial Intelligence in Medicine: Proceedings of the 4th Conference on Artificial Intelligence in Medicine Europe (AIME-93)*, IOS Press, Amsterdam, pp.218-29, 1993.

11. Ramoni, M., Riva A., Stefanelli M., Patel V.L.: Forecasting Glucose Concentration in Diabetic Patients Using Ignorant Belief Networks, in Kohane I.S., et al.(eds.), *AI in Medicine: Interpreting Clinical Data*, AAAI Press, Menlo Park, pp.33-36, 1994.

12. Rutledge G.W., Thomsen G.E., Farr B.R., Tovar M.A., Polaschek J.X., Beinlich I.A., Sheiner L.B., Fagan L.M.: The Design and Implementation of a Ventilator-management Advisor, *Artificial Intelligence in Medicine*, 5(1), pp.67-82, 1993.

13. Shahar Y., Tu S.W., Musen M.A.: Knowledge Acquisition for Temporal Abstraction Mechanisms, Special Issue: Knowledge Acquisition for Therapy-Planning Tasks, *Knowledge Acquisition*, 4(2), 1992.

14. Stefanelli, M.: Therapy Planning and Monitoring, *Artificial Intelligence in Medicine*, 4 (2), pp. 189-90, 1992.

15. Steimann F., Adlassnig K.-P.: Two-Stage Interpretation of ICU Data Based Fuzzy Sets, in Kohane I.S., et al.(eds.), *AI in Medicine: Interpreting Clinical Data*, AAAI Press, Menlo Park, pp.152-6, 1994.

Adaptation and Abstraction in a Case-Based Antibiotics Therapy Adviser

R.Schmidt [a], L.Boscher [b], B.Heindl [b], G.Schmid [a], B.Pollwein [b], L.Gierl [a]

a) Computer Centre of the Medical Faculty
b) Institute for Anaesthesiology
of the Ludwig-Maximilians University of Munich

Abstract

In this paper, we describe an approach to make case-based reasoning methods appropriate for medical problems. From the class of therapeutic problems we have chosen calculated antibiocs therapy advice for patients in an intensive care unit who have developed an infection as an additional complication. As advice is needed quickly and the pathogen is not yet known, we use an expected pathogen spectrum based on medical background knowledge and known resistances, which both will be adapted to the results of the laboratory. Case-based reasoning retrieval methods provide the advice for similar previous patients. The previous solutions are adapted to be applicable to the new medical situation of the current patient. Because of the large and continuously increasing number of cases, we use prototypes as a structural aid. We present some experimental results of our studies on the performance of our prototype design.

1 Introduction

Medicine differs from other knowledge domains by a professional documentation of cases handled in clinical practice. Numerous case collections have been accumulated. However, the intrinsic medical experience of these case bases is not yet fully used in knowledge-based systems. Now, a suitable technique - case-based reasoning, a methodology for reasoning and learning - has reached a state of maturity. The rapidly growing interest of the AI community in case-based reasoning provides an increasing set of methods. Case-based reasoning means to use previous experience to understand and solve new problems. In case-based reasoning, a reasoner remembers a previous situation similar to the current one and uses it to solve the new problem.

At our Faculty, case-based systems had up to now a flat simple prototype (strong abstraction) structure automatically learned while cases were added to the case-base (see for example [1]). Now we investigated the growth of a hierarchical prototype structure built up from a stream of cases. Moreover, we use adaptation to speed up the decision process. The empirical results of this study deal with two problems: (1) the impact of thresholds used within the abstraction process of building the tree of prototypes and cases and (2) the interaction between abstraction and adaptation process. As a first medical domain we have chosen antibiotic therapy.

For patients in an intensive care unit with additional complications, we have developed ICONS, a case-based system that offers antibiotic therapy advice. The goal of ICONS is a quick presentation of appropriate antibiotic therapies considering resistances, the sphere of activity, contraindications and complete coverage of the whole expected pathogen spectrum. Using a potential pathogen spectrum allows us to offer therapy advice without knowing the actual pathogen, which still has to be identified by the laboratory. We attempt to find a calculated therapy deduced from background knowledge on infectious diseases instead of a specific therapy. Only after the laboratory has identified the actual pathogen and has tested its susceptibility against antibiotics defining a specific therapy is possible. The information on the identifications and the susceptibilities of the pathogens is used to update and adapt the knowledge base to the empirical situation of the intensive care unit.

Apart from giving antibiotic therapy advice for unexperienced physicians, the main clinical performance of ICONS is the use of updated information about resistances and expected pathogen spectrums.

2 The Process of Therapeutic Decision Making Selecting Antibiotics

As ICONS is not a diagnostic system, we do not attempt to deduce evidence for the diagnosis of symptoms, frequencies and probabilities, but instead pursue a strategy (shown in figure 1) that can be characterized as follows: find all possible solutions and reduce them using the patient's contraindications and the complete coverage of the calculated pathogen spectrum (establish-refine strategy). First, we distinguish among different groups of patients. For post-operative patients and those patients with community acquired infections, we determine the possible pathogen spectrum according to the affected organ. For patients with nosocomial infections and immuncompromised patients, we use a spectrum deduced from background knowledge. A first list of antibiotics is generated by a susceptibility relation that returns for each group of pathogens all the antibiotics which usually have therapeutic effects. This list contains those anibiotics that can control at least part of the potential pathogen spectrum. We obtain a second list of antibiotics by reducing the first by applying criteria such as the patient's contraindications and the desired sphere of activity. Using the antibiotics of this second list, we try to find antibiotics that cover the whole pathogen spectrum under consideration of the expected susceptibility individually. Except for some community acquired infections, monotherapies have to be combined with synergistic or additive effecting antibiotics. If no adequate single therapy is found, we use combination rules to generate combinations of antibiotics. Each possible combination is tested for the ability to cover the whole expected spectrum. Before the user decides to use one of the presented therapies he or she can investigate potential side effects of the antibiotics. Moreover, he or she may obtain information about the daily costs of each suggested therapy. After the physician has chosen one therapy, ICONS computes the recommended dosage.

3 Adaptation of a Similar Case

One principal argument for case-based reasoning methods [2] is to speed-up the process of finding adequate therapies. Considering the close relation concerning the group of patients and the affected organ, a similar case is retrieved from a hierarchically and generalising storage structure containing prototypes as well as cases. As methods, we apply the similarity measure of Tversky [3] and the Hash-Tree-Retrieval-Algorithm of

Stolter, Henke, King [4]. Furthermore we use a criterion of adaptability during the retrieval, because not very case is adaptable [5]. The adaptation of the previous similar to the current case, is done by a solution transfer that later is reduced by additional contraindications of the current case.

As the number of cases increases continually, the storing of each case would exceed any space limitations. So we decided to structure the case base by prototypes and to store only those cases that differ from their prototype significantly.

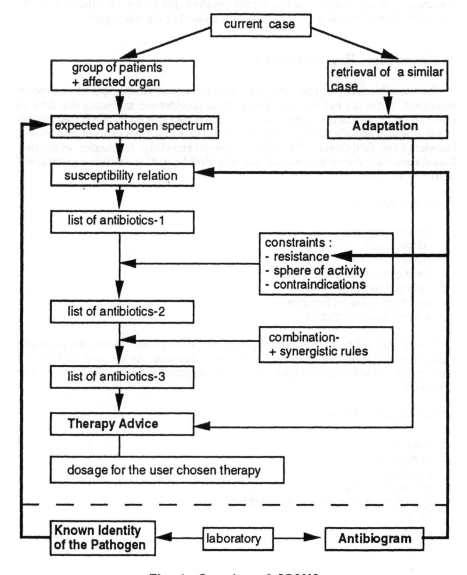

Fig. 1. Overview of ICONS

4 Adaptation to the Results of the Laboratory

The identifications of pathogens and antibiograms are used as control mechanisms. When the pathogen is identified, the flow of control is very similar to the one described above (in section 2). The expected pathogen spectrum is replaced by the identified pathogen. The set of possible antibiotic therapies is reduced by the antibiogram to those antibiotics the pathogens are not resistant to.

In the knowledge base, we supplement each theoretically determined pathogen spectrum with the one that is empirically justified due to the identification of the pathogens. The information about resistances is updated by the antibiogram.

5 Knowledge Representation

We use a hierarchy of pathogens consisting of classes, groups, species and special pathogens. There are two relations, one ot them provides the antibiotics that have an effect on groups of pathogens, the other one supplies the expected pathogen spectrum per group of patient and affected organ. The knowledge base of ICONS exists of case oriented knowledge (medical cases) and medical background knowledge (antibiotics, pathogens). Causal knowledge can be modeled with the latter. Both kinds of knowledge are represented as frame structures.

the case structure:

- group of patients
- affected organ
- contraindications
- some general data on the patient
- pathogen spectrum
- suggested antibiotic therapies
- selected antibiotic therapy

First, a case exists of some general patient data like age, weight etc., the associated group of patients, the affected organ and contraindications. As the program runs, the expected spectrum, the identified pathogen and the suggested and selected therapies will be stored here.

the antibiotic structure:

- side effects
- contraindications
- resistances
- sphere of activity
- synergistic or additiv effecting antibiotics
- dosage
- prices

Besides the procedural knowledge for the handling this knowledge, we use rules for the explicit construction of antibiotics combinations. The implementation of the case-based methods are located in two independent but problem specific modules. For the retrieval and the creation of the prototypes, it is necessary to know the syntactical structure and the important features of the cases and the adaptation criterion. The adaptation modul needs specific information on the kind of reduction of the transferred solutions using the additional contraindications of the current case.

6 Case-Based Reasoning and Prototypes

Empirical research [6] indicates that people consider cases to be more "typical" when the number of features between the presented case and the "normal" case increases. Based on these findings we create prototypes that share most features with most of their cases.

From a medical point of view, prototypes correspond to typical antibiotic treatments associated with typical clinical features of patients. At the highest level, a prototype is created for each affected organ and each group of patients. All cases of a prototype belong to the same group of patients, the same organ is affected and the same pathogen spectrum deduced from background knowledge has to be covered. The cases are discriminated by contraindications. These are antibiotic allergies, reduced organ functions (e.g. kidney and liver), specific diagnosis (e.g. acoustic distortion or diseases of the central nervous system), special blood diseases, pregnancy and the patient's age group (e.g. adult, child, infant). Since each contraindication restricts the solution set of adequate therapies, it is used as a constraint. This results in the condition for the criteria of adaptability, that a similar case can only be adapted if it has no additional contraindications in comparison to the current case. Otherwise, the solution set of the current case would be inadmissibly restricted by the irrelevant constraints of the former similar case.

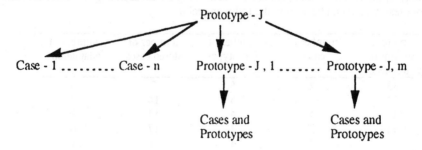

Fig. 2. Possible relationships of a prototype

First, all cases below a prototype are stored. Only upon reaching the threshold "number of cases" the prototype is filled, i.e. the contraindications of the associated cases are inspected, and every contraindication reaching the relative frequency of the second threshold "minimum frequency" are included in the prototype. Subsequently, the prototype is treated like a case and the resulting suggestible antibiotic therapies are stored. Those cases that have no additional contraindications in comparison with the prototype are deleted and their contraindications are included in the frequency table of the prototype.

A new case is incorporated into the case base by updating the frequency table of its associated prototype and recomputing its contraindications. If the contraindications of the prototype change, the suggested antibiotic therapies must be also recomputed. All cases must be inspected again for their need to be stored and we try to create an alternative prototype.

We create an alternative prototype below an existing prototype, if for the latter there exist cases, that have at least one contraindication in common, which the prototype does not include, and which together reach the threshold "number of cases". The alternative prototype is then constructed from the cases deviating from the superior prototype. The difference to the superior prototype consists of other contraindications, other antibiotic therapies and a different position within the prototype hierarchy. A possible prototype hierarchy is shown in figure 2.

7 Results of our Experiments

For testing our prototype design, we have used data of 21 postoperative lung infected patients with on average almost 1.5 contraindications. These cases were entered into ICONS in the same order, which was randomly generated in advance. We have varied the parameters for the two thresholds. After entering these 21 cases, we inspected the current system state. We looked at the number of created prototypes, stored cases and realised adaptations of previous cases or of prototypes. If no such adaptation is possible, we adapt an artificial case without contraindications, which is created after the creation of a prototype. Even the adaptation of a case without contraindications works faster than the normal flow of the program, because the most time-consuming step is the generation of the antibiotic combinations. Though it must be considered, that the state of our system is always subject to the influence of random patient data, the sequence of entry, and the choice of inspection time, it is possible to make some general statements on favourable settings of the two threshold parameters - depending on the desired goal - from the results of our test, which can be seen in the following table.

> relative minimum % frequency	number of cases	number of prototypes	number of stored cases	number of adaptations
33	2	5	7	3
33	3	5	14	7
33	4	4	15	7
33	5	2	17	6
25	2	4	3	1
25	3	3	8	4
25	4	3	15	6
25	5	1	18	4
20	3	2	5	2
20	4	1	10	2
20	5	1	14	4
-	> 21	-	21	6

Table 1: Settings of the threshold parameters and their consequences

Independent of the choice of the two threshold values, the following is evident: the more cases and prototypes are stored, the more adaptations can be executed. To formulate general statements, we will first explain the importance and effects of the two threshold values.

The "minimum frequency" determines from which relative frequency of appearance a contraindication is included in the prototype. Should for instance 3 cases form a prototype, then a "minimum frequency" of 33% means, that only those contraindications are included in the prototype, which exist in more than one case (query for > 33%), while a lower "minimum frequency" would lead to the effect of "initial amnesia": all contraindications of the three cases are included in the prototype, none of the three cases still shows any additional contraindication in comparison with the prototype, and all three cases will therefore be forgotten.

The "number of cases" determines the required number of cases, so that an (alternative) prototype can be created. The lower this threshold is, the more prototypes are created, the number of prototypes increases, and the number of stored cases decreases.

When the prototype incorporates more cases, the frequencies of the contraindications and therefore the filling of the prototype change. The idea is to update the prototype. As a case forgotten once is lost forever, it can happen that a case is assimilated (with the consequence of getting lost) by a prototype, but that this prototype later changes in such a manner, that the case would have at least one additional contraindication and would therefore be kept. Especially in the assignment of parameters of line 5, this effect together with "initial amnesia" can be seen.

For the relationship of both parameters, three different assignments for the "minimum frequency" must be inspected separately. It can be seen, that the goal of maximum number of adaptations is reached, if the product of both parameters is equal to 100, while the goal of minimum storage is reached by the minimum of the product. For the choice of the "minimum frequency" it can be seen, that the number of adaptations increases with the relative frequency, and the number of stored cases falls with relative frequency. Under consideration of realistic boundary conditions (e.g. a prototype should only be created with a minimum of at least two cases) the following target strategies result: (1) to get many adaptations: a high "minimum frequency", and product with the "number of cases" equals 100; (2) minimum storage: low "minimum frequency" and low "number of cases"; (3) balanced compromise: intermediate "minimum frequency" and intermediate "number of cases", and the product is less than 100.

8 Related Work

Methods for realistic applications which provide an automatic incremental refinement and enhancement of a medical knowledge base providing an abstraction mechanism to condense cases are rare. Lebowitz UNIMEM [7] forms a generalised tree of cases. However, he stores all cases. PROTOS [8] uses the "exemplar" approach and stores prototypes and exemplars as more specialised prototypes. However, there is no hierarchy of prototypes as an aid for retrieval and adaptation. Methods for retaining cases described in literature are general enough for realistic medical expert systems. Methods for adaptation, however, usually are very specialised. They can primarily be applied to knowledge rich domains. In CASEY [9], an attempt is made to use more general operators for the adaptation in the domain of heart failure diagnosis, but not all adaptation problems could be solved. In FGP [10] the problem to cope with a large amount of cases is not solved by using prototypes but by retaining only parts of a case in the case base. This cannot be applied as a general strategy in medical domains because it is often not known, which parts of a case are necessary for prospective cases.

9 Conclusions

Because on the one hand, we accumulate the characteristics of several cases into one prototype, and on the other hand, we reduce the number of solutions using constraints, we are confronted with conflicting goals. A case is well adaptable, if it shows as few contraindications as possible. However a case is deleted when it does not show additional contraindications with respect to its prototype, because this is the only possibility to join several cases to one prototype. Both opposing goals can be individually targeted by clear strategies concerning the determination of the threshold values.

The number of more favourable adaptations in relation to the number of cases used seems relatively small. A statement on the representability of the data set we used is still missing. Our investigation only concerns adaptations in relation to contraindications. The spectrum of pathogens and the development of resistances must also be adapted.

We find, that abstractions of individual cases to a hierarchy of prototypes and cases fit the requirements for medical expert systems. What we need in addition are more general adaptation methods for classes of typical medical decision making.

10 Further Work

The aim is to find maximal general adaptation and abstraction methods for decision making in therapy finding. Therefore, we will investigate further medical domains - subsequent the water and electrolyte balance - in generalising our approach. One part is the introduction of an adaptation process concerning the clinic-specific resistance behaviour of pathogens over a long time period. This could be extended by enhancing our approach towards an "early warning" system concerning regional resistance behaviour. Moreover, we are going to integrate the system in our Hospital Information System and to evaluate it on a ward.

Acknowledgements

This research is partly funded by the German Ministry for Research and Technology. It is part of the MEDWIS project for research on medical knowledge bases.

References

[1] Gierl L, Arias-Lewing G, Stengel-Rutkowski S, Jakobeit M, Lohse K (1988): Knowledge Acquisition for Scheme-Based Medical Expert Systems: The Dysmophic Syndrome Example, in: Rienhoff, Piccolo, Schneider (pub.): Expert Systems and Decision Support, Berlin 347-350
[2] Kolodner J. (1993): Case-Based Reasoning, Morgan Kaufmann Publishers, San Mateo
[3] Tversky A. (1977): Features of Similarity, in: Psychological Review 84, 327-352

[4] Stolter R.H., Henke A.L, King J.A. (1989): Rapid Retrieval Algorithms for Case-Based Reasoning, in: International Joint Conference on Artificial Intelligence 11, 233-237

[5] Smyth B, Keane M.T. (1993): Retrieving Adaptable Cases: The Role of Adaptation Knowledge in Case Retrieval, in: First European Workshop on Case-Based Reasoning (EWCBR-93), 76-81

[6] Rosch, E. and Mervis, C.B.(1975): Family resemblances: studies in the structure of categories, in: Cognitive Psychologie 7, 573-605

[7] Lebowitz M. (1987): Experiments with Incremental Concept Formation, Machine Learning, Vol. 2, 103-138

[8] Bareiss R. (1989): Exemplar-based Knowledge Acquisition, Academic Press, San Diego

[9] Koton P. (1988): Reasoning about Evidence in Causal Explanations, in: Proceedings Case-based Reasoning Workshop, Clearwater Beach, Florida, 260-270

[10] Fertig S., Gelernter D. H. (1991): A Virtual Machine for Acquiring Knowledge from Cases, in: IJCAI-91, Sydney, 796-802

Evaluation of Knowledge Based Systems

Field Evaluations of a Knowledge-Based System for Peripheral Blood Interpretation

Diamond LW[a], Nguyen DT[a], Ralph P[b], Sheridan B[c], Bak A[b], Kessler C[d] and Muncer D[d]

[a]Pathology Institute, University of Cologne, Germany
[b]MDS Laboratories, 100 International Boulevard, Etobicoke, Ontario, Canada
[c]Dept. of Pathology, Sunnybrook Health Science Center, University of Toronto, Ontario, Canada
[d]Coulter Technology Center, 11800 SW 147 Avenue, Miami, FL, USA

A medical knowledge-based system for hemogram and peripheral blood smear interpretation was tested in three situations outside of the developers' laboratory: Coulter Corporation's Clinical Evaluations, and two clinical laboratories in the United States and Canada. Rigorous testing of the system's functionality was performed. Hazard testing caught several programming errors which were corrected. The outside evaluators requested a number of changes to the screens and additional functionality. As a result, several user configurable features were added to the system. Medical information collected from 13 of 392 (3.3%) cases effected the knowledge base. Only two of the evaluator's comments concerned the style of the system's displayed output. The system was able to maintain high quality interpretive reports, appropriate to patient care, despite the expected differences in subjective morphologic interpretation between observers.

1. Introduction

An important step in the development of medical knowledge-based systems (KBSs) is evaluation in clinical environments. Once a system has been thoroughly tested and refined in the developers' laboratory, field evaluations must be planned and conducted, to judge the impact of the system on users, patients, and the healthcare system [1,2]. Many parameters can be evaluated, including the reliability of the system's output, and the overall usability of the software.

While the KBS is exclusively in the developers' hands, the bulk of the testing usually consists of carefully controlled and selected cases. When the developers are experienced physicians in the medical specialty, the data that they personally enter into the system has most likely undergone conscientious scrutiny for accuracy. Testing the knowledge base with "perfect" data satisfies the need to examine the system's structure, but it still leaves many unanswered questions about performance in the field, where "real life" data will be entered.

"Professor Petrushka" is a KBS for hemogram and peripheral blood smear interpretation which has been designed as part of a diagnostic workstation for laboratory hematology [3,4]. From its inception in May 1991, the system has been designed to be ultimately incorporated into a hematology laboratory's routine, interfaced to the current series of COULTER® mid-range and high-end hematology analyzers.

As part of the software validation process, "Professor Petrushka" has undergone extensive testing at Coulter Corporation. To establish the feasibility of the product, field evaluations in clinical laboratories in the United States and Canada have been performed. In this report, we present our assessment of the results of this first round of outside testing.

2. Computer Hardware and Software

The hematology workstation consists of an IBM-compatible computer and associated software. The workstation has been tested on two hardware platforms, one for Coulter ONYX sites, and one for STKS sites. The ONYX provides automated three-part leukocyte differentials and is intended for mid-range laboratories with a moderate volume of hematology testing. The STKS is a high performance analyzer with five-part automated leukocyte differentials, suitable for laboratories which process several hundred to several thousand complete blood counts each day.

For the ONYX site, the workstation consisted of an Intel 486-SX 25 MHz computer with 8 MB of RAM, a 170 MB hard disk drive, 15-inch color monitor, and Windows 3.1. The STKS site used an NEC Powermate computer (486-DX2 66 MHz) with 20 MB of RAM, a 540 MB hard disk drive, 17-inch color monitor, and Windows NT 3.1. All systems have been equipped with 150-MB Bernoulli removable cartridge drives for data back-up.

"Professor Petrushka", the KBS for hemogram and peripheral blood smear interpretation, has been previously described in detail [3,4]. Briefly, the system was designed using Turbo Pascal for Windows, version 1.5. Database functions have been programmed with Paradox Engine 3.01. All knowledge engineering and computer programming have been done by a trained hematopathologist.

Both the ONYX and the STKS record flags for suspected morphologic abnormalities, including, but not limited to, circulating blasts, variant ("atypical/reactive") lymphocytes, immature granulocytes, and red blood cell (RBC) agglutination. The hemogram parameters, including suspect flags, are down-loaded and stored in Paradox databases by the interface software.

There is a field in the "users" database which allows the laboratory to set certain configuration parameters using Paradox. Currently, configurable features include: (1) Hematology units (e.g. SI units in Canada); (2) the choice between two available test panels for iron deficiency; and (3) the option for Petrushka to list the suggested antisera panel in the final interpretation, when immunophenotyping is recommended.

Each laboratory can input their own reference ranges by age and sex directly in Paradox. Every technologist or physician with access to the system enters a user name, password, and their preferred keyboard layout for manual differential counting, in the user's database. In this study, the users chose the option of entering manual differential results directly in the dialog box edit fields, rather than using the workstation keyboard as a counting device.

Instructions for the use of the system are available as an on-line Windows help file accessible with a menu selection. At the time of the testing, no printed manual was available.

"Professor Petrushka's " reasoning is based on defined diagnostic patterns [3,4]. When a user requests a specimen for analysis (by date or patient identification number), Petrushka displays the hemogram data. In response to the user's request to proceed, the system determines the predominant preliminary pattern based on the hemogram parameters, age, and sex. When review of a blood film is appropriate, Petrushka recommends scanning at low-power magnification. Based on the predominant pattern and the individual findings of the hemogram, the system suggests a specific approach to blood film review (RBC, WBC and/or platelet morphology). The technologist is free to accept or override any of the system's recommendations according to the impression gained from scanning the blood film. Qualitative morphologic findings are entered using Windows controls (radio buttons and check boxes). When the blood film review is complete, Petrushka determines the final predominant pattern, and generates an editable interpretive report, listing: (1) the pattern; (2) a differential diagnosis; (3) any findings which indicate a more specific diagnosis; and, (4) suggestions for follow-up testing and additional clinical history. When the technologist has completed a case, the data which was entered manually is stored in the database. If a technologist or physician edits Petrushka's final interpretation, the edited report is saved and displayed on review of the case.

An important feature of the program is the use of the database to compare the present specimen with the previous complete blood count from the same patient. The comparison is more than just a delta check of the hemogram data. The final patterns of both specimens are compared, and, if appropriate, the interpretation on the present specimen is based on both sets of findings.

3. Design of the evaluations

In the Coulter Corporation Clinical Evaluations laboratory, testing began in January 1994. The software was put through rigorous usability and hazard testing on workstations interfaced to a STKS and ONYX. Every available function and interface element (menus, dialog boxes, buttons, check boxes, and drop-down lists) was systematically evaluated. Hazard testing included attempts to perform functions not intended by the designers, e.g., entering alphanumeric characters or out-of-range values when editing hemogram or patient demographic data (e.g., age or date of birth). "Action reports", for circumstances which need to be brought to the developer's attention, were generated under the following conditions: (1) for each identified problem; (2) when a question was raised regarding an intended function; or, (3) when the user thought of a possible enhancement to system performance.

Product evaluations in outside clinical laboratories always begin with "off-line" testing on an instrument not being used for routine patient work. In this study, the patient's name and unique identification number were not entered to maintain confidentiality. The specimens were identified only by accession numbers.

Testing at the ONYX site in Temple, Texas, took place in May 1994. STKS cases were collected by a laboratory in Toronto, Canada, between June and August, 1994.

After testing with a minimum number of normal specimens, the outside laboratories were requested to randomly select 10-20 abnormal specimens per day from the patient workload, and re-run the specimens on the instrument interfaced to the KBS. The technologists were asked to interact with the KBS, and enter the peripheral blood morphologic findings, including a manual differential, if performed.

For each specimen, the outside laboratory provided: (1) a copy of the STKS or ONYX printout; (2) a stained peripheral blood film; (3) a printed copy of Petrushka's final display output (using screen capture software resident on the workstation); and, (4) a copy of any laboratory interpretation or pathologist's comments about the specimen.

Clinical information was available on some patients. One laboratory was encouraged to enter clinical information using a "point-and-click" clinical data entry program [5]. The other laboratory made available the results of special hematology testing (e.g., hemoglobin electrophoresis) and chemistry testing on many specimens. These reports were saved for entry by the system developers.

Periodically during the study, the users were asked to fill out "input forms" listing any comments, concerns, recommendations, and suggestions about the workstation.   In particular, the users were asked to express their feelings with respect to the screen layouts, ease of data entry, technical wording, number of screens, and correctness of the logical conclusions of the program. These forms could also be used for the laboratory's "wish lists" for customization and enhancements.

The data, backed up on Bernoulli cartridges, were reviewed by the developers. Based on the original data entered by the laboratory, the developers judged Petrushka's interpretation for correctness and completeness. In addition, all peripheral blood films were reviewed, and a manual differential was performed, when appropriate, by one of the system developers, Dr. D.T. Nguyen (a board-certified hematopathologist). Dr. Nguyen's peripheral blood film findings and manual differential were compared to those reported by the laboratories. When Dr. Nguyen's findings were different, the case was re-played with her morphologic data. Petrushka's interpretive report, based on the new input, was compared to the one generated from the original data. This comparison was done to determine if the system was robust enough to render high quality reports, in the face of the expected differences from one observer to another.

4. Results of the Evaluation

During the period from 1/25/94 to 10/21/94, 104 unique, reproducible action reports were generated, concerning the functionality of the peripheral blood KBS and the clinical data entry program. The general categories of action reports are shown in Table 1. The most common requests were changes to screen displays and requests to change program functionality. In response to user requests, the patient identification number was added to each of the five KBS screens. In the original design, when a previously finalized case was reviewed a second time (e.g., a pathologist reviewing a specimen already completed by the technologist), the default condition was to show the screens as the system had originally displayed them. The reviewer had to click a button to see the technologist's data. At the users request, the current default is to show the technologists data, with comments on the screen to indicate the recommendations that were overridden. In addition, a new dialog box was added to handle comments for the laboratory's own use. These comments do not effect the knowledge base.

Approximately 10% of the action reports reflected either a lack of knowledge about Windows built-in functionality, or were requests for features which already existed in the program. The users were unaware of these features, because they did not read the relevant portions of the on-line instructions, or did not understand the intended functionality. These problems can be addressed by improving the instructions, in order to better educate technologists and physicians as to the intent of the system.

The evaluation, and especially the hazard testing, uncovered six programming errors. For instance, the original program did not trap for the user entering a zero (hematologically, an impossible value) when editing the hematocrit. The system recalculates the MCHC (MCHC=hemoglobin/hematocrit) based on the edited hematocrit. A zero hematocrit would cause a divide-by-zero error and an immediate system crash. The program now traps for impossible values and displays a message box allowing the user to correct their input.

As can be expected, there were several requests to change the terminology used in the system for certain morphologic findings such as "variant lymphocytes." At this point, we have not resolved all of these issues. For a hematology KBS to reason properly, each morphologic term must correspond to a specific cell type, and the meaning should not vary between laboratories. Terms which are potentially ambiguous, even if they are in common usage, impede the transferability of the KBS. There are regional differences in the use of terminology. These issues can best be resolved by standardization and education.

Only two action reports were directed towards Petrushka's final interpretation. One comment was that some of the reports were too "wordy." This reflects the fact that the laboratories normally issued reports containing predominantly short, "canned" comments. There were no serious disagreements as to the medical conclusions in "Professor Petrushka's" reports, however.

Table 1. Number and type of action reports

Reason for action report	Number of Requests
Request to override normal Windows functionality	3
Request to make screen (display) changes	29
User lacked knowledge of program or Windows functionality	14
Request to change Instructions/Help	4
Request to change hematology terminology	6
Request to change program functionality	23
Programming errors	6
Request to change Petrushka's final interpretation	2
Request to improve system performance (e.g., fewer keystrokes)	7
Request to force agreement between analyzer display and KBS	4
Request to improve entry of user configurable data (e.g., reference ranges)	3
Request to change clinical data entry program	3

A total of 392 evaluable cases were submitted by the two laboratories together with all of the required documentation (including peripheral blood films). In 302 of 392 cases (77%), Dr. Nguyen's morphologic interpretation of the peripheral blood film agreed with that of the submitting institution. In 90 cases, Dr. Nguyen's interpretation of quantitative and qualitative features on the peripheral blood smear differed from those reported by the technologists at the submitting laboratory.

The most common morphologic disagreements concerned: (1) intermediate myeloid precursors on the manual differential; (2) low numbers of blasts or lymphoma cells, usually in cases with pancytopenia; (3) a difference in quantitation of red blood cell abnormalities such as the estimate of spherocytes, microcytes, macrocytes, or acanthocytes present; and (4) a difference of opinion regarding hypersegmentation or Howell-Jolly bodies.

A flowchart of the review of the 392 cases is shown in Figure 1. In 65 of the 90 cases with morphologic disagreements, there was no change in Petrushka's final interpretive report. In 25 cases, using Dr. Nguyen's findings resulted in a different interpretive report.

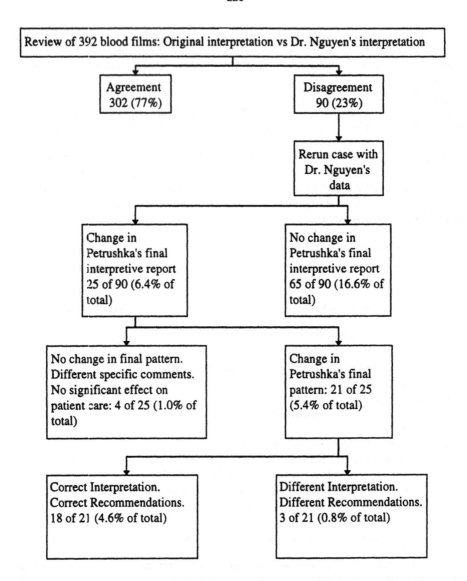

Fig. 1. Flowchart of Evaluable Cases

There were three different scenarios for the 25 cases in which the report changed:

(1) In four cases, the final display output was only slightly different. When specific RBC/WBC abnormalities (e.g., Howell-Jolly bodies, hypersegmentation) were reported, an additional comment was included in the interpretation with regard to those findings;

(2) In 18 cases, the difference of opinion regarding the presence/absence of blasts in low numbers led to two different interpretations: a) "Abnormal Mononuclear Cell Pattern" (typically, bone marrow malignancy); versus b) a different pattern, most often "Pancytopenia" or "Monocytosis" (with bone marrow infiltration by malignant cells included in the differential diagnosis). In both interpretations, the correct follow-up procedure (bone marrow aspirate and biopsy) was recommended. Since the differential diagnosis and recommendations were similar, the differences in the reports were not judged to be significant;

(3) In only 3 of 392 cases (0.8%), the differences in morphologic interpretation resulted in different reports, with no overlap in the differential diagnosis and different sets of recommendations.

As a result of the evaluation, the developers made two significant changes to the knowledge base, and identified 11 other cases where changes to the "rules" will probably be implemented. The two significant changes were implemented at the beginning of the study, to trap for inappropriate RBC morphologic input which grossly conflicted with the RBC histogram and the RBC indices. The other 11 changes will not effect the final pattern, but will trigger specific comments within the report (similar to item number 1 above). Overall, the users were quite pleased with the functionality of the software and the quality of the medical interpretations.

5. Discussion

Testing and evaluation are vital components of the development cycle of medical KBSs. Appropriate review can involve a number of different perspectives such as usefulness (relevance and completeness), and reliability (correctness, robustness, and sensitivity) [6]. Another important aspect of testing is the transferability of a system, i.e., to what degree does the system retain its credibility when it is transferred from one institution to another? Some studies have shown that good performance of a system at one site does not necessarily guarantee equal performance at other sites [7]. Since there are important aspects which need to be included in the evaluation of medical decision-aids, testing guidelines, as part of the European Concerted Action, have been recently proposed [8].

Controlled clinical trials with randomization of patients, doctors or departments are the best way to judge the impact of a decision aid on the quality of medical care [1,2]. Such studies can be difficult to design and require tremendous resources. Preliminary field trials, not conducted as randomized trials, can help answer the important questions about a system's usefulness, reliability, and transferability.

"Professor Petrushka" has been thoroughly tested in the developer's laboratory on more than 4000 cases. On the basis of the field evaluations performed at Coulter Corporation and the two other laboratories, many modifications have been made to the system. The bulk of the changes are related to the user interface and screen layouts. Portions of the screens have been reorganized, including the addition of a new dialog box designed to accept comments such as "few myelocytes seen on scan." Several kinds of user configurable features have been added to the system, such as the ability to turn on or off the immunology panel recommended by Petrushka for FCM. In this field evaluation, only 13 of the 392 cases (3.3%) led to modifications of the knowledge base.

In addition to comments/questions about the user interface, there were requests to modify some medical terminology used by the system, to reflect local laboratory practice. These issues are common to the design of all types of KBSs. Significant work is being done in the field of medical informatics to develop large lexicons of medical terms, to reflect concepts and capture general principles in the usage of medical terminology [9]. With respect to "Professor Petrushka", many of these issues can best be resolved by educating potential users about the system terminology. This information is given in the on-line Windows help. In addition, the medical knowledge contained within the system is addressed in the accompanying electronic textbook [10]. The first chapter of the textbook discusses techniques for preparing reproducible high-quality blood films. The remaining chapters of the textbook are organized by the patterns which Petrushka uses as intermediate conclusions in the reasoning process. Therefore, the textbook, which includes an atlas of digitized photomicrographs, not only serves as an educational tool for peripheral blood interpretation, but as an explanation facility for the interpretations made by the KBS.

Interpretation of morphologic findings is known to be subjective. Hence, there will always be differences of opinion in identifying individual cells in peripheral blood films or in quantitating RBC abnormalities. We were particularly interested in learning how these differences would effect Professor Petrushka's interpretive reports.

Our results confirm the power of heuristic classification using defined diagnostic patterns, which we have applied in the development of all of the modules in the hematology workstation [3,11-13]. Although there were differences in the blood film findings between the users (technologists) and one of the developers in 90 of 392 cases, in 65 cases, the differences did not result in a change in Petrushka's interpretive report. In these 65 cases, the differences were in features which do not contribute to the definition of a pattern (or the criteria of a disease). In the other 25 cases, differences in morphologic interpretation caused Petrushka's report to change. In 22 of the 25 cases, although the reports based on the two sets of findings were different, there was overlap in the differential diagnoses and the resulting recommendations were similar. In three cases, morphologic disagreements in key features led to totally different interpretations of the blood picture.

This study demonstrated that "Professor Petrushka" could function in several environments outside of the developer's laboratory. The hematology knowledge and the overall usability of the system allowed for a high degree of transferability between institutions. In addition, the system was able to deliver high quality interpretive reports, appropriate to patient care, despite the expected differences in subjective morphologic interpretation between observers.

During this study, changes were made to the system's user interface in response to feedback from the users. This method, so-called "user centered design", has been found to be useful in the development of other medical workstations and decision support systems [14].

At this time, several system configurations are actively under consideration. One configuration, for educational use, will contain a database with example cases, the electronic textbook, and multiparameter case studies, but no interface to a hematology instrument. At the other end of the spectrum, the system will be bundled with a hematology communications server capable of handling input from multiple hematology instruments simultaneously. The communications server will provide filtering functions so that normal specimens will be automatically validated and sent to the laboratory information system. In this arrangement, only abnormal specimens requiring technologist review will be sent to the KBS. This configuration will function in a client-server environment with multiple workstations on a network.

Acknowledgments: The authors thank Dr. E.S. Rappaport, Clinical Pathologist, Scott and White Clinic, Temple Texas, for his help in the evaluation.

6. References

[1]   Wyatt J. Field trials of medical decision-aids: potential problems and solutions. In: Fifteenth Annual Symposium on Computer Applications in Medical Care. NewYork: McGraw-Hill 1992;3-7.

[2]   Wyatt J, Spieglehalter D. Evaluating medical expert systems: what to test and how? Med Inf (Lond) 1990;15:205-17.

[3]   Nguyen DT, Diamond LW, Priolet G, Sultan C. A decision support system for diagnostic consultation in laboratory hematology. In: MEDINFO 92. Lun KC, Degoulet P, Piemme TE, Rienhoff O (eds), Amsterdam: North Holland 1992; 591-5.

[4]   Nguyen DT, Diamond LW, Priolet G, Sultan C. Expert system design in hematology diagnosis. Meth Inform Med 1992;31:82-89.

[5]   Diamond LW, Mishka VG, Seal AH, Nguyen DT. A clinical database as a component of a diagnostic hematology workstation. In: Proceedings of the Eighteenth Annual Symposium on Computer Applications in Medical Care. J. Am Med Informatics Assoc. Symposium Supplement. 1994:298-302.

[6]   Barahona P, Durinck J Florin C et al. Knowledge Processing for decision support in the health sector; A perspective for the next decade. In: Barahona P, Christensen JP (eds). Knowledge and decisions in Health Telematics, Amsterdam: IOS Press 1994;3-58.

[7]   Nolan J, McNair P, Brender J. Factors influencing the transferability of medical decision support systems. Int J Biomed Comput 1991;27:7-26.

[8]   Talmon JL. Guidelines for the evaluation of medical decision aids. Technical report, SEM Concerted Action, 1991.

[9]   Rector A. Compositional models of medical concepts: Towards re-usable application-independent medical terminologies. In: Barahona P, Christensen JP (eds). Knowledge and Decisions in Health Telematics, Amsterdam: IOS Press 1994;109-14.

[10]   Diamond LW, Nguyen DT. Linking an electronic textbook to expert systems in haematology. In: Richards B (ed). Healthcare Computing 1994 -- Current Perspectives in Healthcare Computing, Weybridge, UK: BJHC Books, 1994;269-79.

[11]   Diamond LW, Nguyen DT, Andreeff M, Maiese RL, Braylan RC. A knowledge-based system for the interpretation of flow cytometry data in leukemias and lymphomas. Cytometry. 1994;17:266-73.

[12]   Nguyen DT, Cherubino P, Tamino PB, Diamond LW. Computer-assisted bone marrow interpretation: a pattern approach. In: Reichert A, Sadan BA, Bengtsson S, Bryant J, Piccolo U, eds. Proceedings of MIE 93. London: Freund Publishing House, 1993;119-23.

[13]   Nguyen DT, Mishka VG, Moskowitz FB, Diamond LW.   A multiparameter decision support system for lymph node pathology. In: Barahona P, Veloso M, Bryant J, eds. Proceedings of MIE 94. Lisbon: 1994;162-7.

[14]   Engelbrecht R, Fitter M, Rector A. Requirements for a medical workstation using user-centred design. In: Adlassnig K-P, Grabner G, Bengtsson S, Hansen R (eds). Medical Informatics Europe 1991. Lecture Notes in Medical Informatics No. 45. Berlin: Springer-Verlag, 1991;140-4.

Functional Evaluation of Seth:
An Expert System in Clinical Toxicology

Stéfan J. DARMONI, (1), Philippe MASSARI, (2), Jean-Michel DROY, (3), Thierry BLANC, (4), Jacques LEROY, (3)

(1) Information System and Informatics Department, Rouen University Hospital, 1 Rue de Germont, F76031 Rouen Cedex, France;
(2) Medical Informatics Unit, Rouen University Hospital,
(3) Poison Control Centre and Adult Intensive Care Unit, Rouen University Hospital, (4) Child Intensive Care Unit, Rouen University Hospital;

Correspondence and reprints to SJ. Darmoni, Id. Tel: (33) 35.88.49.00 Fax: (33) 35.71.02.21
E-mail: stefan.darmoni@chu-rouen.fr

ACKNOWLEDGEMENTS

This work is supported by a grant from the Conseil de l'Informatique Hospitalière et de Santé (1993).
The authors thank Karen Benattasse for her linguistic assistance.

ABSTRACT

The aim of SETH is to give end-users specific advice concerning treatment and monitoring of drug poisoning. It is developed with an off the shelf expert system shell and runs on a microcomputer. The SETH expert system simulates the expert reasoning, taking into account for each toxicological class delay, signs and dose. The implementation of Seth began in April 1992 in our Poison Control Centre (PCC). SETH is then daily used by residents as telephone response support on drug poisoning. Between April 1992 and October 1994, 2099 cases inputted by residents were analysed by SETH. In October 1994, a functional evaluation of SETH showed that its effect in the daily practise of our PCC is positive: the performance of the residents increased and they would agree to use it outside our University Hospital. An expert system in clinical toxicology is a valuable tool in the daily practise of a Poison Control Centre.

KEY-WORDS:

Evaluation; Expert system; decision making, computer-assisted; drugs poisoning; adult; child.

INTRODUCTION

To support clinicians in the diagnostic and therapeutic process, computer-aided decision support can be used. The aim of SETH is: (a) in a known intoxication to give non-toxicologist physicians better advice for the treatment and monitoring of drug poisoning in adults and children according to clinical manifestations, ingested doses and delay, (b) in an unknown intoxication to identify products according to clinical manifestations and context. After two successful phases of evaluation (1) [phase I: early prototype development, and phase II: evaluation of system's validity (2)] the project group consisting of four experts in toxicology and two medical informaticians from the Rouen University Hospital agreed to install SETH in our Poison Control Centre (PCC) in April 1992.

We are describing below the phase III of Clarke et coll. (2): evaluation of the functionality of the system which concerns user interaction and field trial in a real life situation.

MATERIAL

The use and maintainability of this expert system was foreseen from the start. Analysis of the toxicological reasoning was done before choosing computer programs and hardware (2). Technical choices were made according to this analysis, financial considerations and portability. The domain of this expert system was chosen because drug poisoning is a frequent problem, representing in Rouen University Hospital 8% of adult emergencies and 3% in child emergencies.

SETH is developed using the KBMS expert system shell of Trinzic (3) in the Windows environment. KBMS is an object oriented programming shell. A KBMS application consists of objects and the languages that manipulates them. Objects can be stored in the knowledge base or in an external database. The languages express the instructions that evaluate and manipulate the objects: (a) a rule language in which all rules and inferencing strategies are expressed, and (b) two query languages (Intellect and AISQL). Inferencing strategies are forward chaining, backward chaining, hypothetical reasoning and explicit calls of packets. The query languages allow to manipulate objects in the knowledge base through requests expressed in natural language (English) for Intellect and in SQL for AISQL.

Access is the database management system used to create and easily maintain drugs and toxicological classes tables. The data base contains information on drugs, toxicological classes, potential clinical findings, advice on treatment and monitoring according to severity of poisoning. After each update in the data base, these information are transferred to corresponding objects in the knowledge base. Currently, the data base contains the 1110 most toxic or most frequently ingested French drugs from 76 different toxicological classes. SETH contains also a case database: all data imputed by an end-user, (PCC's resident), such as names of drugs, or generated by SETH such as the conclusions about the intoxication, are stored in the case database.

Hardware is an IBM compatible microcomputer with a 80486 microprocessor and 8 Megabytes of RAM. Hard disk occupation is less than 20 Megabytes.

ANALYSIS

Our cognitive analysis was transposed in the knowledge base. The SETH expert system simulates the expert reasoning, taking into account for each toxicological class delay, signs and dose. SETH describes a level graph, where each level represents a step of the reasoning. The first level contains initial conclusions on delay, dose and signs. These three initial conclusions generate a final conclusion, which represents the second level of the graph. This final conclusion defines for each class accurate monitoring and treatment advice, taking into account drugs interactions (third and last level of the graph, see Fig. 1). All the conclusions are done at the toxicological class level.. Inferencing is used to compute initial conclusions on delay, clinical manifestations and doses, global conclusion, of each ingested class, and to take into account interactions between classes or drugs and treat specific problems. We used, in our application, two inferencing strategies: forward chaining and explicit calls. In November 1994, the SETH knowledge base contains 363 rules in 45 packets, 40 objects and 399 attributes, and 170 AISQL and Intellect requests.

FUNCTIONALITIES OF SETH

The input of the patient and poisoning data is as simple as possible to minimize the time spent to input data. The first input screen includes patient and physician data, the name and quantities of drugs. Drugs data can be inputted by proprietary and non-proprietary names. Quantities of ingested drugs are expressed in tablets or millilitres for proprietary names according to the type of ingestion and in milligrams for non-proprietary names. The second input screen only clinical manifestations chosen from a list of potential symptoms according to the presumed drug ingested. The report includes: a recall of poisoning, an overall conclusion, and advice on emergency actions needed, monitoring, epurative treatments, specific actions, and biological assays. The recall of the poisoning includes all the input data, calculated data such as hypothetically ingested dose for a class, and data from databases such as the name of a toxicological class for a drug. The overall conclusion about the intoxication displays in a textual way, the initial conclusions and the global conclusion for each class, and some details like the list of clinical symptoms which can be explained only by one class. The potential clinical manifestations conclude the report; for each of them, the classes which can explained them are displayed. A print of the main points of the SETH consultation is possible. The report is available for adult and child poisoning, up to 4 drugs or 12 toxicological classes ingested.

RESULTS: FUNCTIONAL EVALUATION

We had already designed two phases to evaluate SETH in our hospital ('internal' evaluation) (2). There is no objective criterion to evaluate SETH, therefore we did use the expert advice as the gold standard. The aim of the first phase was to test respectively the initial conclusions on delay, signs and dose, and the global conclusion (first and second steps of the level graph). The purpose of the second phase was to test the accuracy of monitoring and treatment advice generated by the expert system (third step of the level graph). The aim of the third phase was to evaluate the functionality of SETH, specially the impact of the practical use of SETH (4) using a field trial .

All the errors of phase one and two of the evaluation have been corrected at the beginning of 1992. Then the project group agreed to install SETH in our Poison Centre in April 1992. SETH was then daily used by residents primarily as telephone response support and secondly as an educational tool on drug poisoning. The use of SETH is not mandatory for the PCC's residents. Between April 1992 and October 1994, 2099 drug intoxication cases were inputted, representing 2.1 cases a day, 10% of the Poison Centre's overall activity and 40% of the Poison Control Centre's phone calls regarding drug poisoning. There was an increase in the daily use of SETH: between January and October 1994, 850 cases were analysed by SETH (average of 3.15 cases a day) vs 608 cases (average of 2.25 cases a day) during the same period of 1993. Since the implementation in our Poison Control Centre, there was an increase of phone calls from inside the Rouen University Hospital, specially the residents of the Intensive Care Unit and Emergency Department. Some are coming to obtain a print of the conclusions of SETH.

In October 1994, in order to estimate the impact of SETH on the clinical decision making of residents, a clinical trial was carried out at the PCC of the Rouen University Hospital. The trial has included the nine residents of the PCC who used SETH in the past 30 months. The median of the SETH's utilisation at the time the trial was performed is 24 months (minimum = 10 months and maximum = 30 months). The main objective was to measure whether residents with a computer-aided decision support system changed their management of drug poisoning. The results of the questionnaires are displayed in Table 1.

DISCUSSION

Search in Medline, Toxline and Toxlit shows that less than ten computer-aided decision support systems have been developed in clinical toxicology compared to several dozen in environmental and experimental toxicology (5). Compared with other medical fields, clinical toxicology is probably easier to formalise because few heuristics are used and lots of data can be managed in a data base, such as storing information about drugs and toxicological classes.

From the beginning of the analysis, we have intentionally separated data and knowledge. In SETH, data on drugs, toxicological classes and advice can be updated within the data base application, one line for a new drug and several for a new class describing potential signs and specific advice. The maintenance of data on drugs and toxicological classes is performed with the electronic French drugs dictionary available in our University Hospital and updated every three months (6). When a resident inputs an unknown drug, SETH displays an alarm: its analysis in case of a multiple intoxication does not take into account this unknown drug and SETH stores it in a specific file which is monthly reviewed by the project group to possibly update the drug database. Only reasoning and toxicological classes interaction updates have to be done in the knowledge base. Therefore the overall maintainability of SETH is very easy. The maintenance of the knowledge base implies to review the data stored in the case database, specially the SETH's conclusions, and to confront it with the expert's analysis. The residents can write their remarks about a SETH's analysis in a paper form or in an electronic form according to their preference. We believe that the use of an off the shelf expert system shell, such as KBMS, leads to a better robustness, end-user interface and integration into information system.

The SETH's domain is very precise: exclusively drug poisoning. We defined in the knowledge base some limits of expertise (2), e.g. in case of a drug intoxication with a short delay, without any clinical sign and any information about the ingested dose, the SETH advice is 'Ask a human expert' but it gives the 'maximal' management which is the correct answer in case of doubt in clinical toxicology.

It is also possible to translate SETH in other languages quite easily because of the structure of our model. SETH was definitely not developed for the expert. Anyway they are using part of the system, at least the information stored in the different databases as they could do with an electronic textbook.

The lack of use of expert systems being related to time consumption (2), we have minimised the time spent to input data (2 screens, less than one minute) because it is one of the main reason to explain the lack of use of expert systems in daily practise (2). We have also minimised the overall time of the SETH consultation (less than three minutes) because drug poisoning is an emergency situation. The telephone support must be as quick as possible. Our effort was successful: the PCC's residents have judged that SETH is faster than sources previously used by PCC's residents. The effect of SETH in the daily practise of our PCC is positive: the performance of the residents increased and they would agree to use it outside our University Hospital. SETH does not reduce the time spent to converse with colleagues about drug poisoning which is a very important point to judge the acceptance of the system. The results showing the SETH's acceptance as a useful knowledge source must be underscored because the PCC's residents had a low previous experience with computers (mean score = 2.56). Furthermore, before the use of SETH, the residents thought that CDSS were not really useful in clinical decision making (mean score =

2.71) and after its use, they were willing to use SETH outside the Rouen University Hospital (mean score = 4.44). The relatively poor results of the SETH's interface (mean score of the question 'Easy to use' = 3.44 & mean score of the question 'Clear and convenient presentation' = 3.71) can be explained by the fact that the first version of SETH was developed in the MS-DOS environment. We are now using Windows since the beginning of 1994 but we still have to widely use better interface tools, such as combo boxes and buttons.

Nonetheless, the residents used the expert system in only 40% of the PCC's drug poisoning cases, the most difficult ones, specially multiple drug intoxication. They do not use SETH for a single benzodiazepine intoxication which is very frequent because they already know how to handle it by themselves. Some residents' remarks are astonishing. They assumed that the SETH expert system is faster during the night. Knowing that SETH runs on a stand-alone microcomputer, it has the same time response during the 24 hours. In fact, SETH is not faster but, awaked in the middle of the night, it is the resident's brain which is slower.

The next step of SETH's evaluation will be the external evaluation of SETH in adult and child poisoning, including other French hospitals (PCCs and Emergencies Departments), before it can be transferred to other hospitals. This external evaluation will eliminate the development centre bias: the enthusiasm of the development team may help end-users at the development centre to obtain more benefit from the expert system than elsewhere (7). The transferability of an expert system has been defined as 'the degree of which a system retains its reliability when applied in an another organisational environment (8) In conclusion, we believe than an expert system in clinical toxicology is a valuable tool in the daily practise of a Poison Control Centre.

REFERENCES

1. Darmoni SJ, Massari P, Droy JM, Mahé N, Blanc T, Moirot E, Leroy J. SETH: an expert system for the management on acute drug poisoning in adults. *Comput. Methods Programs Biomed.* 1993; 43: 171-176.
2. Clarke K, O'Moore R, Smeets R, *et al.* A methodology fo evaluation in knowledge-based systems in medicine. *Artificial Intelligence in Medicine* 1994; 6: 107-121.
3. KBMS application development guide. *Trinzic*, Palo Alto, Ca, USA, 1993.
4. Visser, MC. Hasman, A. Van der Linden, CJ. Protocol processing system (ProtoVIEW) to support residents at emergency ward. *Proceedings of MIE 94, Twelfth International Congress of the European Federation for Medical Informatics,* P. Barahona, M. Veloso & J. Bryant, Eds., pp. 138-143.

237

5. Hushon JM: Overview of environmental expert systems. In: *Expert systems in environmental applications*, American Chemical Society, Washington DC: Hushon JM Ed., 1990:1-24.

6. Darmoni SJ, Dufour F, Massari P,. Arnoudts S, Dieu B, Alizon B, Hantute F, Baldenweck M. Consultation of the Electronic Vidal dictionary in the Rouen University Hospital: analysis of the first year of utilisation. In: *Proceedings of MIE 94, Twelfth International Congress of the European Federation for Medical Informatics*, P. Barahona, M. Veloso & J. Bryant, Eds., 1994, pp 384-388.

7.. Wyatt J, Spiegelhalter D: Evaluating medical expert systems: what to test and how? *Med. Inf.* 1990, 15, 208-217.

8. Nolan J, McNair P, Brender J. Factors influenceing transferability of knowledge-based systems. *Int. J. Biomed. Comput.* 1991; 27: 7-26.

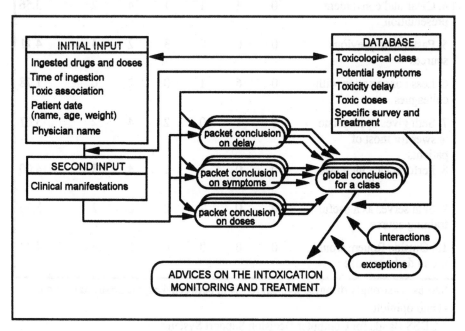

Fig. 1. SETH functional description

Table 1. Acceptance and attitudes towards SETH as a useful knowledge source

Question	-1	1	2	3	4	5	Mean Score
			Frequency of Scores				
1. Previous experience with computers	0	3	3	0	1	2	2.56
2. CDSS* might be useful in clinical decision making	2	1	2	2	2	0	2.71
3. Easy to use	0	1	1	1	5	1	3.44
4. Clear and convenient presentation	0	1	1	1	4	2	3.56
5. Faster than previous sources	0	0	0	3	2	4	4.11
6. Less conversation with colleagues	0	6	1	0	2	0	1.78
7. Seth gives appropriate answer for most of patients	0	1	0	2	4	2	3.67
8. Performance increases	0	0	1	1	4	3	4.00
9. Seth serves as a useful training source	0	0	2	2	3	2	3.56
10. Outside Rouen, would use in daily practise	0	0	0	0	5	4	4.44

Scores: 1=strongly disagree 2=disagree 3=in between 4=agree 5=strongly agree
-1=no opinion
* CDSS stands for Computer Decision Support Systems

Evaluating a Neural Network Decision-Support Tool for the Diagnosis of Breast Cancer

Joseph Downs, Robert F Harrison

Dept. of Automatic Control and Systems Engineering, University of Sheffield
Mappin Street, Sheffield, S1 3JD, UK
e-mail: {downs,r.f.harrison}@acse.shef.ac.uk

Simon S Cross

Dept. of Pathology, University of Sheffield Medical School
Beech Hill Road, Sheffield, S10 2RX, UK
e-mail: s.s.cross@shef.ac.uk

Abstract. This paper describes the evaluation of an application of the ARTMAP neural network model to the diagnosis of cancer from fine-needle aspirates of the breast. The network has previously demonstrated very high performance when used with high-quality data provided by an expert pathologist. New performance results are provided for its use with "noisy" data provided by an inexperienced pathologist. Additionally, ARTMAP supports the extraction of symbolic rules from a trained network and the validity of these autonomously-acquired rules is discussed. It is concluded that the symbolic rules provide an appropriate mapping of input features to category classes in the domain. However, the network in its present form is only suitable for use as a decision-support tool by a senior pathologist, since its performance deteriorated greatly with poor-quality data provided by a junior pathologist. The implications of the findings are discussed.

1 Introduction

Carcinoma of the breast is a common disease which is diagnosed in about 22 000 women in England and Wales each year and is the commonest cause of death amongst the 35–55 age group in same population [23]. Early detection and treatment gives a better prognosis and a Breast Screening Program has been introduced in the National Health Service using mammography as the primary detection modality. The primary method of diagnosis of breast carcinoma, with distinction from benign lesions causing mammographic abnormalities or clinically-detected masses, is cytopathological examination of fine needle aspirates of the breast, FNAB [10].

Large studies of the cytopathologic diagnosis of FNAB have shown a range of specificity of diagnosis of 90–100% with a range of sensitivities from 84–97% [25]. These studies have been produced in centres specializing in the diagnosis of breast disease by pathologists with a special interest in breast cytopathology. In less specialized centres, such as district general hospitals, when a diagnostic FNAB service is being set up the performance is in the lower range of those values with a specificity of 95% and a sensitivity of 87% [21].

The most important performance parameter is the specificity, since a malignant diagnosis on FNAB, combined with clinical assessment, will be the sole diagnostic step before definitive treatment such as mastectomy or wide local excision of the lesion and a false positive result may lead to unnecessary surgery. The acquisition of diagnostic expertise is a relatively slow process in pathology with at least five years study and experience of pathology required. During this period, trainee pathologists are supervised by fully-qualified colleagues but it would be expected that their performance without supervision in FNAB cytodiagnosis would fall well below the figures in published studies. Thus there is scope for an artificial intelligence decision-making tool in the cytodiagnosis of breast FNAB to assist in training junior pathologists and to improve the performance of experienced pathologists.

1.1 Background

In the cytodiagnosis of FNAB there are some observable features which are cited as being important in the recognition of malignant cells, see [24]. Some expert systems have been described which attempt to use human observations of such features in FNAB and then apply computers to process these observations and attach weight to the presence and combination of features, e.g. [11,14,25].

In previous work by the authors [7] we applied a powerful, but little-known, neural network model (termed ARTMAP) to this task. Various configurations of the model gave an accuracy of 94–95%, a sensitivity of 90–96%, and specificity of 92–99% (for full details see [7]). The model was shown to perform at least as well as an expert human pathologist. However, these results were achieved using the high-quality feature assignments provided by the human expert. Less experienced pathologists are more likely to make incorrect feature assignments and thus provide "noisier" input data to the model. This paper provides performance results for ARTMAP under such conditions. Additionally, ARTMAP possesses symbolic rule extraction capabilities which support the validation and justification of its diagnostic predictions. A detailed discussion of the ARTMAP rules used in this domain is provided.

The structure of the remainder of this paper is as follows. Section 2 justifies the selection of ARTMAP for the task. Section 3 describes the data used in the study, and the trials performed with ARTMAP. Section 4 details the results. Section 5 describes and evaluates the symbolic rules extracted from ARTMAP that are used to make diagnoses. Section 6 discusses the findings and suggests directions for further research.

2 Motivation for the Use of ARTMAP

Advances in neurocomputing have opened the way for the establishment of decision support systems which are able to learn complex associations by example. The main thrust of work in this area has been in the use of the so-called feedforward networks to learn the association between evidence and outcome. Examples of such networks include the MultiLayer Perceptron, MLP [19] or the Radial Basis Function networks, RBFN [16].

The MLP or the RBFN have been shown to be rich enough in structure so as to be able to approximate any (sufficiently smooth) function with arbitrary accuracy [6,17]. Thus, given sufficient data, computational resources and time, it is possible to estimate the Bayes-optimal classifier to any desired degree of accuracy, directly and with no prior assumptions on the probabilistic structure of the data. This is an attractive scenario and has been extensively exploited in medical diagnosis.

Nonetheless, the feedback architecture, ARTMAP, [3] possesses a number of attractive features not found in feedforward networks such as: a dynamic architecture which "designs" itself and the ability to distinguish rare from frequent events. More recently it has been demonstrated that, in a modified form, it can also classify data optimally, in a Bayesian sense [15].

For full details of the advantages provided by ARTMAP for medical domains generally see [8,12]. However, in this work, ARTMAP was selected primarily for two reasons. First, it has been demonstrated to provide superior performance to both statistical and rival neural network approaches. With the same data used in [7], logistic regression achieved an accuracy of 92%, sensitivity of 90%, and specificity of 94%; a MLP had accuracy, sensitivity and specificity of 92% [5]. In comparison (see section 1.1), ARTMAP showed both superior sensitivity and specificity. Second, ARTMAP provides explicit symbolic rules which can be easily understood by a human user. This capability will be discussed in detail within this paper.

Owing to space limitations, for an overview of ARTMAP please refer to section 2 of our companion paper in this volume [8]. For the purposes of this paper, three features of ARTMAP are of particular note, the *voting strategy*, *category pruning* and *symbolic rule extraction*. A description of the voting strategy and category pruning is again deferred to [8] (see sections 2.1 and 2.2 respectively). Symbolic rule extraction is detailed in section 5 of this paper.

3 Patients and Methods

3.1 Study Population

The total data set composed cytological specimens from 413 FNAB prepared by a cytocentrifuge method and stained by the Papanicolaou method [9] The final outcome of benign disease or malignancy was confirmed by open biopsy where that result was available. In benign aspirates with no subsequent open biopsy a benign outcome was assessed by clinical details on the request form, mammographic findings (where available) and by absence of further malignant specimens. A malignant outcome was confirmed by histology of open biopsy or clinical details where the primary treatment modality was chemotherapy or hormonal therapy. Idiosyncratic cases were not removed prior to use with the neural network.

3.2 Human Observations

Ten observable features were indicated by binary values. The definitions of these features are given in the Appendix, together with their abbreviated names used for symbolic rule extraction. The observations on the specimens were made independently by a senior pathologist with 10 years experience of interpreting FNAB and a junior pathologist with 18 months experience. The observations were made blind to clinical details or outcome and the pathologists recorded their diagnosis for each case. The interobserver agreement between the two pathologists was assessed using kappa statistics [20].

3.3 Method

Ten ARTMAP networks had been trained previously on 313 data items using the senior pathologist's feature assignments. A severely pruned version of each network had also been derived using the remaining 100 items as a prediction set and a CF threshold for pruning of 0.7. (See [7] for full details.)

An independent test set was derived from the junior pathologist's feature assignments for 82 randomly selected malignant cases and 82 benign cases. Performance results on this test set were recorded for each individual pruned and unpruned network, as well as for the voting strategy using five unpruned nets and five pruned nets. Forced choice prediction was employed in all test cases.

4 Results

Table 1 shows the junior pathologist's performance on the test set in comparison with that of the various ARTMAP networks.

	Accuracy (%)	Sensitivity (%)	Specificity (%)
Junior Pathologist	78.7	57.3	100.0
Unpruned ARTMAP—Individual Mean	73.7	66.7	80.7
Unpruned ARTMAP—Voting Strategy	75.0	74.4	75.6
Pruned ARTMAP—Individual Mean	76.0	57.6	94.5
Pruned ARTMAP—Voting Strategy	75.6	57.3	93.9

Table 1: Relative Performance of Junior Pathologist and Network Types

In [7] it was observed that pruning had the effect of biasing network performance towards increased specificity (an essential requirement for the domain, see section 1), and also that the voting strategy always gave improved performance (albeit slight) over the individual networks. With the data used here, it can be seen that the former effect still occurs, but the latter does not. More importantly, performance of all types of network is not significantly better than that of the junior pathologist. The unpruned networks show better sensitivity but possess unacceptable specificity. The pruned networks achieve higher specificity but at the expense of reducing sensitivity to a very similar level to that of the junior pathologist.

Kappa statistics for the observations of each of the features, reflecting the level of agreement between the senior and junior pathologists, show that for most of the features there was only a moderate level of agreement (about 0.40 for the raw kappa scores, see table 2) and this lack of agreement will be the cause of the reduction in network performance when using the junior pathologist's data. Three of the features, "naked" nuclei, "foamy" macrophages and necrotic epithelial cells, had low levels of agreement that were little better than that expected by chance. Each of these features require a high level of interpretation by the pathologist to identify the cell type ("naked", "foamy" or epithelial) and to assess its apparent biological viability or non-viability at the time of sampling (necrosis). The feature with the highest level of agreement, nuclear size, had the clearest definition requiring least interpretation—the observer simply had to assess whether any epithelial cell nuclei had diameters greater than twice that of adjacent lymphocyte nuclei. The other two features which are prominent in the extracted ARTMAP rules (see section 5) are multiple nucleoli and nuclear pleomorphism and both had reasonable levels of agreement.

Feature	Kappa Statistic
Dyshesion	0.43
Intracytoplasmic lumina	0.34
"3D" epithelial cell clusters	0.38
"Naked" nuclei	0.15
"Foamy" macrophages	0.18
Multiple nucleoli	0.49
Nuclear pleomorphism	0.40
Nuclear size	0.55
Necrotic epithelial cells	0.16
Apocrine change	0.44

Table 2: Kappa Statistics for the Confusion Matrices of Observations of Each Feature by a Senior and Junior Pathologist

5 Symbolic Rules

Most neural networks suffer from the opaqueness of their learned associations [22]. In medical domains, this "black box" nature may make clinicians reluctant to utilise a neural network application, no matter how great the claims made for its performance. Thus, there is a need to supplement neural networks with symbolic rule extraction capabilities. ARTMAP has such an ability [4]. The act of rule extraction is a straightforward procedure in ARTMAP compared with that required for feedforward networks since there are no hidden units with implicit meaning. In essence, each category cluster in ART_a represents a symbolic rule whose antecedent is the category prototype weights and whose consequent is the associated ART_b category.

These rule extraction capabilities provide two advantages. First, a domain expert can examine the complete rule set in order to validate that the network has acquired an appropriate mapping of input features to category classes. Second, the symbolic rules provide explanatory facilities for the network's predictions during on-line operation. In the case of ARTMAP this corresponds to displaying the equivalent rule for the ART_a cluster node that was activated to provide a category decision. (In the case of the voting strategy, a number of such rules, one per voting network, would be displayed.) The diagnosing clinician is then able to decide whether or not to concur with the network's prediction, based upon how valid they believe that rule to be.

Before discussing the specific rules discovered by ARTMAP for this domain, some discussion of the general nature of the rules is needed. These are of a somewhat different nature from those found in conventional rule-based expert systems. Expert system rules are "hard"—an input must match to each and every feature in a rule's antecedent before the consequent will be asserted. In ARTMAP the rules are "soft"—they are derived from prototypical category clusters which are in competition with each other to match closely enough to the input data but do not require perfect matches. This provides greater coverage of the state space for the domain using fewer rules.

Additionally, ARTMAP rules are self-discovered though exposure to domain exemplars, rather than having been externally provided by a human expert. ARTMAP is thus able to bypass the difficult and time-consuming knowledge-acquisition process found with rule-based expert systems [13]. (However, collection of the data may itself be a non-trivial task in many medical domains.) A drawback of this approach is that the rules are "correlational" rather than causal, since ARTMAP possesses no underlying theory of the domain but simply associates conjunctions of input features with category classes. (This occurs with neural networks generally.) However, this difficulty is probably not of great importance from an applications viewpoint since useful diagnostic performance can often be achieved from correlational features without recourse to any "deep" knowledge of the domain.

A final general point concerns the learning rule in ARTMAP which governs the formation of category clusters, and hence the rules that will be derived from these clusters. Under the "fast-learning" conditions used in this application, whenever an input is successfully matched to an existing category cluster node the new weights for that node are formed by taking the logical AND of the input pattern and the existing weights for that cluster. This has the effect of deleting all features from the category cluster weights that are not also present in the input pattern. Hence, the weights tend to denote progressively more general clusters as they encode more input patterns and more features are deleted. Additionally, all features that are still present in the weights for a cluster once training ceases are known to have been present in all input vectors encoded by that cluster.

Rule extraction from the 10 pruned nets used in this domain yielded 14 distinct rules, 12 for malignant outcomes and 2 for benign. The full list of rules is shown in table 3, ranked by how many of the 10 pruned networks each rule occurred in. No single rule in the set should be taken as canonical, since each is derived from a node which covers only a portion (albeit an important one) of the overall state space covered by each diagnostic category. However, taking the rules as a whole, a picture of a typical benign or malignant case can be constructed.

Benign cases are likely to display either no features, or the FOAMY feature in isolation. Malignant cases are almost certain to display a combination of NUCLEOLI, PLEOMORPH and SIZE. The 3D feature is also strongly implicated in malignancy. FOAMY, ICL, NECROTIC, and DYS may further be present, although with a lower likelihood. The senior pathologist in this study confirmed the validity of these rules and the relative importance of the features, with the exception that he places no value on the presence or absence of the FOAMY feature. This matter will be discussed later in this section.

Wells et al. [24] provide a canonical list of diagnostic criteria for FNAB which includes all features used in this study, although no assessment of their relative importance or likelihood is given. In summary, they cite FOAMY, APOCRINE and NAKED as indicators of benignancy, and all other features used here as indicators of malignancy. The self-discovered rules of ARTMAP show good overall agreement with these criteria apart from two notable exceptions. First, APOCRINE and NAKED are conspicuous by their absence from any of the ARTMAP rules. Second, FOAMY has an ambiguous status, being present in rules for both benign and malignant outcomes.

The first discrepancy can be explained by reference to the way in which CFs are calculated for nodes in ARTMAP based equally upon both usage and accuracy. The high CF threshold for pruning in this application requires a node to be both highly accurate and to encode a large proportion of exemplars of a particular category in order to remain unpruned. It is thus possible for a node with very good predictive accuracy but low usage to be pruned. This indeed happens in the case of nodes containing the APOCRINE and NAKED features, which both occur rarely in the data. Examination of the unpruned networks revealed the frequent occurrence of nodes where these features, in isolation or conjunction with the FOAMY feature, indicate a benign diagnosis. Although such nodes usually have a perfect accuracy score, they also have a very low usage score and hence their overall CF value falls below the threshold for pruning.

In future work this anomaly might be corrected by using a different weighting for the CF calculation, so as to bias the overall CF score more towards accuracy than usage. However, this has the risk that the resultant networks will possess incomplete coverage of all possible cases in the domain owing to the absence of high usage nodes encoding general cases.

Rule 1 (10 Occurrences)
IF
  NO-SYMPTOMS
THEN
  BENIGN

Rule 2 (8 Occurrences)
IF
  3D=TRUE
  NUCLEOLI=TRUE
  PLEOMORPH=TRUE
  SIZE=TRUE
THEN
  MALIGNANT

Rule 3 (8 Occurrences)
IF
  3D=TRUE
  FOAMY=TRUE
  NUCLEOLI=TRUE
  PLEOMORPH=TRUE
  SIZE=TRUE
THEN
  MALIGNANT

Rule 4 (7 Occurrences)
IF
  FOAMY=TRUE
THEN
  BENIGN

Rule 5 (4 Occurrences)
IF
  ICL=TRUE
  3D=TRUE
  NUCLEOLI=TRUE
  PLEOMORPH=TRUE
  SIZE=TRUE
THEN
  MALIGNANT

Rule 6 (4 Occurrences)
IF
  DYS=TRUE
  NUCLEOLI=TRUE
  PLEOMORPH=TRUE
  SIZE=TRUE
THEN
  MALIGNANT

Rule 7 (3 Occurrences)
IF
  FOAMY=TRUE
  NUCLEOLI=TRUE
  PLEOMORPH=TRUE
  SIZE=TRUE
THEN
  MALIGNANT

Rule 8 (3 Occurrences)
IF
  NUCLEOLI=TRUE
  PLEOMORPH=TRUE
  SIZE=TRUE
THEN
  MALIGNANT

Rule 9 (2 Occurrences)
IF
  3D=TRUE
  FOAMY=TRUE
  NUCLEOLI=TRUE
  PLEOMORPH=TRUE
  SIZE=TRUE
  NECROTIC=TRUE
THEN
  MALIGNANT

Rule 10 (2 Occurrences)
IF
  3D=TRUE
  FOAMY=TRUE
  PLEOMORPH=TRUE
  SIZE=TRUE
  NECROTIC=TRUE
THEN
  MALIGNANT

Rule 11 (2 Occurrences)
IF
  DYS=TRUE
  ICL=TRUE
  NUCLEOLI=TRUE
  PLEOMORPH=TRUE
  SIZE=TRUE
THEN
  MALIGNANT

Rule 12(1 Occurrence)
IF
  FOAMY=TRUE
  NUCLEOLI=TRUE
  PLEOMORPH=TRUE
  SIZE=TRUE
  NECROTIC=TRUE
THEN
  MALIGNANT

Rule 13 (1 Occurrence)
IF
  ICL=TRUE
  NUCLEOLI=TRUE
  PLEOMORPH=TRUE
  SIZE=TRUE
THEN
  MALIGNANT

Rule 14 (1 Occurrence)
IF
  ICL=TRUE
  3D=TRUE
  PLEOMORPH=TRUE
  SIZE=TRUE
THEN
  MALIGNANT

Table 3: Symbolic Rules Extracted from Pruned ARTMAP Networks

The status of the FOAMY feature is more problematic. In [24] it is classified as an indicator of benignancy. However, the senior pathologist in this study regards its occurrence as little more than "background noise" which is as likely to be found in malignant cases as benign. Its status in the ARTMAP rules is certainly ambiguous. In isolation, the FOAMY feature frequently indicates a benign outcome. However, it is also present, in conjunction with other features, in a number of rules with malignant outcomes. The frequent occurrence of this feature in the rules as a whole indicates that it is present in a large proportion of the data, regardless of outcome. If the relative frequency of occurrence is considered, the feature can be seen to be present in 1 of the 2 distinct rules for benignancy, and 5 of the 12 for malignancy. Alternatively, if occurrence without regard for distinctiveness is considered, it occurs in 7 of the 17 benign rules and 16 of the 39 malignant rules. By either calculation the proportions between outcomes are very similar. We therefore conclude that, at least for this particular data set, the FOAMY feature tends more towards being "background noise" than a useful indicator of benignancy. This conclusion may be tested in future work by training new networks which omit the FOAMY feature from the inputs and observing whether performance is subsequently degraded.

6 Discussion

The findings in section 4 indicate that although the existing ARTMAP application should prove useful as decision-support tool for senior pathologists [7], its performance is inadequate with poor-quality input data provided by a junior pathologist. The results further suggest that initial feature identification rather than subsequent diagnostic decision-making is the key criterion which distinguishes expert and neophyte performance in this domain. If this hypothesis is verified by further research, it obviously has implications for the training of junior pathologists in this field. Further studies are required however with a larger number of pathologists to evaluate the levels of agreement in identification of the observed features used by the network.

The symbolic rule extraction process described in section 5 provides more positive immediate results. ARTMAP has been shown to have acquired autonomously a valid mapping from input features to category classifications for the domain. This mapping is made explicitly available by means of the symbolic rules, and thus the "black box" criticism common to neural networks is alleviated.

From a purely AI viewpoint, we would further hope that ARTMAP's symbolic rules might serve another purpose beyond validation and justification of predictions—the discovery of novel information about the domain and/or the resolution of disagreements between domain experts about diagnostic criteria. For example, the ARTMAP rules provide an indication of the relative importance of different indicators of malignancy, based upon both frequency of occurrence and predictive accuracy, a matter on which no canonical information seems to be available. Furthermore, the rules suggest that the FOAMY feature should not perhaps be regarded as an important indicator of benignancy.

Future work might also include an empirical comparison of ARTMAP's performance with "pure" symbolic machine learning methods. (As noted earlier in section 2.1, ARTMAP has already been shown to have superior performance to the MLP and logistic regression in this domain.) In particular, the ID3 algorithm and its variants [18] seem most suitable for comparison with ARTMAP since it is well-used and known to be the equal or better of other symbolic learning algorithms in many domains. A preliminary qualitative comparison can be given here. Both ARTMAP and ID3 are supervised learning methods based on induction of concepts from exposure to domain exemplars. Additionally, variants of both ID3 and ARTMAP have been developed to handle data that has continuous values, is noisy, or has missing items. The decision-tree representation of ID3 provides a compact, hierarchical structuring of data features that is not possible in ARTMAP. However, the same representation has been criticized as being often unintelligible to domain experts. Furthermore, unlike ARTMAP, ID3 is not suitable for incremental learning in non-stationary environments.

Acknowledgement

This research was supported by the Science and Engineering Research Council (SERC) of the UK, grant number GR/J/43233.

References

[1] G.A. Carpenter and S. Grossberg (1987) A Massively Parallel Architecture for a Self-Organizing Neural Pattern Recognition Machine, *Computer Vision, Graphics and Image Processing*, 37, 54–115.

[2] G.A. Carpenter, S. Grossberg, N. Markuzon, J.H. Reynolds and D.B. Rosen (1992) Fuzzy ARTMAP: A Neural Network Architecture for Incremental Supervised Learning of Analog Multidimensional Maps, *IEEE Transactions on Neural Networks*, 3(5), 698–712.

[3] G.A. Carpenter, S. Grossberg and J.H. Reynolds (1991) ARTMAP: Supervised Real-Time Learning and Classification of Nonstationary Data by a Self-Organizing Neural Network, *Neural Networks*, 4(5), 565–588.

[4] G.A. Carpenter and A.H. Tan (1993) Rule Extraction, Fuzzy ARTMAP, and Medical Databases, *Proceedings of the World Congress on Neural Networks*, Volume I, 501–506.

[5] S.S. Cross, T.J. Stephenson, Y. Diez, R.F. Harrison, J.C.E. Underwood and J. Downs (In Press) Diagnosis of Breast Fine Needle Aspirates Using Human Observations and a Multi-Layer Perceptron Neural Network, to appear in *J. Pathol.*.

[6] G. Cybenko (1989) Approximations by Superpositions of a Sigmoidal Function, *Mathematics of Control, Signals and Systems*, 2, 303–314.

[7] J. Downs, R.F. Harrison and S.S. Cross (In Press) A Neural Network Decision Support Tool for the Diagnosis of Breast Cancer, to appear in *Proceedings of the 10th Conference of the Society for the Study of Artificial Intelligence and Simulation of Behaviour* (AISB-95). Amsterdam: IOS Press.

[8] J. Downs, R.F. Harrison and R.L. Kennedy (1995) A Prototype Neural Network Decision Support Tool for the Diagnosis of Acute Myocardial Infarction, this volume.

[9] S.A.C. Dundas, P.R. Sanderson, H. Matta and A.J. Shorthouse (1988) Fine Needle Aspiration of Palpable Breast Lesions: Results Obtained with Cytocentrifuge Preparation of Aspirates, *Acta Cytologica*, 32, 202–206.

[10] C.W. Elston and I.O. Ellis (1990) Pathology and Breast Screening, *Histopathology*, 16, 109–118.

[11] P.W. Hamilton, N. Anderson, P.H. Bartels and D. Thompson (1994) Expert System Support Using Bayesian Belief Networks in the Diagnosis of Fine Needle Aspiration Biopsy Specimens of the Breast, *J. Clin. Pathol.*, 47, 329–336.

[12] R.F. Harrison, C.P. Lim and R.L. Kennedy (1994) Autonomously Learning Neural Networks for Clinical Decision Support. In: E.C. Ifeachor and K.G. Rosen, (Eds.) *Proceedings of the International Conference on Neural Networks and Expert Systems in Medicine and Healthcare* (NNESMED–94), Plymouth, UK, 15–22.

[13] F. Hayes-Roth, D.A. Waterman and D.B. Lenat (1983) *Building Expert Systems*. London: Addison-Wesley.

[14] H.A. Heathfield, N. Kirkham, I.O. Ellis and G. Winstanley (1990) Computer Assisted Diagnosis of Fine Needle Aspirates of the Breast, *J. Clin. Pathol.*, 43, 168–170.

[15] C.P. Lim and R.F. Harrison (In Press) Modified Fuzzy ARTMAP Approaches Bayes Optimal Classification Rates: An Empirical Demonstration, to appear in *Neural Networks*.

[16] J. Moody and C. Darken (1989) Fast Learning in Networks of Locally-Tuned Processing Units, *Neural Computation*, 1, 281–294.

[17] J. Park and I. Sandberg (1991) Universal Approximation Using Radial Basis Function Networks, *Neural Computation*, 3, 246–257.

[18] J.R. Quinlan (1986) Induction of Decision Trees, *Machine Learning*, 1, 81–106.

[19] D. Rumelhart, G. Hinton and R. Williams (1986) Learning Representations by Back-Propagating Errors, *Nature*, 323, 533–536.

[20] P.B. Silcocks (1983) Measuring Repeatability and Validity of Histological Diagnosis—A Brief Review with Some Practical Examples, *J. Clin. Pathol.*, 36, 1269–1275.

[21] R.D. Start, P.B. Silcocks, S.S. Cross and J.H.F. Smith (1992) Problems with Audit of a New Fine-Needle Aspiration Service in a District General Hospital, *J. Pathol.*, 167, 141A.

[22] G.Towell and J.W. Shavlik (1993) Extracting Refined Rules from Knowledge-Based Neural Networks, *Machine Learning*, 13(1), 71–101.

[23] J.C.E. Underwood (1992) Tumours: Benign and Malignant. In: J.C.E. Underwood, (Ed.) *General and Systematic Pathology*, 223–246. Edinburgh: Churchill Livingstone.

[24] C.A.Wells, I.O. Ellis, H.D. Zakhour and A.R. Wilson (1994) Guidelines for Cytology Procedures and Reporting on Fine Needle Aspirates of the Breast, *Cytopathology*, 5, 316–334.

[25] W.H. Wolberg and O.L. Mangasarian (1993) Computer-Designed Expert Systems for Breast Cytology Diagnosis, *Anal. Quant. Cytol. Histol.*, 15, 67–74.

Appendix: Definition of Input Features

DYS: True if majority of epithelial cells are dyshesive, false if majority of epithelial cells are in cohesive groups.

ICL: True if intracytoplasmic lumina are present, false if absent.

3D: True if some clusters of epithelial cells are not flat (more than two nuclei thick) and this is not due to artefactual folding, false if all clusters of epithelial cells are flat.

NAKED: True if bipolar "naked" nuclei in background, false if absent.

FOAMY: True if "foamy" macrophages present in background, false if absent.

NUCLEOLI: True if more than three easily visible nucleoli in some epithelial cells, false if three or fewer easily visible nucleoli in epithelial cells.

PLEOMORPH: True if some epithelial cell nuclei with diameters twice that of other epithelial cell nuclei, false if no epithelial cell nuclei twice the diameter of other epithelial cell nuclei.

SIZE: True if some epithelial cells with nuclear diameters at least twice that of lymphocyte nuclei, false if all epithelial cell nuclei with nuclear diameters less than twice that of lymphocyte nuclei.

NECROTIC: True if necrotic epithelial cells present, false if absent.

APOCRINE: True if apocrine change present in all epithelial cells, false if not present in all epithelial cells.

Knowledge-Based Systems for Lymph Node Pathology: A Comparison of Two Approaches

Nguyen DT, Park IA, Cherubino P, Tamino PB and Diamond LW

Pathology Institute, University of Cologne, Germany

A formal evaluation was performed between two knowledge-based systems for lymph node pathology: 1) Intellipath, Kiel edition, a Bayesian system; and, 2) "Professor Amadeus", which uses categorical reasoning and defined diagnostic patterns. The evaluation, involving three pathologists, was based on 57 lymph node biopsies. Intellipath demonstrated satisfactory performance, with from 63.2% to 71.9% correct answers, depending on the experience level of the pathologist. "Professor Amadeus" achieved better results, with accuracy rates ranging from 93% to 96.5%. In this study, the better performance of "Professor Amadeus" could be attributed to a more effective multiparameter approach, fewer errors in the knowledge base, and better handling of input parameters.

1. Introduction

One of the challenges in knowledge-based system (KBS) development is the choice of appropriate methods for knowledge representation and reasoning. Deber and Baumann point out that medical reasoning consists of two separate processes, problem-solving and decision-making [1]. Treatment selection is an example of decision making, since it involves a choice between several prospective alternatives [1]. Diagnosis, on the other hand, is an example of problem solving, the search for the cause of an event that has already taken place. It has been suggested that categorical reasoning, which can be based on pattern recognition, is ideally suited for problem-solving, while probabilistic reasoning is appropriate for decision-making [1-5]. Other researchers view medical problem-solving and decision-making as being intertwined. They regard the physician solely as a decision maker, and model medical reasoning as a series of lotteries or gambles [6].

The probabilistic, decision-making model, has been applied to diagnostic hematology in the development of normative expert systems for peripheral blood smear interpretation [7] and lymph node (LN) pathology [8-10]. In these systems, knowledge is represented by thousands of conditional probabilities.

We share the opinion that medical problem-solving and decision-making are distinct and require different approaches [1-5]. We have developed a diagnostic hematology workstation with modules for peripheral blood interpretation [11,12], flow cytometry (FCM) immunophenotyping and DNA content analysis [13,14], bone marrow morphology [15,16] and LN pathology [17]. Our systems, based entirely on categorical reasoning, use defined diagnostic patterns to subdivide the task into more manageable steps. The modules communicate with each other through a set of relational databases [18].

One of the best ways to evaluate the strengths and weaknesses of different KBSs is by direct comparison of two or more systems on the same set of cases. Several evaluations of this type have been reported in the medical literature [19-21]. In this study, we performed a formal comparison of the commercially available probabilistic LN expert system, Intellipath (Kiel Version), and our hematology workstation module for LN pathology, "Professor Amadeus." In order to overcome the difficulties of choosing the appropriate "gold standard" to compare the system's performances, we utilized three pathologists with varying degrees of experience in LN pathology. Only cases in which all three pathologists independently made the same diagnosis were included in the study. This diagnosis was used as the "gold standard." Since the pathologists correctly diagnosed all of the cases, it can be inferred that each was familiar with appropriate morphologic criteria for those diseases, and was correctly identifying the features according to their own view of the criteria. The evaluation tested the ability of the two systems to reach the correct diagnosis when given the features identified by each of the pathologists.

2. Description of the Systems

Intellipath (LN module), the commercially available version of the Pathfinder project, has been described in detail elsewhere [8-10]. Briefly, Intellipath consists of a probabilistic expert system based on Bayesian belief networks, and an accompanying analog videodisc with didactic material and photomicrographs. The LN module exists in two versions, an American version which utilizes the LN classification system favored by the primary domain expert, and a Kiel version which follows the non-Hodgkin's lymphoma (NHL) classification popular in Europe [22]. In this study, the Kiel edition was tested (version 5.02 with the second release of the videodisc).

In Intellipath, the pathologist enters observations using a mouse under the MS-DOS operating system. The 146 features known to the system include routine morphology, immunologic findings (mostly immunohistochemistry on paraffin sections), and clinical features such as age, HIV status, or a history of mycosis fungoides (MF). Each feature is organized as a set of two or more mutually exclusive values. For example, the feature "Medium lymphoid cell number" can assume the values absent, sparse (1-10%), moderate (11-50%), numerous (51-90%) and striking (>90%). After each feature is entered, the system lists a differential diagnosis with disease probabilities.

Our diagnostic LN module, "Professor Amadeus", has also been previously described [17]. The system is designed for use in the Microsoft Windows environment. All knowledge engineering has been performed by a trained hematopathologist. "Professor Amadeus" utilizes the heuristic classification method as defined by Clancey [23]. The core of the system is a set of defined diagnostic patterns (Table 1) based on immunologic, cell kinetic, and morphologic features. Amadeus is designed to be used in conjunction with our KBS for FCM analysis, "Professor Fidelio" [13,14]. If FCM results are available in the database, "Professor

Amadeus" utilizes that information in the interpretation. Since the cases in this study were received in consultation, without FCM data, this feature of the system was not tested.

The user interface in "Professor Amadeus" consists of five screens: (1) clinical data, FCM and/or frozen section immunology; (2) low-magnification LN examination; (3) high-magnification cytology on sections and touch imprints; (4) special stains and paraffin tissue immunohistochemistry results; and (5) final interpretation. Microscopic observations are input using Windows objects including dialog boxes, drop-down lists, check boxes, and radio buttons.

Table 1. Lymph Node Patterns

1.	Low-grade B-cell NHL	7.	Histiocytic/dendritic neoplasms
2.	High-grade B-cell NHL	8.	Non-hematopoietic elements
3.	T-cell NHL	9.	Hematopoietic neoplasm, lineage or grade undetermined
4.	Hodgkin's disease	10.	"Undifferentiated" malignant neoplasm
5.	Reactive conditions	11.	Non-specific
6.	Leukemic infiltrates		

Morphologic evaluation of LN sections begins with low-magnification findings. In our systematic approach, we use a subset of the histologic architectures described by Nathwani [24]. To describe the LN architecture, the pathologist can select one of five choices (diffuse, follicular, interfollicular, sinus, or focal) from a drop-down list. Alternatively, the system will determine the architecture based on the input of individual low-power features (e.g., "germinal centers increased"). Separate screens are available to enter cytologic details and the results of special stains, including immunohistochemistry on paraffin sections.

There are 70 morphologic features in this version of "Professor Amadeus", including a much larger panel of immunologic markers than that available in Intellipath. The overall number of morphologic features in Amadeus is fewer than that of Intellipath because the architectural and cytologic features are concisely packaged to describe the limited repertoire of lymphoid cell morphology. For example, to describe the large cells in B-immunoblastic lymphoma in Intellipath, requires the input of four separate features: "Large lymphoid cell (LLC) number", "LLC cytoplasm", "LLC nuclear shape", and "LLC nucleolar features." In Amadeus, these four features can be input by a single selection from a drop-down list, i.e., "large transformed cells with eccentric nuclei and basophilic cytoplasm."

Based on the features entered, the system determines the final pattern by comparing the data to the pattern definitions. After determining the predominant pattern, the program invokes a procedure containing "diagnostic rules" specific to that pattern. These rules are used to generate the text of the report, which includes: 1) a summary of the available clinical information and FCM findings (if any); 2) the relevant LN morphologic findings including immunohistochemistry and special stains; 3) the final pattern; and 4) the diagnosis or differential diagnoses. When appropriate, the system recommends obtaining additional data (e.g., clinical history and/or laboratory tests such as molecular genetics) to establish the final diagnosis.

3. Case Selection and Design of the Study

Seventy-six LN cases were selected from the consultation files of the Pathology Institute, University of Cologne, based on the quality of the material. Three pathologists participated in the study. At the time of the evaluation, pathologist A was finishing a year of concentrated fellowship in LN morphology. Pathologists B and C are board certified hematopathologists, with 8 and 16 years of experience in LN pathology, respectively. The two more experienced pathologists were involved in both an initial evaluation of Pathfinder/Intellipath [25] and the development of "Professor Amadeus."

Each pathologist independently reviewed all 76 cases with knowledge of the patient' s age, sex and any clinical information (e.g. HIV infection, history of MF) available from the records. Complete interobserver agreement between the three pathologists was achieved in 57 of the 76 cases. This subgroup of 57 cases formed the basis for the evaluation of the KBSs. The agreed-upon diagnosis served as the "gold standard".

The 57 cases encompassed a variety of diagnoses, including: 1) seven different forms of reactive lymphadenopathy (Kikuchi's lymphadenitis, florid follicular hyperplasia with and without progressive transformation of germinal centers, toxoplasmosis, granulomatous reactions, Castleman's disease, florid and involutionary reactions in AIDS patients, and dermatopathic lymphadenopathy in a patient with cutaneous MF); 2) all subtypes of Hodgkin's disease; 3) low-grade B-cell NHL, including involvement by chronic lymphocytic leukemia, Centroblastic/Centrocytic (CB/CC) lymphoma (follicular), Centrocytic lymphoma, and Immunocytoma; 4) high-grade NHL of both the B- and T-cell types (Lymphoblastic, Burkitt's, Immunoblastic, Centroblastic, Ki-1+ large cell lymphoma, and peripheral T-cell lymphomas of the Lennert's and AILD-like subtypes); 5) Histiocytic proliferations (Histiocytosis-X, Follicular Dendritic Cell Sarcoma); and 6) Miscellaneous conditions (Metastatic Carcinoma, Plasmacytoma, involvement of LNs by MF, acute myelomonocytic leukemia, and extramedullary hematopoiesis).

Each of the three pathologists was familiar with the use of both the Intellipath and the "Professor Amadeus" systems. Each pathologist was given at least three days to study the Intellipath didactic material and videodisc and had access to textbook material on LN diagnosis written by Dr. Nathwani [24,26].

The goal of the study was to test each system's ability to match the "gold standard" diagnosis when given the features observed by each pathologist. In order to prevent the participants from being influenced by either system's interpretation, each pathologist entered their microscopic observations on hard-copy printouts which listed the histologic features of each system. The age, sex and clinical history (if any) of the patient, were entered into the respective computer systems along with the morphologic observations of each pathologist.

The report generated by "Professor Amadeus" was considered correct only if the diagnosis matched the "gold standard" diagnosis. Intellipath's interpretation was considered correct if the disease with the highest probability, based on the features selected by the pathologist, closely matched the reference diagnosis. In some cases, because of the limited repertoire of final diagnoses in Intellipath, and the limited set of immunologic markers available, less specific diagnoses were accepted as correct. For example, in the case of follicular dendritic cell sarcoma, the generic interpretation "sarcoma" (a disease category) was accepted as a correct diagnosis.

When the Intellipath system indicated that no diagnosis was compatible with the features input, the observations of that pathologist were reviewed by at least two pathologists to determine if the findings were valid. Nuclear diameters were measured with optical micrometers, when necessary, and extensive counts were made of cellular components (e.g., the number of mast cells in the sections) to validate the pathologist's input. In every instance, the pathologist's input was verified as acceptable.

When the Intellipath system gave no diagnosis, and when Amadeus listed a descriptive interpretation with a differential diagnosis that included the reference diagnosis, it was considered a minor error. Misinterpretations which would have no effect on therapy (e.g., one type of low-grade NHL misclassified as a different type of low-grade NHL) were also considered minor errors.

Incorrect diagnoses which would result in inappropriate therapy were regarded as major errors. Examples of such major mistakes include malignant lymphomas misclassified as benign diseases, or a NHL diagnosed as Hodgkin's disease.

4. Results

The number of histologic features entered per case is listed in Table 2 by pathologist and system. Overall, the more experienced pathologists entered fewer features. Fewer features were required for Amadeus than Intellipath.

Table 2. Number of histologic features input per case

Pathologist	Amadeus (# of features)		Intellipath (# of features)	
	Range	Mean	Range	Mean
A	3-14	8.7	4-19	11.2
B	3-15	8.6	3-18	10.6
C	3-14	7.8	4-18	10.4

The performance of Intellipath varied with the pathologist's level of expertise. The highest number of correct answers was obtained when the system was given the features selected by the most experienced of the three pathologists.

Based on the features selected by pathologist A (least experienced), the top-ranked diagnosis in Intellipath closely matched the "gold standard" diagnosis in 36 cases (63.2%). With one exception (probability of 0.44), the probabilities listed for these diagnoses ranged from 0.75 to 1.0. The system was able to reach the correct diagnosis with absolute certainty (i.e. with no other competing diseases listed as differential diagnoses) in nine cases. Three of the 21 errors were major errors: 1) Kikuchi's lymphadenitis (self-limiting and requiring no therapy) diagnosed as a mycobacterial histiocytosis (which requires treatment); 2) Hodgkin's disease, mixed cellularity misdiagnosed as metastatic malignant melanoma; and 3) Dermatopathic lymphadenopathy misinterpreted as involvement by MF/Sezary syndrome.

Using the features entered by pathologist B, Intellipath was able to diagnose 38 cases correctly (66.7%), including ten cases in which a single diagnosis was achieved. The probabilities associated with the correct answers ranged from 0.75 to 1.0. Again, there were three major errors: 1) a T-cell NHL misclassified as Hodgkin's disease; 2) Immunocytoma misinterpreted as AIDS involutionary; and 3) the dermatopathic lymphadenopathy case noted above.

With the features selected by pathologist C (most experienced), the Intellipath system was able to diagnose 41 cases (71.9%) correctly. The probabilities associated with these correct answers ranged from 0.75 to 1.0, except in one case where the top-ranked diagnosis had a probability of only 0.52. A single correct diagnosis was reached in eight cases. There were two major errors (immunocytoma and dermatopathic lymphadenopathy cases).

The performance of "Professor Amadeus" varied little with the pathologist's level of expertise. The system achieved a correct diagnosis in 53 of 57 cases (93%) based on the features selected by pathologist A, and 55 of 57 cases (96.5%) for both pathologists B and C. In all cases, the misinterpretations were minor errors. For example, Amadeus misdiagnosed one case of "Ki-1+ large cell lymphoma" as "centroblastic lymphoma, diffuse" (pathologist A), but both diagnoses are high-grade NHL. In one case (pathologist A) follicular CB/CC lymphoma with few blasts was diagnosed as follicular CB/CC with many blasts (both low-grade NHL). In two cases (all three pathologists), the system rendered the interpretation "Polymorphous infiltrate, positive for UCHL-1." Amadeus added the comment that the "differential diagnosis includes peripheral T-cell lymphoma and abnormal immune response" and suggested FCM on fresh tissue to determine a final diagnosis. The correct answer was peripheral T-cell lymphoma and the response was therefore graded as a minor mistake.

5. Discussion

In this study, we tested the ability of two KBSs, one Bayesian and one utilizing categorical reasoning with defined diagnostic patterns, to make correct LN diagnoses based on the input of three pathologists. The Bayesian system, Intellipath (Kiel Edition), showed good performance (63-72% correct responses). Intellipath's accuracy improved with the experience level of the pathologist entering the histopathologic observations. One reason to account for this, is the natural tendency of experienced pathologists to only report features which support their already arrived-at diagnosis.

This phenomenon has also been observed by the developers of Intellipath [25]. Experienced pathologists enter features which reflect their mental image of a typical example of the diagnosis.

The mental image of a given diagnosis may differ among pathologists, however. In one study, eight well-known American and European experts in ovarian pathology were given the name of an ovarian disease and asked to list all of the diagnoses they considered to be morphologically similar to the given disease [27]. There was very little consensus among the experts. The study concluded that after leaving a training program, each pathologist develops his or her own diagnostic criteria through experience. The limited consensus in differential diagnosis can be traced back to a limited consensus in diagnostic criteria.

The diagnostic accuracy rate achieved by Intellipath in this study is comparable to other normative diagnostic systems. When Professor Sultan's Bayesian system for evaluating complete blood counts was tested with 180 cases of anemia, the correct diagnosis was made in 64.5% of cases [7]. Furthermore, when a probabilistic system for diagnosing renal masses was evaluated by comparing three different inference strategies, the three strategies had accuracies from 61-64% when highly specific tests such as angiograms, computerized tomography, and magnetic resonance imaging were not considered, and accuracies ranging from 69%-76% when all available information was considered [28]. The renal mass study also looked at the reliability and discriminating power of the diagnoses reached by the systems. The authors concluded that for the probabilistic models that they studied, "perfect performance in all of these parameters may not be easily attainable" [28].

One study of the American version of Pathfinder/Intellipath system concluded that the diagnostic performance was "at least as good as that of the Pathfinder/Intellipath expert" [25]. In that study, however, the diagnosis of record was not used as the "gold standard." A subjective method was employed in the evaluation. The domain expert was shown the values input into the system and was asked to rate the output of the system on a scale of one to ten [25].

Our choice of the "correct diagnosis" as the "gold standard" in this study can be justified by the fact that all three pathologists independently agreed to the same diagnosis. We were only testing the ability of the systems to reach the "gold standard" diagnosis using the features reported by each pathologist. We did not utilize Intellipath's ability to recommend to the pathologist which features to look for at each stage of the consultation.

Undoubtedly, the diagnostic accuracy of the Kiel edition of Intellipath can be improved by correcting several errors in the knowledge base that this study uncovered. This version does not recognize the small cell variant of lymphoblastic lymphoma, even though the authors of the Intellipath knowledge base illustrated this variant in a recent publication [26]. Furthermore, the system cannot diagnose LN involvement by MF when the cytologic input is "small, slightly irregular lymphocytes" (instead of "cerebriform"), despite the fact that many well documented cases of MF do not contain typical cerebriform cells [29].

Several of the major errors made by Intellipath are more disturbing. In particular, the Kiel version of Intellipath will diagnose MF, solely on the basis of clinical history, even in LNs which show only the morphologic features of benign dermatopathic lymphadenopathy [29]. Similarly, the system ignored monoclonality (diagnostic of B-cell malignancy in a LN) and misdiagnosed a malignant lymphoplasmacytic infiltrate as a benign condition.

In this study, the experience level of the pathologist did not have a large effect on the performance of "Professor Amadeus." One explanation for this, is that Amadeus has fewer cytologic descriptors. In Intellipath, cytologic observations are segregated into cell size (small, medium, large) and multiple other components (e.g., percentage of the cells, nuclear shape, nucleolar features, chromatin structure). To fully describe the predominant cell population, a feature-value needs to be entered for each component. If the separate descriptors are not selected carefully, "nonsense" combinations may be generated, which not only misleads the program, but can confuse physicians with little training in LN pathology. In Amadeus, each of the choices is a valid description.

We believe that for problem-solving systems, defined diagnostic patterns are a better method of knowledge representation than conditional probabilities [3]. The methods for morphologic LN interpretation advocated by the Intellipath domain expert, based on low-power LN architectural patterns, are well documented in the literature [24]. Each morphologic pattern corresponds to a specific set of differential diagnoses. The Intellipath domain expert's knowledge of LN morphology is largely contained in these patterns, which he has carefully defined [24]. The pattern approach to LN diagnosis is well represented in the didactic material distributed with Intellipath, but is not well reflected in the expert system itself.

In this study, knowledge representation by defined diagnostic patterns, allowed Amadeus to focus on pertinent features and ignore extraneous ones. As a result, the system could better accommodate the differences among the three pathologist's observations. Only differences in key diagnostic features will cause "Professor Amadeus" to change an interpretation. A further advantage of our system is its speed. On a 486SX-25 MHz computer, it took an average of only 45 seconds to run a case on Amadeus, as opposed to an average of 5.5 minutes per case for Intellipath.

The histopathological analysis of LN sections is only one of the important steps in the proper diagnosis and treatment of patients with hematologic diseases. Previous large scale multi-institutional studies on the reproducibility of LN histologic diagnosis have yielded uniformly discouraging results, with overall agreement rates less than 60% [30,31]. Similarly, poor reproducibility (64.8% on 105 cases) has been encountered in the diagnosis of leukemias on bone marrow aspirates when only routine Wright-stained films were evaluated [32]. When the bone marrow aspirates were evaluated together with the available cytochemical stains, however, a higher number of diagnostic agreements was obtained. The maximum degree of reproducibility for leukemia diagnoses (99% agreement between two observers on 93 cases), was attained when the FCM immunologic parameters are included in the evaluation [32]. These results confirm that a multiparameter approach is extremely helpful for the accurate diagnosis of hematologic disorders.

In this study, paraffin immunohistochemical stains were available in most of the cases, resulting in 75% reproducibility (57 of 76 cases) which, though still unsatisfactory, is much better than that in earlier reports [30,31]. Reproducibility can be expected to markedly improve, when FCM immunology findings, clinical history, and other pertinent laboratory values are used in the definitions of malignant lymphomas [33]. Better reproducibility of diagnoses improves the value of comparative clinical and survival studies among different institutions. KBSs based on a multiparameter approach should improve the reproducibility of hematologic diagnoses, and thus contribute to patient care in hematology.

Intellipath functions as a stand-alone system and concentrates mainly on routine LN morphology. The current repertoire of immunologic parameters and clinical features is limited. For example, the system cannot handle many of the newer immunology markers performed on fresh tissue. To reflect our multiparameter approach to LN diagnosis, "Professor Amadeus" is designed as one module in a comprehensive hematopathology workstation. The system can function as a stand alone, already equipped to handle a large panel of immunohistochemical stains including frozen section immunology. The full potential of the system is attained, however, when it is used in conjunction with the other workstation modules. In this mode, "Professor Amadeus" can retrieve the clinical data stored in the patient database, as well as the interpretation of peripheral blood, bone marrow and FCM studies, if these tests have been already performed.

Intellipath ships with an analog videodisc containing thousands of LN images and on-line educational material which most people find to be an extremely useful resource. The next phase in the development "Professor Amadeus" will include further testing and validation, along with implementation of an explanation facility in the form of an electronic textbook with an atlas of digitized photomicrographs. Our LN textbook will be organized by patterns, similar to our existing textbook for the peripheral blood module [34].

6. References

[1]   Deber RB, Baumann AO.   Clinical reasoning in medicine and nursing: Decision making versus problem solving. Teach Learn Med. 1992;4:140-6.

[2]   Szolovits P, Pauker SG. Categorical and probabilistic reasoning in medical diagnosis. Artif Intell. 1978;11:115-44.

[3]   Diamond LW, Mishka VG, Seal AH, Nguyen DT.   Are normative expert systems appropriate for diagnostic pathology? J Am Med Informatics Assoc. 1995; (in press).

[4]   Feinstein AR. An analysis of diagnostic reasoning I. The domains and disorders of clinical macrobiology. Yale J Bio Med. 1973;46:212-32.

[5]   Feinstein AR.   An analysis of diagnostic reasoning II. The strategy of intermediate decisions. Yale J Bio Med. 1973;46:264-83.

[6]   Elstein AS.   Paradigms for research on clinical reasoning: A researcher's commentary. Teach Learn Med. 1992;4:147-9.

[7]   Sultan C, Imbert M, Priolet G.   Decision-making system (DMS) applied to hematology. Diagnosis of 180 cases of anemia secondary to a variety of hematologic disorders. Hematol Pathol. 1988;2:221-8.

[8]   Heckerman DE, Horvitz, EJ, Nathwani BN.   Toward normative expert systems: Part I The pathfinder project. Meth Inf Med. 1992;31:90-105.

[9]   Heckerman DE, Nathwani BN. Toward normative expert systems: Part II Probability-Based Representations for efficient knowledge acquisition and inference. Meth Inf Med. 1992;31:106-16.

[10] Nathwani BN, Heckerman DE, Horvitz EK, Lincoln TL. Integrated expert systems and videodisc in surgical pathology: An overview. Hum Path. 1990;21:11-27.

[11] Nguyen DT, Diamond LW, Priolet G, Sultan C. A decision support system for diagnostic consultation in laboratory hematology. In: Lun KC, Degoulet P, Piemme TE, Rienhoff O, eds. MEDINFO 92, Amsterdam: North Holland, 1992;591-95.

[12] Diamond LW, Nguyen DT, Sheridan BL, Strul M, Bailey K, Bak A. An automated hematology laboratory with computer-controlled robotics. In: Greenes RA et al. MEDINFO 95 (in press).

[13] Diamond LW, Nguyen DT, Jouault H, Imbert M. Evaluation of a knowledge-based system for interpreting flow cytometric immunophenotyping data. In: Reichert A, Sadan BA, Bengtsson S, Bryant J, Piccolo U (eds). Proceedings of MIE 93 London: Freund Publishing House, 1993;124-8.

[14] Diamond LW, Nguyen DT, Andreeff M, Maiese RL, Braylan RC. A knowledge-based system for the interpretation of flow cytometry data in leukemias and lymphomas. Cytometry. 1994;17:266-73.

[15] Nguyen DT, Cherubino P, Tamino PB, Diamond LW. Computer-assisted bone marrow interpretation: a pattern approach. In: Reichert A, Sadan BA, Bengtsson S, Bryant J, Piccolo U (eds). Proceedings of MIE 93. London: Freund Publishing House, 1993;119-23.

[16] Nguyen DT, Diamond LW, Cherubino P, Koala WBJ, Imbert M, Andreeff M. A diagnostic workstation for neoplastic bone marrow diseases: Evaluation on 526 cases. In: Greenes RA et al. MEDINFO 95 (in press).

[17] Nguyen DT, Mishka VG, Moskowitz FB, Diamond LW. A multiparameter decision support system for lymph node pathology. In: Barahona P, Veloso M, Bryant J (eds). Proceedings of MIE 94. Lisbon:1994;162-7.

[18] Diamond LW, Nguyen DT. Communication between expert systems in haematology. In: Richards B (ed). Healthcare Computing 1993 -- Current Perspectives in Healthcare Computing, Weybridge, UK: BJHC Books, 1993;111-19.

[19] Middleton B, Shwe MA, Heckerman DE, Henrion M, Horvitz EJ, Lehmann HP, Cooper GF. Probabilistic diagnosis using a reformulation of the INTERNIST-1/QMR knowledge base II. Evaluation of diagnostic performance. Meth Inform Med 1991;30:256-67.

[20] Li Y-C, Haug PJ. Evaluating the quality of a probabilistic diagnostic system using different inferencing strategies. In: Safran C, ed. Proceedings of the Seventeenth Annual Symposium on Computer Applications in Medical Care. New York: McGraw-Hill, 1993;471-7.

[21] Berner ES, Webster GD, Shugerman AA, Jackson JR, Algina J, Baker AL, Ball EV, Cobbs G, Dennis VW, Frenkel EP, Hudson LD, Mangall EL, Rackley CE, Taunton OD. Performance of four computer-based diagnostic systems. N Engl J Med 1994;330:1792-6.

[22]   Lennert K. Malignant Lymphomas other than Hodgkin's Disease. New York: Springer-Verlag, 1978.

[23]   Clancey WJ. Heuristic classification. Artif Intell 1985; 27: 289-350.

[24]   Nathwani BN. Diagnostic significance of morphologic patterns in lymph node proliferations. In: Knowles DM, ed. Neoplastic Hematopathology. Baltimore: Williams & Wilkins, 1992;407-25.

[25]   Heckerman DE, Nathwani BN. An evaluation of the diagnostic accuracy of pathfinder. Comput Biomed Res. 1992;25:56-74.

[26]   Nathwani BN, Brynes RK, Lincoln T, Hansmann ML. Classifications of Non-Hodgkin's lymphomas. In: Knowles DM, ed. Neoplastic Hematopathology. Baltimore: Williams & Wilkins, 1992:555-601.

[27]   van Ginneken, van der Lei J. Understanding differential diagnostic disagreement in pathology. In: Proceedings of the Fifteenth Annual Symposium on Computer Applications in Medical Care. New York: McGraw-Hill, 1991:99-103.

[28]   Li Y-C, Haug PJ. Evaluating the quality of a probabilistic diagnostic system using difference inference strategies. In: Proceedings of the Seventeenth Annual Symposium on Computer Applications in Medical Care. New York: McGraw-Hill, 1993:471-477.

[29]   Vonderheid EC, Diamond LW, Lai SM, Au F, Dellavechia MA. Lymph node histopathologic findings in cutaneous T-cell lymphoma. A prognostic classification based on morphologic assessment. Am J Clin Pathol 1992;97:121-9.

[30]   Kim H, Zelma RJ, Fox M, Bennett JM, Berard CW, Butler JJ, Byrne GE Jr., Dorfman RF, Hartsock RJ, Mann RB, Neiman RS, Rebuck JW, Sheehan WW, Variakojis D, Wilson JF, Rappaport H. Pathology panel for lymphoma clinical studies: a comprehensive analysis of cases accumulated since its inception. J Nat Cancer Inst 1982;68:43-67.

[31]   Velez-Garcia E, Durant J, Guams R, Bartolucci A. Results of a uniform histopathologic review system of lymphoma cases. Cancer 1983;52:675-9.

[32]   Browman GP, Neame PB, Soamboonsrup P. The contribution of cytochemistry and immunophenotyping to the reproducibility of the FAB classification in acute leukemia. Blood 1986;68:900-5.

[33]   Harris NL, Jaffe ES, Stein H, Banks PM, Chan JKC, Cleary ML, Delsol G, Wolf-Peeters CD, Falini B, Gatter KC, Grogan TM, Isaacson PG, Knowles DM, Mason DY, Muller-Hermelink HK, Pileri SA, Piris MA, Ralfkiaer E, Warnke RA. A revised European-American classification of lymphoid neoplasms: A proposal from the international lymphoma study group. Blood 1994;84:1326-92.

[34]   Diamond LW, Nguyen DT. Linking an electronic textbook to expert systems in haematology. In: Richards B, ed. Healthcare Computing 1994 -- Current Perspectives in Healthcare Computing, Weybridge, UK:BJHC Books, 1993;269-79.

Diagnostic Support Systems

Mapping Laboratory Medicine onto the Select and Test Model
to Facilitate Knowledge-Based Report Generation in Laboratory Medicine

Kindler H.*, Densow D.', Fischer B.', Fliedner T. M.'

*) Institute for Applied Knowledge Processing, University of Ulm
Postfach 2060, D-89010 Ulm, kindler@faw.uni-ulm.de
') Department of Clinical Physiology and Occupational Health, University of Ulm

This research applies results of the project GAMES-II, which was partially funded by the AIM Programme of the Commission of the European Union. The partners of this project have been SAGO (Florence, Italy), Foundation of Research and Technology (Heraklion, Greece), Geneva University Hospital (Switzerland), University of Amsterdam (The Netherlands), University College of London (UK), the University of Pavia (Italy), and the University of Ulm (Germany).

Many of the designations used by manufacturers and vendors to distinguish their products are claimed as trademarks. Where those designations appear in this publication, and the authors were aware of a trademark claim, the designations have hereinafter been printed in initial caps or all caps. In cases the authors were not aware of any such claims which were not mentioned therefore, this does not imply that they do not exist.

Abstract

A prototype of a knowledge-based system in laboratory medicine that produces report proposals for haematology is presented. The medical problem-solving process in laboratory medicine can be modelled with the ST-model (select and test). This decreases the complexity of the inference mechanisms and allows the construction of knowledge-acquisition tools that are easy to use for laboratory physicians. The possibility of how to connect the knowledge-based system with a LIS (laboratory information system) is described.

1. Introduction

After receiving the specimen and the request in laboratory medicine tests are performed resulting in wet data. These wet data will be compared with age and sex dependent reference intervals. In large laboratories the request and specimen management, the performance of tests, and the comparison of the results with reference intervals are automatically done by LISs. The automation leads to an increase in productivity, standardisation, and a reduction of common place errors. The final report generation is mostly not supported by LISs and has to be done by the laboratory physicians. Evaluating a sample of laboratory requests and the respective results shows that about 80% of the results are either within the normal or minuscule deviations from normal put in words. Routine diagnoses and the clue for further investigations can be found in about 15% of the reports. About 5% of the reports require further consideration by the laboratory physician only. Supporting the laboratory physician in report generation by a knowledge-based system may allow to produce 95% of the reports automatically to be authorised by the laboratory physician. This will save time and standardise a laboratory's output. For processing the remaining 5% a physician's work place integrating a text processor and access to information systems, e. g., MEDLINE, case collections, email, will be provided.

A demonstrator of such a knowledge-based system has been realised for a commercial laboratory. For reasons of confidentiality its name will not be mentioned. The employment of a knowledge-based system will lead to a restructuring of the working processes within the laboratory.

For practical purposes a circumscribed domain, i. e., haematology has been chosen to commence with. At the moment the knowledge-based report generation is still limited on wet data received from automatic analysers, e. g., flow-cytometry. This covers a third of the daily requests of the above mentioned laboratory.

The ST-model has been applied for the development of the knowledge-based report generation system. Within the premises of the GAMES-II project the ST-model has been utilised for many medical domains (Kindler 1993, Leaning 1993, Safran 1993) and has proven to be an appropriate means to model medical decision making processes. The GAMES-II methodology allows for the easy integration of knowledge-acquisition tools and the generation of knowledge-based systems.

For the sake of clarity the German of the application has been translated into English.

2. Modelling the Laboratory Medicine Work Flow with the ST-Model

The ontology underlying the ST-model serves to describe and structure the facts belonging to a laboratory request and its result values. The facts are depicted as white ovals in fig. 1. Arrows on the white background in fig. 1 symbolise tasks to be performed in the clinical laboratory, whereas, arrows on the grey background indicate tasks of the requesting physician. The white ovals on the borderline between light and dark grey indicate communication between the physician and the laboratory. The sequence of tasks starts with the request for laboratory analyses by the physician. After submitting the request and the specimen to the laboratory the specimen is processed leading to results. The results which are of quantitative or qualitative nature are transferred on-line to the LIS.

For typical values see column *result* in fig. 2. These wet data are compared with age and sex dependent reference intervals. Examples of reference intervals can be found in fig. 2 in the column *reference interval*. This task is called abstraction (Ramoni 1992). The output of the abstraction are the values *very low*, *low*, *normal*, *high*, and *very high*. In column *assessment* in fig. 2 these values are indicated by --, -, *o*, +, ++. In general, nowadays abstraction is performed by an LIS. The results and assessments in fig. 2 form the basis for the diagnosis and the suggestion for further diagnostic investigations by the laboratory physician. This task is called abduction (Ramoni 1992). In a deduction step it is decided whether a diagnosis can be ascertained, further clinical investigations will be required, or supplementary laboratory tests will have to be performed. Abduction and deduction are up to now rarely taken over by an LIS. When performing the abduction and deduction by a knowledge-based system most of the reports (in the chosen scenario over 95%) can be generated automatically.

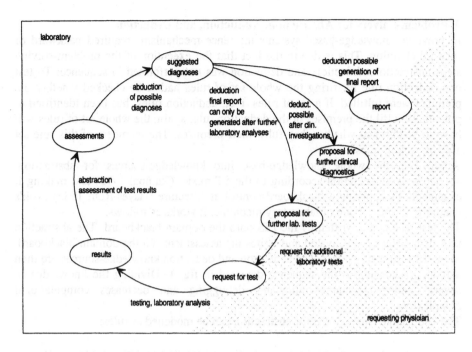

Fig. 1: ST-model of the laboratory medicine problem solving process.

According to the ST-model problem-solvers can be related to each of the above mentioned tasks. The tests are performed by technicians and automatic analysers. The abstraction is mostly performed by the LIS using decision tables. Problem-solvers for abduction and deduction can be both laboratory physicians and knowledge-bases.

parameter	reference interval	SI unit	result	assessment
platelets	187,00-406,00	Giga/l	122,00	−
leukocytes	4,50-11,00	Giga/l	30,00	++
neutrophils	1,80-7,70	Giga/l	28,00	++
myelocytes	-	% NEU	5,00	
metamyelocytes	-	% NEU	4,00	
band neutrophils	0,00-0,70	% NEU	20,00	++
segmented neutrophils	1,80-7,00	% NEU	60,00	++
lymphocytes	1,00-4,80	Giga/l	2,30	o
monocytes	0,00-0,80	Giga/l	0,40	o
eosinophils	0,00-0,45	Giga/l	0,20	o
basophils	0,00-0,20	Giga/l	0,00	-
erythrocytes	3,80-5,20	Tera/l	5,08	+
haemoglobin	117,00-157,00	g/l	158,00	++
haematocrite	0,35-0,47	l/l	0,46	+
MCV	80,80-100,00	fl	90,55	o
MCH	26,40-34,00	pg	31,10	o
MCHC	314,00-358,00	g/l	343,48	o

Fig. 2: Hypothetic example of wet and abstracted data in laboratory medicine.

3. Problem-Solvers for Abstraction, Abduction, and Deduction

In previous knowledge-based systems inference mechanisms required backward or forward chaining. This is due to the fact that the execution of the problem-solving steps abstraction, abduction, and deduction was not performed in sequence. To test one deduction rule for firing the whole set of rules had to be checked whether the premises were fulfilled. If a set of rules, i. e., abduction rules, has been identified to potentially fulfil the premises of the deduction rule, again, the whole set of rules will have to be checked whether they fulfil their premises. The members of this rule set are abstraction rules.

Subdividing the whole knowledge-base into knowledge-sources for abstraction, abduction, and deduction according to the ST-model (Ramoni 1992) and making it executable by applying a blackboard-control architecture (Hayes-Roth 1985) avoids chaining for the domain of laboratory medicine. It works as follows.

The test results are written by the LIS onto the domain blackboard. The abstraction knowledge-source is activated and writes the assessments on the domain blackboard. Many LISs already do this. The abduction and deduction knowledge-sources are then activated according to the sequence depicted in fig. 1. Dividing the knowledge in several knowledge-sources alleviates debugging and decreases computational complexity.

The knowledge chunks used for abstraction can be modelled as rules:

$$\underset{c,r}{\forall}\ \text{parameter}(c,r) \wedge \text{sex}(c,r) \wedge \text{age\_interval}(c,r) \wedge \text{reference\_interval}(c,r) \rightarrow \text{abstracted\_value}(c,r) \ (1)$$

The inference performed on these rules is simple pattern matching and performed for all results r of all cases c. Given a laboratory result with its parameter name, the patient's sex, the patient's age, and the result value all abstraction rules will be compared whether they match this result. Matching means that the parameter name and sex are identical, and both the age and the result value lie within the respective intervals. In case that the premises of a rule match its *abstracted_value* for the appropriate case is abstracted and written on the domain blackboard. The knowledge in the abstraction rules as described above can be acquired from reference value tables.

The knowledge chunks used for abduction can be modelled as rules:

$$\underset{c}{\forall}\ \text{condition\_for\_parameter\_1}(c) \wedge ... \wedge \text{condition\_for\_parameter\_n}(c) \rightarrow \text{potential\_diagnosis} \ (2)$$

The predicate condition_for_parameter_i is given in (3). It is complex for reasons of dealing with open intervals. (3) can be transformed in an internal representation with only 4 premises.

The inference performed on these rules is a complex pattern matching. The rule format allows to take the quantitative results and the qualitative assessments into account. Each of the laboratory parameters in the premises has to be checked whether its value and its abstracted value are lying within the qualitative and quantitative intervals given in the rule. In case the premises are fulfilled for all the laboratory parameters the rule is fired and the potential diagnosis established.

The predicate condition_for_parameter_i:

$$
\begin{aligned}
&\text{(qt} \geqq \text{lower quant. border} \land \text{qt} < \text{upper quant. border)} \\
\lor\quad &\text{(qt} \geqq \text{lower quant. border} \land \text{qt upper quant. border not given)} \\
\lor\quad &\text{(qt} \geqq \text{lower quant. border not given} \land \text{qt} < \text{upper quant. border)} \\
\lor\quad &\text{(ql} \geqq \text{lower qual. border} \land \text{ql} \leqq \text{upper qual. border)} \\
\lor\quad &\text{(ql} \geqq \text{lower qual. border} \land \text{upper qual. border not given)} \\
\lor\quad &\text{(lower qual. border not given} \land \text{ql} \leqq \text{upper qual. border)} \\
\lor\quad &\quad\text{(lower qual. border not given} \land \text{upper quant. border not given} \\
&\land\quad \text{lower qual. border not given} \land \text{upper qual. border not given)}
\end{aligned}
\tag{3}
$$

qt is the case's result for the parameter i and ql is the assessment of the case's result for parameter i.

The knowledge to define the potential diagnosis *shift-to-left*, given below, can be easily expressed with the above structural description.

> **If**
> the leukocyte and neutrophil concentrations are at least increased, the myelocytes are below 7%, the metamyelocytes are between 3.5% and 13.5%, the band neutrophils are between 15% and 35%, and the segmented neutrophils are between 55% and 85% of the neutrophils
> **then**
> shift-to-left is a potential diagnosis.

Fig. 3: Typical abduction rule in laboratory medicine.

This definition is prototypical for most of the abduction knowledge in laboratory medicine.

The deduction knowledge is not formalised. The deduction step is mostly performed by the general practitioner or clinician, since in general, supplementary clinical information is required. Consequently, these "problem-solvers" can best be served by plain text describing what further clinical exams or laboratory tests are required to confirm a particular diagnosis.

> **If**
> first examination
> **then**
> check acute phase protein, send-in control within 10 days.

Fig. 4: Typical deduction rule in laboratory medicine.

4. Knowledge-Acquisition-Tools

Knowledge develops dynamically. Reference intervals and methods may differ from laboratory to laboratory. Even abductive and deductive knowledge may differ between laboratories. This is despite the fact that medical bodies are active in the field of standardisation, especially including clinical chemistry.

One design requirement demanded by the commercial laboratory was easy knowledge editing by laboratory physicians. Also mechanisms were requested to define access restrictions for knowledge modification. On the one hand it could be interesting to start with a standard knowledge-base and let each user modify it for himself, on the other hand a standard kernel could be delivered and each user may add additional knowledge to make the knowledge-base more sensitive and specific. The syntax of a rule description language for medical knowledge in the domain of clinical chemistry as proposed by Pohl (1988) is too intricate to be used by a physician and the task concept is incorporated implicitly only. Therefore, the laboratory physician needs a knowledge-engineer to formalise his knowledge.

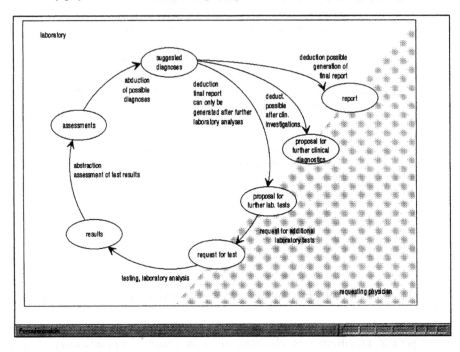

Fig. 5: KA-tools selection menu adopting the problem solving model of laboratory medicine.

The KADS approach (Wielinga 1992) was intended to provide easy to use knowledge-acquisition (KA) tools on the basis of problem-solving processes. It is, however, disadvantageous that knowledge-based systems cannot be automatically produced out of the knowledge acquisition results applying the KADS methodology. A major goal of the GAMES-II project (Schreiber 1993) was to produce KA tools based on the KADS methodology to automatically generate executable knowledge-based systems. A first prototype of a model-driven KA tool allowing to generate knowledge-based systems is M-KAT (Lanzola 1993). A wide variety of different kinds of knowledge can be acquired with this system. Powerful, but, expensive workstations are required for the use of M-KAT.

According to the philosophy of the GAMES-II project, a prototype of a knowledge-based system combined with a KA component has been built for laboratory medicine. For the task-oriented main KA menu, a visualisation of the ST-model of laboratory medicine, see fig. 5. The ST-model allows to access the KA editors for the different knowledge-sources by a pointing device. The editor for the abstraction knowledge-source is shown in fig. 6. The input into one sheet of the abstraction editor is translated in a set of rules as depicted above. Each row of the platelet reference intervals in fig. 6 is transferred into one rule.

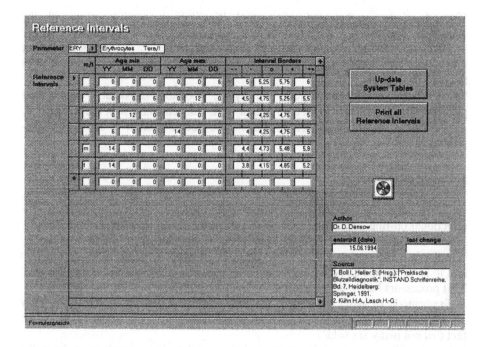

Fig. 6: Editor for the acquisition of abstraction knowledge.

For the editor of the abduction knowledge-source see fig. 7, which is a formalisation of the conditional clause defining *shift-to-left* in fig. 3.

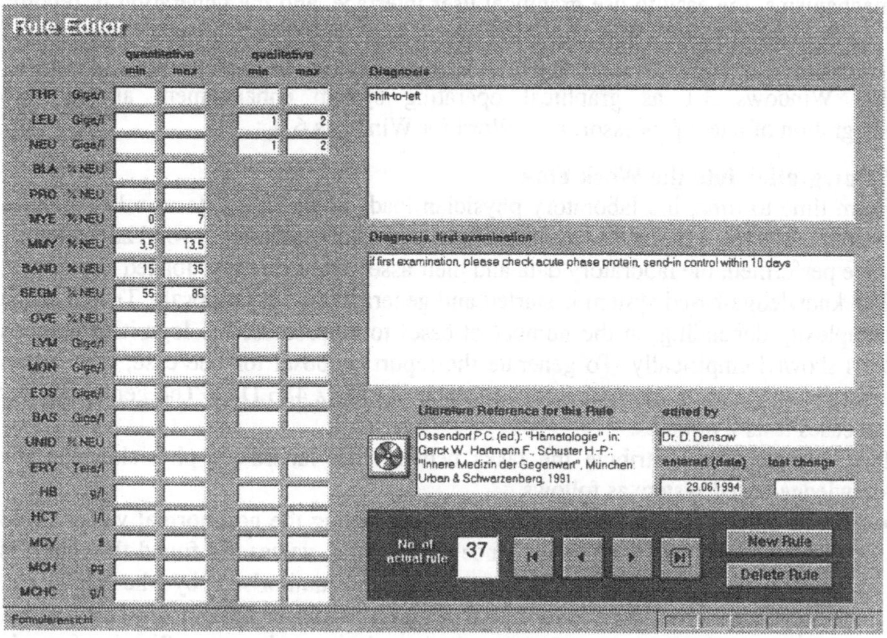

Fig. 7: Editor for the acquisition of abductive knowledge.

Since the deduction knowledge is textual, it will be inserted together with the abduction knowledge. The conditional clause of fig. 4 proposing further investigations can be found on the right-hand side in fig. 7. The physician's input into these editors is automatically transferred into knowledge-chunks as described in the previous chapter.

The ST-model is understood quite well by the laboratory physicians. Only a four page manual is necessary to let a physician work independently with the KA editors. A print-out of the editor sheets provides good documentation of the knowledge-sources.

5. Design Decisions

The commercial laboratory already had a LIS running on a mainframe. Therefore, the knowledge-based system could be attached as a satellite to the LIS only. Bi-directional on-line connection to the LIS was necessary to read down the laboratory results and assessments and to write back the reports for documentation purposes. The single work-station for the knowledge-based support of the report generation had to be rather inexpensive and reliable. Easy to use interfaces had to be provided. Text processors had to be integrable.

A disadvantage of many AI-shells is that they need quite performant hardware. They are rather expensive and the number of bugs is relatively high which is due to the fact that not many are sold.

Using the ST-model for laboratory medicine renders the advantage to keep the inference mechanisms relatively simple as explained above.

The solution we chose to meet the above requirements and to rapidly produce a running system was a combination of software components running under Microsoft Windows 3.1. The database system Access 2.0 allowed to realise the inference mechanisms, the easy to use graphical user interface, and the generation of reports. Due to password protection multiple roles of users could be realised. The access to the mainframe could be provided by a terminal emulation running under Windows 3.1. Windows 3.1 as graphical operating system enhancement allowed the integration of a text processor, i. e., Word for Windows 6.0.

6. Integration into the Work Flow

From time to time, the laboratory physician loads down case data of a list of cases from the LIS main frame. In case that laboratory tests extending beyond haematology were performed, the laboratory data and their assessments are also loaded down.

The knowledge-based system is started and generates report proposals. The run-time complexity depending on the number of cases to be processed is logarithmically as tests showed empirically. To generate the report proposal for one case, e. g., in an emergency situation, takes about 40 seconds (50 MHz 486 DX). The generation for ten cases lasts 2 minutes.

It is envisaged to distribute the tasks between the laboratory physician and the knowledge-based system as follows.

- In case that an unambiguous diagnosis explaining the non-normal values or no diagnosis could be established since only normal values are found the report is printed out, put into a routine stack, and authorised by the physician. Nevertheless, the knowledge-based report generation system produces a text processor file for eventual corrections and an explanation file for forensic reasons.

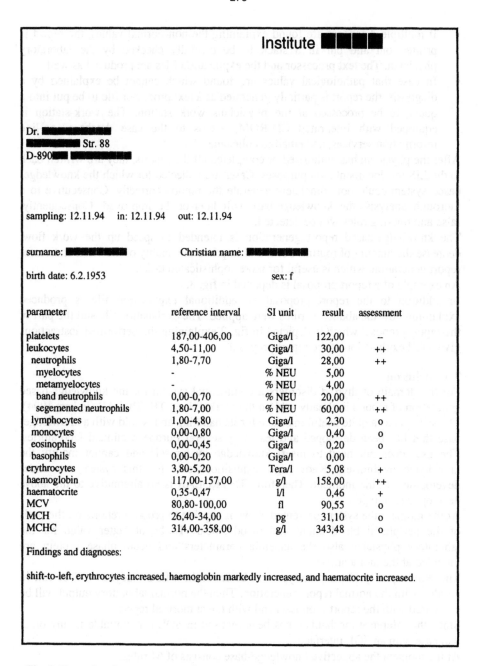

Fig. 8: Example of a report proposal generated by a knowledge-based system..

- If multiple diagnoses are found explaining the non-normal values, the report is printed out and put in a stack to be carefully checked by the laboratory physician. The text processor and the explanation files are produced as well.
- In case that pathological values are found which cannot be explained by a diagnosis, the report is partially generated as a text processor file to be put into a queue to be processed at the physicians work station. The work-station is equipped with integrated CD-ROM, access to the case database, on-line information services, and email to colleagues.

After the physician has authorised or completed all the reports, they are written back to the LIS for documentation purposes. Cases are collected for which the knowledge-based system could not completely generate the reports correctly. Consecutive to a thorough analysis, the knowledge-base will later-on be improved. Consequently, false and missing rules will be detected.

The knowledge-based report generation is intended to speed up the work flow, decrease the amount of routine work, and improve the quality of the work-station for report generation, which is useful for more sophisticated tasks.

An example of a report proposal is depicted in fig. 8.

In addition to the report proposal an additional explanation file is produced explaining how the abductive rules were applied. The explanation file and the part of the report proposal which is depicted in fig. 2 explaining the performed abstractions give a full explanation of the report proposal.

7. Conclusion

For the domain of thyroid disorders a commercial system for the knowledge-based generation of reports is already sold on the market, i. e., THYROLAB (Eckert 1994). The system has a quite useful interface for its handling and is sold with a knowledge-base that has been developed and tested by several European clinical laboratories. The user does not need to insert knowledge by himself and cannot modify the knowledge by himself. Knowledge acquisition tools for this system are under development at the moment. Therefore, THYROLAB is no alternative to the above developed prototype.

At the moment the system presented above is capable to propose reports on the basis of the peripheral blood count and blood smear. To be of better value for the laboratory physician also the anaemia parameters and acute phase protein are included at the moment.

The report quality of the prototype will be tested in a field study. It will be run in parallel with the normal report generation. Then the normal laboratory output will be compared with the report proposals and with the authorised reports.

Since the inference mechanism has been realised in SQL it is portable to any other database with an SQL interface.

At the moment the abductive knowledge-base consists of 64 rules.

8. Literature

1. Barosi G., Magnani L., Stefanelli M.: "Medical diagnostic reasoning: epistemological modeling as a strategy for design of computer-based consultation programs", Theoretical Medicine 14, 1993, 43-55
2. Causer M. B., Findlay G. A., Hawes C. R., Boswell D. R.: "Assessment of a computerized system for the diagnosis of iron deficiency", Pathology 26, 1994, 37-39

3. Eckert M., Bepperling C., Pohl B. Trendelenburg C.: "Thyrolab - Technische Anforderungen an ein wissensbasiertes Befundungssystem für den Routineeinsatz", Laboratoriumsmedizin 18, 1994, 572-576

4. Edwards G., Compton P., Malor R., Srinivasan A., Lazarus L.: "PEIRS: a pathologist-maintained expert system for the interpretation of chemical pathology reports", Pathology 25, 1993, 27-34

5. Grimson J. B.: "Integrating knowledge-based systems and databases", Clinica Chimica Acta 222, 1993, 101-115

6. Kindler H., Densow D., Fliedner T. M.: "A knowledge-based advisor supporting to deal with rare diseases", in: Andreassen S., Engelbrecht R., Wyatt J.: "Artificial Intelligence in Medicine", IOS Press, 1993, Amsterdam, 129-133

7. Hayes-Roth B.: "A blackboard architecture for control", Artificial Intelligence 26, 1985, 251-321

8. Lanzola G., Stefanelli M.,: "Inferential knowledge acquisition", Artificial Intelligence in Medicine 5, 1993, 253-268

9. Lanzola G., Stefanelli M.: "A specialized framework for medical diagnostic knowledge-based systems", Computers in Biomedical Research 25, 1992, 351-365

10. Leaning M. S., Samuel P., Austin T., Hassan T., Modell M., Patterson D., Rubens R.: "Clinical application of a general methodology for decision-support systems", in: Andreassen S., Engelbrecht R., Wyatt J.: "Artificial Intelligence in Medicine", IOS Press, 1993, Amsterdam, 146-152

11. Pohl B., Trendelenburg C.: Pro. M. D. - a diagnostic expert system shell for clinical chemistry test result interpretation, Methods of Information in Medicine 27, 1988

12. Ramoni M., Stefanelli M., Magnani L., Barosi G.: "An epistemological framework for medical knowledge-based systems", IEEE Transactions on Systems, Man, and Cybernetics 22, 1992, 1-14

13. Safran E., Borst F., Thurler G., Pittet D., Lovis Ch., Rohner P., Auckenthaler R., Scherrer J.: "An evolutive alert system for diagnosis of nosocomial infections based on archiving databases", in: Andreassen S., Engelbrecht R., Wyatt J.: "Artificial Intelligence in Medicine", IOS Press, 1993, Amsterdam, 180-184

14. Schreiber A. T., van Heijst G., Lanzola G., Stefanelli M.: "Knowledge Organisation in medical KBS construction", in: Andreassen S., Engelbrecht R., Wyatt J.: "Artificial Intelligence in Medicine", IOS Press, 1993, Amsterdam, 394-405

15. Wells I. G., Cartwright R. Y., Farnan L. P.: "Practical experience with graphical user interfaces and object-oriented design in the clinical laboratory", Clinica Chimica Acta 222, 1993, 13-18

16. Wielinga B. J., Schreiber A. T., Breuker J. A.: "KADS: A modelling approach to knowledge engineering", Knowledge Acquisition 4, 1992, 5-53

Machine Learning Techniques Applied to the Diagnosis of Acute Abdominal Pain

C. Ohmann[1], Q. Yang[1], V. Moustakis[2], K. Lang[1], P.J. van Elk[3]

[1] Theoretical Surgery Unit, Department of General and Trauma Surgery, Heinrich-Heine-University, Germany

[2] Institute of Computer Science, Foundation of Research and Technology, University of Crete, Greece

[3] Stichting Deventer Ziekenhuizen, Deventer, Netherlands

1. Introduction

The diagnosis of acute abdominal pain is still a major problem despite considerable improvements by using imaging technology (e.g. ultrasound) and special laboratory investigations (1). Recent studies have reported error rates of 40% for the initial examiner and of 20% for the final examiner (2). Computer-aided diagnosis in acute abdominal pain has been propagated for more than 25 years (3). The results of several evaluation studies have been convincing, however existing systems are rarely used in clinical routine and the diagnostic accuracy does not exceed 60% (4).

Existing programmes are mainly based on the Independence Bayes model. Promissing techniques of artifical intelligence have been used not very often (5). In order to improve the diagnostic accuracy, we tested several automatic rule induction techniques on two databases and compared the results with the standard model. Special attention was given to the problems of geographical variation and training sample-size.

2. Patients and methods

2.1. Data collection

The investigation was based on a prospective clinical data base collected in the framework of a Concerted Action of the European Community (COMAC-BME-European Community Concerted Action on Objective Medical Decision Making in Patients with Acute Abdominal Pain; project leader: F.T. de Dombal (Leeds, UK)) (6). Six surgical departments in Germany participated in the study (University Surgical Department: Düsseldorf, Köln-Merheim, Marburg, Homburg/Saar and surgical department: Bürgerhospital Frankfurt, Hospital Siegburg (2)). Included in the study were all patients with acute abdominal pain of less than one week duration. A structured and standardized history and clinical examination were performed in every patient and the data were documented prospectively using a form suitable for computer use. This form was based on the original abdominal pain chart of the World Organisation of Gastroenterology (OMGE). Terminology and definitions were taken from the European Community Conerted Action (6). Final diagnosis was based on operative findings, special investigations and the course of the disease during hospital stay. 46 parameters from the history and clinical examination were used for prediction of 15 diagnoses. In addition, a dutch database with 1350 cases from 3 hospitals was used for evaluation. This database was collected according to the same criteria as the german database with some minor differences. 47 parameters from history and clinical examination were available for prediction of 14 diagnoses. In order to make the results comparable between the databases 3 parameters from the german database and 4 parameters from the dutch database had to be excluded for the analysis of geographical variation and sample size.

2.2.   Rule induction techniques

Six different techniques of automatic rule induction were applied to the two databases and compared with the standard statistical model, Bayes theorem with conditional independence (table 1).

technique	reference	implementation	conditions
ID3	Quinlan (7)	Knowledgeseeker™	create size: 10 stop size: 50 default rule
C4.5	Quinlan (8)	original	-
NewId	Turing Institute (9)	original	no pruning, default rule
PRISM	Cendrowska (10)	own	stop size: 20 default rule
ITRULE	Smyth (11)	own	K = 500 default rule
CN2	Clark (12)	original	no threshold significance default rule star size: 10

Table 1:   **Automatic rule induction techniques**

For evaluation the german database was randomly splitted into a training set (n = 839) and a test set (n = 415). The same procedure was performed for the dutch database (training set: n = 902, test set: n = 448). All techniques were trained on the training set and evaluated on the test set. To achieve unbiased estimates of the diagnostic accuracy only the results with the test set were analysed.

Geographical variation was investigated by training of ID3 and PRISM on the german training set (n = 839) and comparing the test results on the german test set (n = 415) and a random sample of the dutch database with the same sample size. The same procedure was done with training on the dutch training set (n = 902) and testing on the dutch test set (n = 448) and a german random sample of the same size. The role of the **sample size** of the training database was studied by training of ID3 and PRISM on the german training set (n = 839) compared to training on the full german database (n = 1254). Testing was performed with the random test sample of the dutch database. The same was done vice versa.

3. Results

The diagnostic accuracies for the techniques applied to the original data are presented in table 2. No differences in overall accuracy were observed, except for NewId, which was about 5% less accurate compared to the other techniques. None of the algorithms did exceed an overall accuracy of 50%.

technique	Diagnostic accuracy in the test set (%)	
	german data[1]	dutch data[2]
Bayes	45.1	-[3]
ID3	47.5	48.4
NewID	40.2	-
PRISM	44.6	46.0
CN2	47.4	-
C4.5	46.3	-
ITRULE	42.7	-

Table 2: **Overall accuracy of the learning systems**
[1]training set: n = 839, test set: n = 415
[2]training set: n = 902, test set: n = 448
[3]investigation not performed

Minor differences in diagnostic accuracy were observed, when two of the techniques were applied to the reduced parameter set. ID3 showed no statistical differences with respect to geographical variation (table 3). Slightly better results were achieved with ID3 trained on german cases and tested on german versus dutch cases (+1.9%) and vice versa (+ 3.6%). For PRISM the results were different, with a higher diagnostic accuracy for PRISM trained on the dutch database and tested on dutch versus german cases (+7.8%).

A small, however not statistically significant effect of the sample size of the training set could be demonstrated for ID3. Training on the full german database (n = 1254) versus the smaller training set (n = 839) resulted in a 3.4% higher diagnostic accuracy when testing on dutch data. With training on dutch data and testing on german data the difference was smaller (+1.6%). PRISM did not result in

algorithm	training database	testing database	accuracy (%)	database	accuracy (%)	p-value[1]
ID3	G (839)[2]	G (415)	46.0	N (415)	44.1	n.s.
	N (902)	G (448)	43.5	N (448)	47.1	n.s.
PRISM	G (839)	G (415)	47.7	N (415)	48.4	n.s.
	N (902)	G (448)	42.0	N (448)	49.8	p<0.05

Table 3: **Effect of geographical variation**
for description of databases and algorithms see text
[1]chi-square test, n.s. = not significant
[2]G = Germany, N = Netherlands, (sample size)

algorithm	training database	testing database	accuracy (%)	p-value[1]
ID3	G (839)[2]	N (415)	44.1	n.s.
	G (1254)	N (415)	47.5	
	N (902)	G (448)	43.5	n.s.
	N (1350)	G (448)	45.1	
PRISM	G (839)	N (415)	48.4	n.s.
	G (1254)	N(415)	47.2	
	N (902)	G (448)	42.0	n.s.
	N (1350)	G (448)	42.2	

Table 4: **Effect of training sample size**
for description of databases and algorithms see text
[1]chi-quare test, n.s. = not significant
[2]G = Germany, N = Netherlands, (sample size)

any improvement when trained on the full database compared to the smaller training set (differences: -1.2%, +0.2%).

4. Discussion

Machine learning techniques of automatic rule induction did not improve the results of the standard model (Independence Bayes). The following reasons may be responsible. Studies demonstrating high predictive accuracy with these techniques in other areas were mainly based on low dimensional problems, whereas in our study many diagnoses had to be differentiated. Compared to published data, the size of our training sets was high. Due to a skew distribution with a very low prior probability for the majority of diagnoses, the ability of efficient learning in these subgroups was considerably reduced. The differences between the rule induction techniques were small, except for NewId. For this technique no pruning was applied, thus leading to a model of high complexity. Many rules and a high number of conjunctions in a rule are responsible for overspecialization and overfitting.

The analysis on geographical variation shows that despite common and agreed terminology significant deteriorations in performance of rule-induction techniques may occur, if a system developed in one country or region is applied in another country or region. Consequently, computer-aided diagnostic systems in acute abdominal pain should be at least partially based on local data if there is geographical variation. An increase of the training set by about 50% resulted in a slightly higher diagnostic accuracy, as expected, however, not statistically significant. An ongoing study, in which a prospective and quality controlled database with 3000 cases in acute abdominal pain is built up, will again focus on this aspect with more data.

Acknowledgements

The work was supported by grant of the German Ministry of Research and Technology (MEDWIS-program A 70). The authors thank Ursula Willems for typing the manuscript.

References

1. J. Hoffmann, O. Rasmussen, Aids in the diagnosis of acute appendicitis, Br. J. Surg. 76 (1989) 774-779

2. C. Ohmann, M. Kraemer, S. Jäger, H. Sitter, C. Pohl, B. Stadelmayer, P. Vietmeier, J. Wickers, L. Latzke, B. Koch, K. Thon, Akuter Bauchschmerz - Standardisierte Befundung als Diagnoseunterstützung. Ergebnisse einer prospektiven multizentrischen Interventionsstudie und Testung eines computerunterstützten Diagnosesystems, Chirurg 63 (1992) 113-123.

3. F.T. de Dombal, D.J. Leaper, J.R. Staniland, A.P. McCann, J.C. Horrocks, Computer-aided diagnosis of acute abdominal pain, British Medical Journal II (1972) 9-13.

4. C. Ohmann, M. Kraemer, Evaluierung von Entscheidungsunterstützungssystemen bei der Diagnose von akuten Bauchschmerzen - Eine Analyse publizierter Systeme, Biometrie und Informatik in Medizin und Biologie 23 (1992) 107-111.

5. C. Ohmann, V. Moustakis, Q. Yang, K. Lang, Evaluation of automatic knowledge acquisition techniques in the diagnosis of acute abdominal pain Artifical Intelligence in Medicine (in press)

6. F.T. de Dombal, H. de Baere, P.J. van Elk, A. Fingerhut, J. Henriques, S.M. Lavelle, G. Malizia, C. Ohmann, C. Pera, H. Sitter, D. Tsiftsis, Objective medical decision making acute abdominal pain in: J.E.W. Benken and V. Thevenin, eds., Advances in Biomedical Engineering (IOS Press, 1993) 65-87.

7. J.R. Quinlan, Induction of decision trees, Machine Learning, 1 (1986) 81-106.

8. J.R. Quinlan, C4.5: Programs for machine learning, Morgan Kaufman Publishers. Inc. San Mateo, California (1993)

9. The Turing Institute, NewID Software system, George House, 36 North Hanover St. Glasgow G1 2 AD, Scotland (1993)

10. J. Cendrowska, PRISM: An algorithm for inducing modular rules, Knowledge-Based Systems 1 (1988) 255-276.

11. P. Smyth, R. Goodman, An information theoretic approach to rule induction from databases. IEEE transactions on knowledge and data engineering 4 (1992) 301-316

12. P. Clark, T. Niblett, The CN2 induction algorithm, Machine Learning 3 (1989) 261-283.

Reflections on Building Medical Decision Support Systems and Corresponding Implementation in Diagnostics Shell D3

Bernhard Puppe, University of Wuerzburg, Department of Medicine,
Allesgrundweg 12, D-97218 Wuerzburg, Germany
e-mail: bpuppe@informatik.uni-wuerzburg.de

Abstract

We take a closer look at the medical environment in which decision support systems will have to operate and which ultimately determine their success of failure. Based on the experience accumulated in ten years of active involvement into research revolving around the construction of expert systems, we put forward for discussion a couple of judgements including representation of symptomatic detail, temporal reasoning, case data validation, rule syntax, intermediate conclusions, modularity, test indication/sequence and domain choice. For each of the items touched we describe how our preferences and conclusions have been implemented in the diagnostics shell D3 and thereupon-based diagnostic systems.

0   Introduction

Anybody involved in research directed toward development of decision support systems in medicine must have been surprised to read in a news magazine with as high a reputation as "The Economist" that "the future of doctors looks bleak" because "much diagnosis and even some surgical procedures are being automated", combined with the recommendation that far-sighted physicians should hurry to concentrate on the psychosocial dimension of disease (1). The assessment is all the more surprising because it has become standard practice in research cycles to unfavorably compare the optimistic predictions of the discipline´s youth to the less-than-overwhelming achievements after more than two decades of world-wide research (2).

Automation of diagnosis is certainly not even the goal of Artificial Intelligence in Medicine (AIM), let only its actual consequence. However, this is not to say, that The Economist´s writer´s opinion is completely beside the point. When insiders are bound to loose confidence because progress is incremental and frequently prophesied breakthroughs fail to materialize, a fresh look from the outside can be helpful to acknowledge achieved results and put them into proper perspective.

While leaning toward the optimistic side, we think it useful to take a closer look at the medical environment in which decision support systems will have to operate and which ultimately determine their success of failure. Based on the experience accumulated in ten years of active involvement into research revolving around the construction of expert systems, we would like to put forward for discussion a couple of judgements concerning the successful development of decision support systems in medicine. It is our interpretation of the lesson taught by the past two decades of numerous more or less successful contributions toward the common goal of assisting the practicing physician in his responsible task of treating people suffering from disease. By forwarding our view, we hope to induce a discussion leading to sharper recognition of the actual requirements decision support systems will have to fulfil to contribute to better patient care.

Our reflections will touch on the subjects listed in table 1. For each of them, we explain how we implemented our judgements and choices in D3, a large and powerful diagnostics shell (3), and in thereupon-based diagnosics systems (4,5).

1. Symptomatic detail
2. Temporal reasoning
3. Rule syntax power
4. Intermediate conclusions
5. Knowledge base modularity
6. Test indication and sequence
7. Choice of Task and Domain

Table 1: Subjects covered in the article

1 Symptomatic Detail

The question of how much symptomatic detail should be represented in a knowledge base applies mainly to history, physical examination and radiology, areas relatively rich in symptomatic detail. It is reasonable to restict representation to items necessary for drawing conclusions. However, while satisfying present needs, such a representation could be hard to extend to accomodate future difficult cases the resolution of which requires more detailed data. On the other hand, it is neither possible nor desirable to represent the entire spectrum of complaints and their nuances and shades as revealed by detailed history. To cut them down to a reasonable and manageable number, abstractions are inevitable. For example, instead of listing a thoracic pain as pressing, crushing, squeezing, tight, constricting, elephant-sitting-on-chest-like etc, all these descriptors could be summarized as angina pectoris-like.

For the PARACELSUS knowledge base (4), we use a rather uniform set of descriptors to characterize complaints, including intensity, temporal type/duration, mode of onset, progress, location, quality, aggravating factors, relieving factors, accompanying symptoms, external circumstances, specific drug history and prior treatment. Together with symptom-specific desciptors, they seem capable of catching most of the information relevant for diagnosis.

The more specific and detailed findings are, the more specific the conclusions they can support. However, it often requires a specialist to recognize the distinguishing features with the necessary precision. There are immediate implications for finding representation:

For example, cardiac murmurs can be characterized as to their intensity, punctum maximum (best heard at), radiation, cardiac phase (e.g. diastolic), shape (e.g. crescendo), pitch/frequency and quality (e.g. blowing). However, not all examiners are able to accurately and correctly analyze and describe all these aspects of a murmur. If a system draws diagnostic conclusions only from sophisticated auscultation data, it will often draw no conclusions at all because ordinary users not specialized in cardiology won't provide the necessary data. However, each examiner can forward some information, albeit less specific. Just as a consultant adapts to the level of sophistication of diagnostic evidence provided by a consult-seeking colleague, so should an expert system. Hence, when auscultation data are insufficient for a system to classify murmurs expertly (e.g. as "ejection", "holosystolic", "late systolic", "diastolic decrescendo" etc), it should be able to make the best out of simple data (such as "systolic" or "diastolic") even the least sophisticated user can be expected to provide.

2 Temporal Reasoning

Already in its beginnnings, temporal reasoning was recognized as one of AIM´s biggest problems, and it has remained so ever since. The question bears some resemblance to the preceding one of how much symptomatic detail should be represented in a knowledge base. Of course, it would be possible to give every single finding a time tag specifying when it began, was first observed, how long it lasted etc. However, this approach could easily double or even multiply a knowledge base´s volume with only marginal gains in diagnostic accuracy. Accurate temporal information is, after all, not required to solve every question. The problem therefore is to find a way to exploit time not in a brute force fashion, but more selectively and intelligently, i.e. to include it in the inferential process only when necessary.

In medicine, temporal reasoning can be divided into two categories: anamnestic/historical and monitored/present. The course of disease in the past is accessible mostly over the patient´s history. Therefore, accurate information is often missing. Also characteristic for historical temporal reasoning is that it has to cover a great time span, sometimes from birth up to the presence. All this is in sharp contrast to temporal reasoning in a monitored environment in which the patient enters after consulting a physician, especially after hospitalization. Here, temporal reasoning usually spans only a short time interval in which changes of findings are documented meticulously. Since both types are so different, it is difficult to accomodate them efficiently with a single model. One way of doing it would be to routinely record the time of onset of every symptom in absolute terms (year, day, hour etc), similarly the time of observation of every sign and of every test or procedure done. Such a system could always determine the sequence of appearance or observation of all represented findings.

In D3, we explore a different model of anamnestic temporal reasoning. Rather than uniformly assigning every finding its place in the temporal dimension, D3 provides the possibility to characterize findings by qualitative descriptors. For example, complaints are usually characterized by temporal type (acute, intermediate, chronic), mode of onset (paroxysmal, abrupt, gradual), presence (continuous, recurrent, intermittent, episodic) and progress (unchanging, better, worse, variable). Further temporal qualifiers can be added as needed. The same applies for quantitative temporal information. For example, the duration of a complaint can be accurately registered in hours or days etc (rather than simply as acute or chronic). Experience however showed that qualitative representation of time is most often sufficient for drawing conclusions. In the historical context, registration of exact years, months, days, hours etc more often than not fails to improve diagnostic accuracy, but costs computational resources resulting in longer response times. Also, some temporal categories of enormous diagnostic importance - such as "episodic", "intermittant", "paroxysmal", "recurrent" - are hard to extract from quantitative data. Even a scheme registrating quantitative data extensively would have to ask for these categories explicitly.

However, the qualitative approach does not always allow to compute temporal relations between two or more findings which is often important to understand the underlying disease process. For example, in a patient with dependant edema and ascites, it is useful to know which appeared first because this information can help solve the differential diagnosis of cardiac disease (edema first) vs hepatic disease (ascites first). If the temporal type of both edema and ascites is given as "chronic", there is no way to decide which came first.

Fortunately, this is not as big a disadvantage as it might seem. When the sequence of appearance of symptoms matters, we prefer asking the user explicitly. This has the additional advantage of providing more reliable information because patients tend to remember better the sequence of appearance of their complaints than their exact dates. Also, our approach spares us the need to define somewhat arbitrarily the relevant temporal relationships such as prior, simultaneous, after etc (what is the *minimal* time period for concluding that ascites appeared *before* dependant edema?).

Once the patient is hospitalized and monitored for his disease, the nature of temporal reasoning changes. All data collected und recorded under these circumstances invariably carry a temporal tag, and a very exact one since timing no longer depends upon the patient's memory. Also, disease progress tends to be much faster. In such a context, our model for anamnestic reasoning no longer applies. So far, D3 does not posess a model for temporal reasoning in this environment.

3   Rule Syntax Power

Even simple syntactic schemes can produce astonishing results. For example, QMR/Internist does not allow combinations of findings as precondition in the rules it uses to quantify diagnostic evidence, but restricts itself to scoring two relationships (evoking strength and frequency) between diagnoses and particular findings (6). For classical Bayes, the situation is similar since conditional probabilities are usually given only for single findings, not for combinations of them. Since the results produced by such simple conclusional schemes have been sometimes quite impressive, one has to conclude that they can approximate the complexity of relationships between findings and diseases.

However, this is not sufficient for systems striving to simulate the physician's line of thinking or way of drawing conclusions. We feel that a general decision support system should be built upon a rule syntax powerful and flexible enough to exactly express any diagnostic "rules" or definitions stated in the medical literature. These are sometimes complex logical combinations of major and minor criteria Even if it is possible to achieve satisfactory results with simple conclusional schemes, they might still be inadequate because they complicate the knowledge acquisition process: Instead of directly entering a chunk of knowledge from the medical literature or as specified by an expert into the program, a knowledge engineer has to translate it into the system's syntax.

This is not to say that diagnostic inference should follow medical school teaching under all circumstances. Bayes' theorem does not loose its attractiveness for automatic decision support because the human mind is not very good at doing the related computing. However, the appeal of a diagnostic shell to system builders increases with the number and variety of mechanisms it offers to express knowledge. Shell architects should try to incorporate as many scoring schemes as possible although the resulting heterogeneity considerably complicates the program.

D3 has a remarkably flexible and powerful rule syntax. It allows almost all combinations of findings (e.g. connected by and, or, n from m) as preconditions for its if-then inferences. Furthermore, each element of a combination in a precondition can itself be a combination. Fig. 1 shows a rule with such a complex precondition.

If	or	fatigue = present (1)
	or	dyspnea = exertional (2)
and	or	fever = presently (3)
	or	fever = by history (4)
	or	infection = present (5)
and	or	bleeding = pathologic (6)
	or	easy bruising = present (7)
then suspect		bone marrow insufficiency

Fig. 1: Rule precondition with complex syntax. On the upper level, three preconditions are combined by "and"; on the lower level, each of these can be inferred from sub-preconditions connected by "or". (conditions 1/2 point to anemia, 2-4 to neutropenia, 6/7 to thrombocytopenia)

Incorporation of sensitivity and specificity measures into non-statistical systems is especially difficult because it implies crossing over from one world of diagnostic inference into another. While statistical approaches (such as Bayes) are often non-applicable in the early stages of the diagnostic process (because of lacking formal preconditions to apply the method or because the statistical coefficients are unknown), they can be used in the end game when sensitivity and specificity measures are available for the decisive tests. To exploit them, it is necessary to calculate the pre-test probability of the hypothesis under consideration. This cannot be done exactly when the inference process up to this point has been non-statistical or pseudo-statistical. The obvious alternative - that we favor in D3 - is to convert the twin-pair sensitivity and specificity into pseudo-probabilities compatible with the system´s main scoring mechanism.

4   Intermediate Conclusions

Diagnosis often does not jump from observations directly to the final specific diagosis, but often gets there by a series of smaller steps aggregating several lower-level findings into higher-level abstractions (cp fig. 2). As the diagnostician moves through the hierarchy or network of syndroms - they might be called "intermediate" or "auxillary" concepts of diagnosis - the focus of his search sharpens while the breath and number of hypotheses and differential diagnoses contemplated decreases.

While many inference methods, e.g. Bayes´ statistics, do not readily support simulation of this process, it can be easily modeled by a heuristic system allowing to draw conclusions not only from findings, but also from their derivates (syndroms, intermediate concepts). The knowledge base of HEMATO-PARACELSUS is constructed along these lines (4). 3 advantages of the resulting inferential structure deserve special emphasis:

First, the transparency of the symptomatic profile of each diagnostic concept - and hence, of the entire knowledge base - is increased because the number of inferences per concept tends to be low. In the example of fig. 2, the syndrom "bone marrow failure" is derived from 3 symptoms (fatigue, infection, bleeding) plus the syndrom "pancytopenia" which is itself derived from 3 laboratory findings. Without the mediation of "pancytopenia", its successor "bone marrow failure" would have 6 contributing factors. Since important syndroms usually have much more - fig. 2 shows only a fraction of the whole story - the resulting picture can get uncomfortably complex if not counter-measures are taken. Especially probabilistic reasoning - combining partial evidence from many findings - can get out of control if there are too many factors to integrate. Sequential aggregation of findings into higher-level concepts decreases complexity by reducing the number of inferences per concept. The resulting transparency of the knowledge base allows its builder to improve its quality and helps the user to understand its design.

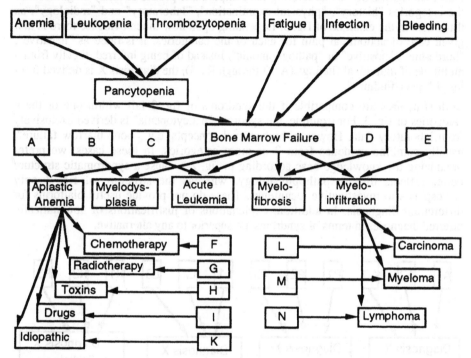

Fig. 2: Incremental diagnosis by repeated aggragation of findings into higer-level concepts. The letters A through N represent nonspecified findings

The second advantage of the proposed structure is a considerable overall reduction of inferences. In fig. 2, it takes 27 rules to infer the 8 specific diagnoses (the "leaves" at the bottom) with the help of 4 intermediate syndroms. Without them, 64 inferences would be necessary to achieve the same result.

While the first 2 stated advantages are mainly technical - facilitating knowledge representation - the last one relates to system performance: Each intermediate concept can serve as the basis for differential diagnosis. Instead of directly suggesting specific diseases as diagnostic hypotheses - which might not be possible in complex cases - such a system indicates the differential diagnoses of its highest-level syndrom and suggests tests for differentiating among them, a process that helps the diagnostic search to stay focused and is especially useful in solving tough cases.

5   Knowledge Base Modularity

Another consequence of a knowledge base structure similar to fig. 2 is increased modularity. Since the diagnostic task is divided in multiple steps, each one can be modified without affecting the overall action. For example, if the knowledge base developer decides to change the way "bone marrow failure" is inferred, he can do so without bothering to adjust inferences based on this syndrom. Otherwise, he would probably have to change the symptomatic profile of all the 8 differential diagnoses of "bone marrow failure".

An even higher degree of modularity has been achieved in the system "Abdominal Pain-PARACELSUS" by structuring the symptomatic profile of many diagnoses into the categories "history", "physical examination", "laboratory" and "radiology", as illustrated in the example of fig. 3. Depending upon how much evidence there is in a given case of abdominal pain for each of the categories, it is rated as "negative", "borderline", "positive" or "pathognomonic". Instead of being inferred directly from a multitude of individual findings (A1-3 through R1-3), the diagnosis X is derived from just 4 "super-findings".

Both structures are compatible if the syndroms of fig. 2 fall within one of the 4 categories of fig. 3. For example, the syndrom "pancytopenia" is derived exclusively from laboratory data. However, higher livel concepts, e.g. "bone marrow failure", usually combine evidence from 2 or more categories. In these cases, we prefer organizing the knowledge base according to fig. 2 because a syndromatic structure better reflects disease pathophysiology, whereas the structure of fig. 3 only corresponds to investigative techniques. These do not provide useful platforms for differential diagnosis. In addition, explanations or justifications of automatically inferred diagnoses in terms of syndroms are superior to any alternative.

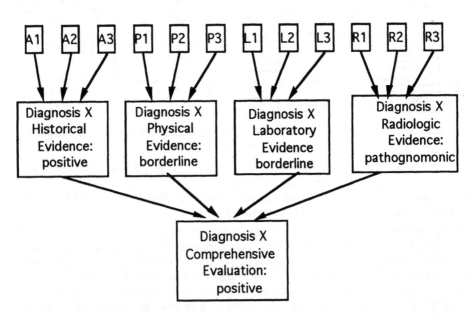

Fig.3: Modular structure of diagnostic knowledge by aggregation of evidence into categories corresponding to investigative techniques. A1-3 through R1-3 represent nonspecified findings

A big advantage of a structure like fig. 3 is the possibility to divide knowledge base building among several specialists. This not only tends to make knowledge base building faster, but might also improve its quality. The division of labor is obvious: A clinician specializing in the domain under consideration works out the categories "history" and "physical examination", while a laboratory specialist and a radiologist cover their respective domains. The participating specialists do not even have to work in the same place. Since all subtasks are well-defined and completely independant from one another, they can be carried out in complete independance. Coordination becomes necessary only in the final stage when the aggragated clinical, laboratory and radiologic evidence has to be combined for a comprehensive evaluation of each diagnosis (which corresponds to the lower half of fig. 3). Here, clinicians and radiologists - a purely fictitious example - might well argue over the relative contributions of "their" evidence toward diagnosing certain diseases.

We are presently in the process of reimplementing "Abdominal Pain/PARACELSUS", a tutorial system for the differential diagnosis of acute abdominal pain, in the described labor-dividing fashing, but have not yet reached the stage where the speculated differences of opinion might emerge.

6   Test Indication and Sequence

Apart from interpreting findings and test results correctly, a diagnostic decision support system should also suggest a reasonable sequence of tests to find the diagnosis in as short, simple and cheap a way as possible.

Many diagnostic systems, including D3´s heuristic module, use hypothesize-and-test as their main problem solving strategy. While this method has been proven to be widely used by physicians (7), it is not first choice to guide the work-up of frequently occuring diagnostic problems. For them, standardized schemes (e.g. flowcharts) are available in the literature (8). They have been developed for nearly all important findings or syndroms (e.g. edema, hepatomegaly, vertigo, hematuria etc). In decision support systems, it is desirable to integrate both methods in a way that gives priority to standard procedures when available and resorts to free hypothesize-and-test when they are not.

For example (cp. fig. 2), after recognizing "pancytopenia", a physician might immediately (in the way of hypothesize-and-test) suspect a myeloma and order specific tests to confirm his hypothesis. This would certainly not comply with recommendations for a standardized work-up which includes as first and immediate measures a differential, red blood cell morphology and reticulocyte count.

Systems including both types of procedural knowledge need a control structure to assure a reasonable sequence of self-indicated tests. It should suppress free-climbing hypothesize-and-test as long as standard schemes for the diagnostic work-up are available.

In D3, the goal of assuring a reasonable test sequence is achieved by assigning priority to tests indicated by special indication rules (e.g. if finding F is observed, then indicate tests A,B,C) over tests indicated by disease hypotheses (e.g. if diagnosis X is suspected, then indicate tests D,E,F). Inside both subgroups, further measures apply to secure proper sequence. In the first (forward-indicated), tests are arranged in a sequence with historical questions at the top, followed by physical examination, laboratory tests and technical investigations. In the second subgroup (backward-indicated), test sequence depends on special scores assigned to every test and investigation reflecting their cost (in terms of risk and pain for the patient and resources required) and usefulness to confirm or rule-out a disease hypothesis. Together, the 3 mechanisms assure that tests self-indicated by the system are suggested to the user in proper sequence.

7   Choice of Task and Domain

Since few of the many good systems developed over the years actually found their way into practice, a mismatch between supply and demand can be safely assumed as a contributing factor. To test this hypothesis, it would be rewarding to conduct a classical market analysis (what sort of system doctors really need?). To do it however, is hard because the majority of doctors still remain sceptical toward the concept of a physician being supported by an intelligent decision making tool. Consequently most projects develop in cooperation with a single medical partner. While this scenario assures that the effort will not be competely in vain, it does not guarantee acceptance of the product by a significant section of the medical community.

Easier than conducting a veritable market analysis is to count how much for what sorts of questions and problems physicians consult each other. Combined with reflections, this might be helpful to identify rewarding scenarios for expert system use. Generally speaking, medical decision support can be addressed to one of the following categories: diagnosis, therapy, and day-to-day management. The latter comprises the vast majority of medical decision making and hence, expert consulting. For example, once a diagnosis of hypertension or diabetes mellitus has been made, a chain of innumerable management decisions streching over several decades is likely to follow. Since each of these decisions tends to be relatively simple and straightforward, there is probably little incentive for consulting an expert system, especially when taking into consideration its disproportionately high time requirements.

The second category, therapy, will profit from automatic decision support when it is sufficiently complex and not primarily dependant in its quality on manual dexterity, psychosomatic impact or verbal skills of the physician (e.g. chemotherapy). Obviously these restrictions apply to many treatment modalities. When coupled with diagnostic decision support, automatic therapy recommendations can be provided cheaply, since most necessary data would be known to the machine already. Hence, without adding much additional input, the user could be rewarded for diagnostic consulation by an automatically generated treatment plan.

Since diagnostic problems can be very demanding intellectually - probably the most demanding in medicine altogether - they constitute classical and frequent consultation reasons and are likely to profit most from automatic decision support. While it is relatively easy to determine the cause of localized disease causing local manifestations, recognition of systemic diseases with diffuse findings can be quite complicated. By and large, the difficulty of the diagnostic task is maximal at its start when the number of disease hypotheses to be contemplated is big and heterogeneous. This has implications for building diagnostic systems:

Before arguing the point for more comprehensive programs, we acknowledge that important and useful systems can also be small. There a hundreds of niches - in diagnostics, but also in therapy and patient management - waiting to be occupied by smart focused programs. They are the first to enjoy routine consultation if the task is not surrendered completely (9). However, if our analysis is correct, the biggest impact on medicine will come from general diagnostic systems.

Building systems to tackle difficult problems of differential diagnosis (e.g. subclassification of acute leukemia, differentiation of inflammory bowel disease) can be rewarding if the scope of differential diagnoses to be considered is limited and hence resolution does not require represensation of hundreds of findings. However, use of such narrow systems is restricted to the relatively unfrequent occasions when these problems come up, and they rely on the physician to narrow the diagnostic problem down to a point where the specialized system can take over. Often this narrowing is more difficult than the following refinement. Under these circumstances, consulting the computer might not be worth while.

The potential usefulness of a diagnostic system rises with the scope of its domain. However, construction and maintanance of big medical knowledge bases is such an awesome task that few projects can command the necessary resources. In the field of internal medicine, only a handful of general systems emerged over the last 20 years including QMR, DXplain, Meditel, Iliad (10). The majority of projects lack the necessary resources.

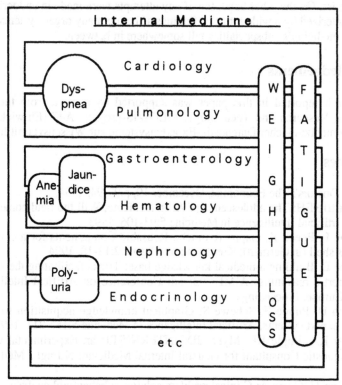

Fig. 4: Internal Medicine, its subspecialties and some major findings

Figure 4 suggests possible domains for diagnostic systems in the intermediate space between small niches (specialized differential diagnosis) and complete comprehensiveness. They fall into 2 categories: the first, smaller one comprises major complains, signs and findings (e.g. dyspnea, hepatomegaly, anemia), whereas the second, bigger one contains officially recognized subspecialties (in our example from internal medicine, they include cardiology, hepatology etc).

Major findings make suitable candidates for diagnostic expert systems if their differentials are reasonably narrow. For example, fatigue or weight loss obviously do not fulfil this criterium (cp. fig. 4). Since they accompany almost every significant disease process, a system offering advice on their diagnosis would have to be quite comprehensive and hence overstretch most projects´ resources. By contrast, complains like dyspnea or polyuria are much more specific. A system helping in the search for their etiology could restrict itself to representation of a limited number of findings and (differential) diagnoses.

A subspecialty is a good candidate for a diagnostic system if cases can be easily assigned on the basis of presenting complaints or other straightforward findings. If this assignment requires considerable expertise (e.g. oncology), a significant part of the entire diagnostic task has already been done by the time the diagnostic system comes into play; hence its potential usefulness would be limited.

Domains showing little interdependance with other medical fields also qualify for diagnostic systems. Among internal medicine´s subspecialties, rheumatology seems to be the best in this regard. Not surprisingly, several rheumatological systems emerged over the years. On the other hand, few if any attempts were made in endocrinology, a field characterized by a wide range of findings affecting many organ systems. The rest of internal medicine´s subspecialties fall somewhere in between.

Acknowledgements

The research reported in this paper was supported by a grant from the German Ministry of Research and Technology, MEDWIS-project A47 (Entwicklung und Evaluation medizini-scher Diagnostik-Expertensysteme zur Wissensvermittlung).

References

(1)   Why Doctors? The Economist, December 10, p.93-94, 1994,

(2)   Shortliffe EH: The adolescence of AI in medicine: Will the field come of age in the ´90s? Artificial Intelligence in Medicine 5:93-106, 1993

(3)   Puppe F, Poeck K, Gappa & others: Reusable Components for a configurable diagnostics shell (in German); Künstliche Intelligenz 2:13-18, 1994

(4)   Puppe B: Building a medical knowledge base: Tricks facilitating the simulation of the expert´s reasoning. AIME ´93: 5th Conference on Artifical Intelligence in Medicine Europe, Proceedings

(5)   Gappa U, Puppe F, Schewe S: Graphical knowledge acquisition for medical diagnostic expert systems. Artificial Intelligence in Medicine 5:185-211, 1993

(6)   Miller RA, Pople HE, Myers JD: INTERNIST1: an Experimental Computer Based Diagnostic Consultant for General Internal Medicine; N Engl J Med 307:468-476, 1982

(7)   Kassirer U, Gorry A: Clinical problem solving: a behavioral analysis; Annals of Int. Me. 89:245-255, 1978

(8)   Puppe B, Puppe F: Standardized forward and hypothetico-deductive reasoning in medical diagnosis; Proc. of MEDINFO-86, p 199-201, 1986

(9)   Aikins JS, Kunz JC, Shortliffe EH, Fallat RJ: PUFF: An exert system for interpretation of pulmonary function data. Computers in Biomedical Research, 16: 199-208; 1983

(10) Berner E, Webster G, Shugerman A: Performance of four computer-based diagnostic systems; N Engl J Med 330:1792-6), 1994

Models for Clinical Information Systems

Decision models for cost-effectiveness analysis: a means for knowledge sharing and quality control in health care multidisciplinary tasks

S. Quaglini[a], M. Stefanelli[a], F. Locatelli[b]

[a]Dpt. Informatica e Sistemistica - Università di Pavia, Pavia, Italy
[b]Clinica Pediatrica - IRCCS Policlinico S. Matteo, Pavia, Italy

Abstract. This study presents a methodology for creating and employing decision-analytic models in order to solve problems involving different health care expertises. Several agents may concur in making a choice, each of them having a different role both in building and using the model. As a matter of fact, each agent can provide specialized knowledge for the creation of parts of this model. Operators may use different sets of data for exploring possible model improvement, and may manipulate different variables influencing final decision. The training example given by the paper will be a model for a therapeutic choice based on the cost/effectiveness ratio. Two agents will be taken into consideration: a physician, able to predict health outcomes and a hospital manager, able to compute costs. They both have access to the hospital data base, but with different logical views, that allow them to validate and improve the relative part of the model. The aim of this work was directed towards demonstrating that the use of explicit knowledge representation concerning budget-related choices in health care, can improve communication among different operators, and eventually ameliorate the quality control of health care services.

1 Introduction

How best to allocate limited economic resources is becoming an ever increasing problem in health care. Two factors mainly influence the increased demand in this field: the technological progress, that has allowed industry to offer more and more sophisticated instruments to health care operators, and the increased life expectancy. On the other hand, the scarcity of resources makes satisfaction of requests impossible, and imposes controls for avoiding or limiting waste of resources. In the last years an exponential growth in the literature on economic evaluation of health care programs has been observed and the common denominator of these papers is *cost-effectiveness analysis*. However, several problems have been highlighted about the interpretation of results reported in the studies mentioned above. In their book [2], Drummond, Stoddart and Torrance proposed a checklist for assessing economic evaluations in health care. Each item in this checklist concerned a particular aspect of the evaluation that can be faced in different ways. In summary, studies may greatly differ for *cost evaluation*, *point of view*, and *evaluation of consequences* of the programs to be compared. It is true that a complete standardization is impossible in this field because of differences among health care organizations in different countries, and because some aspects of evaluation are still very controversial (i.e. the discount of future life years, the use of different utility measures and the inclusion of indirect costs and benefits, that can be difficult to measure). However, in a recent paper [3], Drummond discusses about the need of *some* standardization, mainly based on three motivations: 1) the maintenance of methodological standards, 2) the facilitation of the comparison of the results of economic evaluations for different health care interventions, and 3) the transferability of study results among different settings.

We underline two additional points. The first one is the multidisciplinarity of these studies. Typically, physicians and health economists are involved, but also statisticians, in particular epidemiologists, might play a role in assessing the efficacy of health programs, while psycologists and sociologists are involved in trying to assess scales for the evaluation of life quality. Most papers do not highlight the roles of the different professional figures, and do not describe the basic tools used for their inferences (i.e. data base, statistical packages, source of cost description, etc.). We think that making these aspects more explicit could help the communication among the different health care professionals.

The second point is the high rate of obsolescence of these studies. As a matter of fact, the rapid progresses in medicine and technology often produce variations in clinical practice that invalidate the hypotheses on which studies were previously based. Then, it is essential to ground these analyses on a general architecture incorporating specific models that are allowed to change, both in structure and quantification, so as to follow *natural* evolutions of the real setting. The architecture, which is fixed for a certain class of problems, is the basis for standardization. It is essential to establish both general principles, such as *who is responsible for which models or parts of a model*, and general model-related questions, such as *which information is needed* to estimate probabilistic parameters, and where this information is stored (i.e. data base tables or views, literature, personal communications, etc.). Within this frame, specific models will reflect specific realities and will *learn* from experience on such realities: clarity in both model building methods and aims will allow an easier comprehension and exploitation of results.

Cost/effectiveness (CE) studies concern a wide range of problems: screenings, technological park assessment, prevention programs, etc. The present study addressed the assessment of a general architecture for the CE analysis of therapeutic choice. Our opinion is that, through this architecture, a closest collaboration may be reached among the different health care operators, namely physicians and hospital managers, in order to solve some difficult problems. Our main purpose was to avoid a passive acceptance of solution proposals by operators, thus permitting a constant control on the quality of the current policy by using normative methodologies and objective data.

2 The architecture for dynamic Cost-Effectiveness analysis

Figure 1 shows both human and software agents involved in what we call *dynamic CE analysis*.

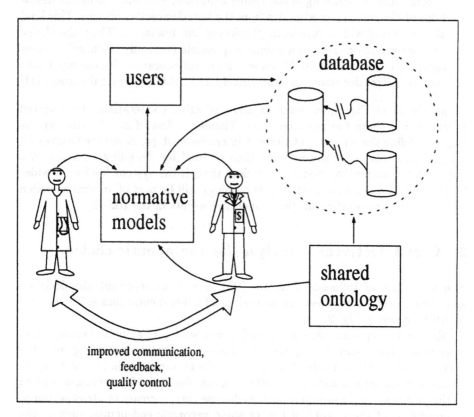

Fig. 1. The architecture of a system for performing dynamic cost/effectiveness analysis. Both human and software agents are shown

- The expert physician(s) (i.e. the team responsible for the knowledge about the available clinical treatments and outcomes) and the hospital manager(s) (i.e. the team responsible for knowledge about budget resources and management) provide the structure of normative models, such as influence diagrams or decision trees, and provide initial estimates of probabilistic parameters, costs, and utilities.
- The model is adopted for clinical routine. It may be a model that suggests strategies for the management of a problem in the whole patient population (policy-oriented model), in subsets of patients, or in the specific patient. The

suggestions coming from the model may or not be followed in the practice; in any case data about patients' outcome and follow-up are stored into local and/or remote data bases. These latter can be useful for collecting a sufficient amount of data in order to perform the requested statistics.

- The model parameters are updated by using learning algorithms exploiting the growing data base [10].
- A continuous monitoring of the model behaviour will allow to detect deviations of the parameter estimates from the initial expert's opinions. This kind of monitoring will involve both physicians and managers. Thus, the choice on costs would not rest on a unique responsible (typically the hospital manager), but this system could allow a more collaborative clinical input into capital budget decisions, an issue already highlighted by several authors [11]

As shown in Fig. 1, in order to allow effective cooperation, the involved agents should share the same ontology. Therefore, first of all, domain experts ought to define the relevant entities and interrelationships, necessary to describe the problem. These definitions could then be used for designing both the data base and the normative models. It is clear that, if the system fits into a wider hospital information system (HIS), its ontology will be part of, or shared with a wider ontology supporting all the cooperative agents in the HIS [8].

3 Cost-Effectiveness analysis for therapeutic choice

The formalism of Influence Diagram (ID) is used to represent the decision-analytic model for cost/effectiveness analysis of a therapeutic choice. The general model is shown in Fig. 2.

Rectangles represent decisions, circles represent probabilistic variables and diamonds value nodes. Firstly we can consider the network starting from the decision node in the left side of the picture. The therapeutic choice involves prediction of both desired and adverse effects. Examples of the former are, according to the problem, *final* end-points such as disease cure, increase of life expectancy, improvement of the quality of life, or some *surrogate* end-points, such as the normalization of some indices, like blood pressure or serum cholesterol, that are thought to improve the patient's health status. Examples of adverse effects are risk of death (i.e. surgical intervention mortality), acute toxicity, early and late therapy-induced morbidities. In the last cases, additional treatments are normally needed to cure side effects of the primary therapy. The knowledge for building this part of the model is normally provided by physicians, who also decide how detailed the model should be, i.e. the complexity of the causal paths from the decision node to the value node. In order to clarify this point, let us consider Fig. 3, that shows two detailed levels of knowledge representation.

From part (a) we are simply informed that a certain therapy may cause infections as an adverse effect. From part (b) we can understand *i)* how infection may be generated (due to immunodeficiency related to primary therapy), *ii)* what kind of infections can develop and *iii)* what are the signs and symptoms

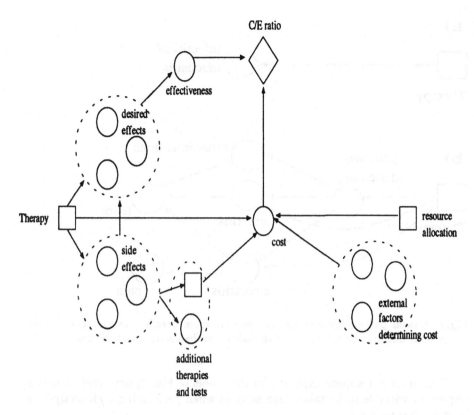

Fig. 2. The influence diagram formalism is used to represent the general task of performing cost/effectiveness analysis for a therapeutic choice; decision makers are both physicians (left side of the picture) and hospital managers (right side of the picture)

of infection. Thus, when knowledge is represented as in (b), identification of detailed diagnostic and therapeutic approaches is more straightforward, and this simplifies the process of costs determination. The drawback is that probabilistic quantification of the model becomes more cumbersome. As usual, a compromise must be reached between complexity and computability. In any case, for each ID node representing a pathological state, we can identify some of the possible cost components:

- Laboratory tests
- Drugs
- Medical Interventions
- Surgery
- Length of hospitalization

that may be convenient to consider separately, in order to obtain detailed cost analysis (see as an example the description of a system for calculating antibiotic cost [7]).

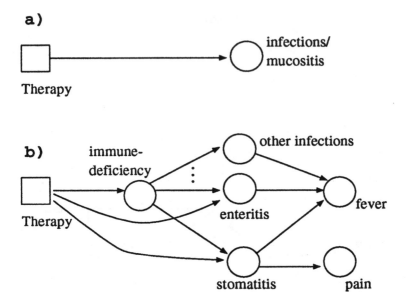

Fig. 3. Causal paths with different complexity: more details may simplify the cost assessment, but may be associated with difficult probabilistic quantification.

Time does not appear explicitly in the model of Fig. 2; nevertheless it is an important variable to be taken into account when performing C/E analysis, at least for two reasons:

1. the well-known discounting process, that allows to differently weight future costs and benefits with respect to actual costs and benefits. In fact it must be considered that if S ECU will be payed n years from now, and r is the discount rate, the corresponding actual amount is $A = \frac{S}{(1+r)^n}$. In the same way, future life years may be discounted [6];

2. to distinguish whether a pathological event is a side effect of a certain therapy or whether it is independent from it. We already mentioned that a data base will be exploited for updating the model probabilistic parameters. This implies the existence of automatic procedures able to extract from the data base both rough data and data abstractions that label the nodes of the ID. An example of these abstractions is *conditioning-related infections* (conditioning regimen is the pre-transplant treatment). Probably many infectious events are stored in the data base for a particular transplanted patient, related or not to the conditioning regimen. In a simple situation, a rule could take into account the timing of the event and establish the *conditioning-related infection* as *true* only if time interval from conditioning does not exceed, for example, 20 days.

Let us now look at Fig. 2 from the rigth side. Another decision node is shown, namely the manager competence. Costs reported in hospital price lists are normally computed by considering several factors, mainly grouped in two categories:

1. direct costs, i.e. salary of involved persons, with different professional competence (physicians, nurses, technicians, etc.), cost of consumer goods (X-ray plates, barium, radioisotopes, etc.), capital expenditure (buildings , big instruments, etc.);
2. general management cost (electric power, water, telephone, etc.)

Other categories, like indirect costs (time loss by patients or relatives, salary losses, etc.) and intangible costs (pain, anxiety, etc.) are normally not considered, due to the difficulty in their quantification.

Some of these cost components belong to the manager. He could decide, for example, to invest an amount of money either for buying a therapeutic instrument, or for leasing it, in the case it does not already exist, or for stipulating conventions with other health organizations that may furnish cheaper remote services. He could also decide to move some persons from a job to another, or reduce some working areas.

Other factors are outside the decisional power of the manager, being dependent on the economic situation, such as interest and inflation rate.

4   An example: the choice of the conditioning regimen for bone marrow transplantation

Figure 4 shows the ID created for performing cost/effectiveness analysis on the choice of conditioning regimen for bone marrow transplantation (BMT). In the case of a patient with acute myeloid leukemia, busulfan and total body irradiation (TBI) can be considered alternatives. The trade off can be summarized as follows: busulfan is very much cheaper than TBI (about 100 ECU vs 800-1300 ECU), but it can produce early adverse effects, whose treatment is very expensive. The efficacy of the two treatments can be considered similar. Let us describe the problem in detail, by using as a guideline the structure of the general model shown in Fig. 4. Effectiveness will be expressed in Quality Adjusted Life Years (QALYs).

One of the goals of the conditioning treatment is to prepare the recipient for accepting the donor's marrow. The desired effect is, therefore, the *engraftment* of donor hematopoiesis, that depends mainly on the histocompatibility between patient and donor (HLA-identity). Failure to engraft may either cause death or require a second BMT. Although previous reports suggested an higher efficacy of TBI in comparison to busulfan, no clear evidence exists at present on a difference in the engraftment probability in patients transplanted with an HLA-identical sibling after different conditioning regimens. As regards adverse effects, it is convenient to consider separately early and late morbidities. Among the former, infections, mucositis (mainly stomatitis and enteritis), interstitial pneumonitis (IP) mainly due to cytomegalovirus (CMV) infection, and venocclusive disease (VOD) should be included. Each of them requires additional tests

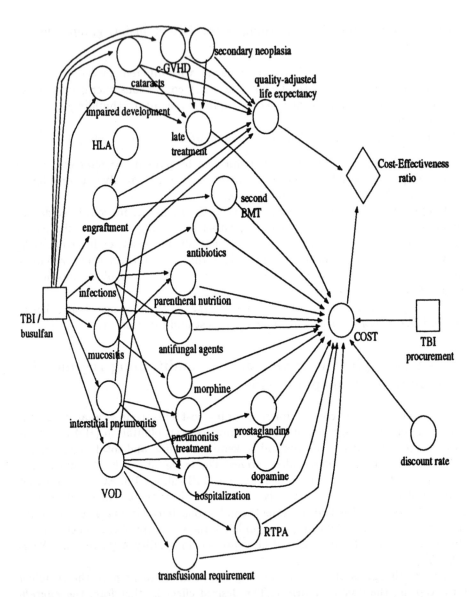

secondary neoplasia

c-GVHD

cataracts

quality-adjusted
life expectancy

impaired development

HLA

late
treatment

Cost-Effectiveness
ratio

second
BMT

engraftment

antibiotics

infections

parentheral nutrition

TBI /
busulfan

mucositis

antifungal agents

COST

TBI
procurement

morphine

interstitial pneumonitis

pneumonitis
treatment

prostaglandins

VOD

dopamine

hospitalization

discount rate

RTPA

transfusional requirement

Fig. 4. The Influence diagram for deciding about the conditioning regimen for a patient affected by acute myeloid leukemia and undergoing bone marrow transplantation. The choice will take into account early and late morbidities, survival, and costs of both primary treatment and side effect related treatment

and treatments, i.e. antibiotics and antifungal agents for infections, antigenemia monitoring for CMV infection, prostaglandin for hemorrhagic cystitis and
dopamine, prostaglandin and recombinant tissue plasminogen activator (RTPA)
for VOD. Treatment of VOD increases transfusional requirement. Some of these
side effects, namely VOD and IP, may be very severe and sometimes fatal.

Late morbidities are represented by growth failure, delayed pubertal development, cataracts, chronic graft-vs-host disease (c-GVHD), that mainly impair
quality of life, and secondary malignancies, that can also decrease life expectancy.
Some of these morbidities require specific treatments such as hormonal replacement therapy for growth failure and surgical repair for cataracts, while other
morbidities such as c-GVHD may have a wide range of organ involvement, and,
thus, the required treatments are hardly predictable.

The decision node on the right side of Fig.4 represents a choice for the manager: a linear accelerator or a Cobalt source for TBI are very expensive instruments, that may not be available in the center where BMT is performed. Thus,
patients have often to be transferred to the nearest equipped center, with additional discomfort and cost. For this reason, it is worth of investigation how the
acquisition or the leasing of the instruments above mentioned would affect the
C/E ratio.

Of course, the network with tables of conditional probability values, mainly
for late morbidities and cost of their treatment is difficult to assess. Again, we
underscore the opportunity of having a learning system. As long as data are
collected about patients' follow-up and real costs, outcome forecasting for new
patients becomes more and more reliable.

4.1 The system advice

In this section preliminary results obtained for the problem described above
are presented. Tools for implementation are GAMEES [1], as the editor for the
ID, and NBS (Net Base System, a public domain software) as the relational
data base. The link between the ID and the data base is performed through an
interface that exploits the shared ontology in order to use appropriate data or
to transform rough data into the abstractions represented into the ID, with the
final goal of updating the model parameters.

The following results have been obtained by using a set of simulated data,
since the size of the actual database does not allow sensible statistics to be
performed.

Prior estimates provided by the domain experts have led to the following
result:

Perform conditioning with busulfan, independently of TBI cost.

This advice is illustrated in Fig. 5a, which shows that busulfan has an expected C/E ratio lower than that of TBI, for a sensible range of TBI cost. Actual
cost of TBI is about 900 ECU. Cost in case of leasing or whole capital investment
are calculated by suitable formulas. In order to verify how the model can learn
from the acquired information on patients, we simulated data from probabilistic
distributions similar to the prior, with the exception of development of VOD,

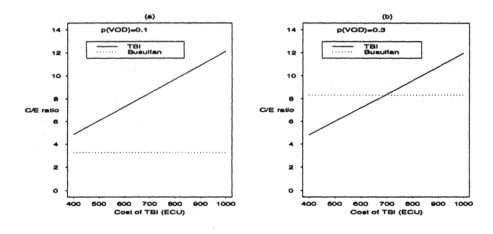

Fig. 5. Expected C/E ratio of the two conditioning regimens as a function of the TBI cost, a) before learning from data b) after learning form data

for which we increased the probability in case of busulfan conditioning. In other words, let us imagine that many patients undergo busulfan treatment, and more patients than expected developed VOD as a side effect. After learning, the estimate of the VOD development probability raised from 10% to 30% for patients treated with busulfan, while it did not vary (from 8% to 7.8%) for patients receiving TBI. As a result of this learning process, the sensitivity analysis (Fig. 5b) shows that a slight decrease in TBI cost could lead to the suggestion of TBI as the best cost/effective choice. As a further support of this observation, inspection of marginal probabilities has been performed. As a result, hospitalization and probability of death represented the most important factors in determining the increase in the C/E ratio for patients given busulfan.

5 Discussion

According to Paul Ellwood *"the health care system is an organism desperately in need of a central nervous system that can help it to cope with the complexities of modern medicine"* [4]. Using this metaphor, one of the major aims of the nervous system is to collect information and to synthesize it in order to perform a task in the most effective way. The system proposed in this study is intended as a mean for gathering and elaborating data on patients' outcomes and costs with the aim of providing physicians and hospital managers with a synthetic information that can influence the decisional process among alternative health programs or strategies.

This work motivates from the observation that hospital information systems are more and more diffused, and the potential of large data repositories is only partially exploited. It is rare that existing information systems are used to perform data retrieval other than those for routinary tasks (i.e. reports print, patient's observations display, etc). Coupling data bases with decisional models may represent a progress towards a more intelligent use of information systems.

A great amount of literature on decision psychology, trying to highlight factors affecting the *practice styles* or the *clinical decision rules* has been published: in particular Poses et al. [9] showed that improving physician's knowledge of probabilities may not affect their decisions. These observations could belittle the use of quantitative models, such as the one proposed in our study, for medical decision making. Nevertheless, issues such as quality control, cost-effectiveness verification, local responsibility for the hospital health policy are becoming mandatory within many health care systems and thus both physicians and managers will be somehow *forced* to consider them. Therefore, they will need systems helping them in these tasks, in particular when the latter imply long-term predictions about patients' outcomes and costs.

Of course future work is needed, in order to make computerized architectures such as that described here useful not only for cooperation among health care operators of the same hospital, but also for *world-wide exchange* of ideas and results. Different groups could gain access to different models through telematic networks, and consult them, after opportune tailoring towards their specific context. Both consultation and tailoring implies that these architectures must incorporate *i)* tools for sharing ontologies among groups with common interests *ii)* tools for performing different kinds of analysis on the same issue and for comparing results (for example, authors use different measures for the quality of life, thus it could be useful to compare the same data translated into risk-neutral QALYs, risk-adverse QALYs, health years equivalent, etc.), iii) tools for interpreting the different data from which to compute life expectancy (survival time, percent survival, death risk, etc.), and for managing different data base from which to learn.

Acknowledgments

This work is partially supported by the European Community, through the AIM project 2034 GAMES II - A General Architecture for Medical Expert Systems. We thank Dr. Patrizia Comoli for her valuable suggestions and comments.

References

1. Bellazzi R, Quaglini S, Berzuini C, Stefanelli M. GAMEES : a probabilistic environment for expert systems. Computer Methods and Programs in Biomedicine 1991, 35:177-191.

2. Drummond MF, Stoddart GL, Torrance GW. Methods for the Economic Evaluation of Health Care Programmes. Oxford: University Press, 1987.
3. Drummond M, Brandt A, Luce B, Rovira J. Standardizing methodologies for economic evaluation in health care, Practice, problems and potential. International Journal of Technology Assessment in Health Care, 9:1 (1993), 26-36.
4. Ellwood PM. Outcome management: a technology of patient experience. New England Journal of Medicine, 1988; 318:1549-1556
5. Horngren CT. Cost accounting: a managerial emphasis. Prentice Hall, Englewood Cliffs, N.J., 1982.
6. Keeler E, Cretin S. Discounting of life savings and other nonmonetary effects. Management Science 29, 3:300-306.
7. Kerr J. The antibiotic cost calculator; an expert system for global antibiotic cost calculation Computer Methods and Programs in Biomedicine 1995, 46:13-21.
8. Lanzola G, Falasconi S, Stefanelli M. Cooperative Software Agents for Patient Management Proc. of the AIME95 Conference, 1995.
9. Poses RM, Cebul RD, Wigton RS. You Can Lead a Horse to Water-Improving Physicians' Knowledge of Probabilities May Not Affect Their Decisions Medical Decision Making 1995, 15:65-75.
10. Spiegelhalter DJ, Dawid AP, Lauritzen SL, Cowell RG. Bayesian Analysis in Expert Systems. Research report 92-6, MRC Biostatistics Unit, Cambridge (U.K.), 1992.
11. Stetler C.B., Fagan J., Hanson M., Biancamano J., Curry S. Patient Centered Redesign. International Journal of Technology Assessment in Health Care, 10:2 (1994), 227-234.

Model-Based Application: The Galen Structured Clinical User Interface

Laurence Alpay[1], Anthony Nowlan[3], Danny Solomon[4], Christian Lovis[2], Robert Baud, Tony Rush[3], Jean-Raoul Scherrer[1]

[1] Medical Informatics Centre, Geneva Hospital, Geneva, Switzerland

[2] Medicine Department, Geneva Hospital, Geneva, Switzerland

[3] Medical Products Group, Hewlett Packard Ltd, Bristol, U.K

[4] Medical Informatics Group, University of Manchester, Manchester, U.K

Researchers in the Artificial Intelligence in Medicine community are facing the challenge to design and develop clinical systems which are intuitive to use and adequately expressive. This paper reports on a Structured Clinical User Interface (SCUI) which achieves this goal by using technologies developed in the Galen project: a new formalism to represent models of terminology coupled with a server to access and use this knowledge. This approach opens the way to the development of clinical systems which use conceptual knowledge, driven by models of terminology, in a dynamic and flexible way. Furthermore, the overall task of building a clinical application is separated into a terminological part handled by the server and an application part handled by application developers without needing to worry about implementing the terminology. The SCUI is specifically developed and tested in the context of infectious diseases to satisfy the demands made by the medical intensive care unit to the Geneva Hospital's microbiology laboratory.

1 Introduction

The SCUI is developed in the context of the European AIM project Galen. The purpose of Galen is to establish the foundations for the next generation of coding systems to support integrated clinical information systems. The heart of Galen is a Semantical Encyclopaedia of Terminology (SET). The SET consists of a generative Master Notation (also referred to as GRAIL for Galen Representation and Integration Language) and a Coding Reference (CoRe) model for medical terminology formulated in that notation [1], [2] and information modules providing multilingual lexicons, conversions to and from existing coding systems. The SET is encapsulated by an application programming interface to form a general Terminology Server (TeS) for use by other applications in a client-server environment [3], [4]. The SCUI is a test bed for the Galen technologies; and as such it is aimed to provide input (to the requirements) for the bottom up development of the Master Notation and the CoRe model, and to evaluate the effectiveness of this notation within a clinical setting. Furthermore, the SCUI [5] is initially implemented as a bilingual system (French and English) to provide a means of testing the multilingual module of the TeS [6].

2   Context of the SCUI

2.1   Clinical Setting

The SCUI is intended to collect clinical information of the infectious disease domain in a large university hospital like the Geneva Hospital. We have specified *nosocomial infections* within the infection diseases specialty as the domain of application. Nosocomial infections or hospital infections are best defined as infections which are neither present nor incubating at the time of the admission to the hospital, and which are usually developed at least 48 hours following the patient's admission [7], [8]. Detailed definitions for various types of nosocomial infections have been established by the Centers for Disease Control (CDC) in Atlanta in 1988 [9]. Such infections are relevant since they increase patient morbidity and mortality as well as hospital costs [8], [10]. Being acquired in the hospital, they represent a major objective for the development of surveillance systems [11]. Within the domain of infectious diseases, we have focused on the subdomain of urinary infections. There are a number of reasons for choosing this infectious subdomain. First, there is a strong interest from clinicians for this type of infection. In addition, urinary infectious diseases are very frequent and easy to detect. Furthermore, urinary infections are typical nosocomial infections and represent an additional problem to the patient's morbidity. Finally, the symptomatology of urinary infections, as well as the necessary laboratory tests, are well known.

We have chosen the *medical intensive care unit* (MICU) as the target ward for the development of the demonstrator. Patients hospitalised in intensive care units are at a much higher risk of developing nosocomial infections than in more general wards: infections acquired in these wards account for more than 20% of nosocomial infections, although these care for less than 10% of hospitalised patients [12]. The urinary tract as an infection site counts for 24% and reasons for higher infection rates in intensive care units include, among other things, urinary bladder catheterization [13].

2.2   Design Considerations

The development of the SCUI is centered around requests made to the microbiology laboratory by the MICU. The initial aim of the SCUI is to improve requests made by the MICU by asking significant and relevant clinical information about the patients [14]. The current procedure at the Geneva Hospital for ordering a microbiology examination is as follows: 1) the health care provider in the MICU makes a microbiology request; 2) the request goes through the Geneva Information System for laboratory exams called UNILAB [24] (a part of the DIOGENE Hospital Information System [15]); 3) the request arrives at the laboratory, is dispatched, and investigations are made until results are obtained; 4) results are collected, validated, and returned to the clinical service; 5) the clinical service interprets the results, takes action following the test results (e.g. treatment, no treatment, other investigations) and may make a new request.

This current procedure has some drawbacks in that some clinical information (like state of immunity, state of illness, presence of fever) which would be useful to the laboratory is usually missing. Morever, the data collected during a microbiology laboratory request are not analyzed further, so that the clinical staff can ask for additional information. With the SCUI, the requests made to the microbiology laboratory should be improved: On one hand, the microbiology laboratory needs to gather relevant clinical information with the test requested in order to increase the chance of identifying an infectious agent. On the other hand, the clinician aims at receiving a better and faster service from the laboratory in the process of microbiological test requests.

2.3 Clinical Information Requirements

The SCUI is applicable whenever a urinary infection is suspected in the clinical ward. The types of data to be collected while making a request to the laboratory have been grouped into the following sections. It is much more detailed than the information usually presented and is an important aim to make detailed patients' information available to the user where necessary:

1) GENERAL INFORMATION e.g. the patient's mobility, consciouness, dependency;

2) URINARY HISTORY e.g. the patient's past urinary infections and any related treatment, as well as information about any event of lithiasis and nephrectomy;

3) SIGNS AND SYMPTOMS e.g. general findings such as fever, chills; specific findings such as alguria, dysuria, haematuria;

4) DIAGNOSTIC INFORMATION e.g. information on the diseases related to urinary infections such as cystitis, cystopyelitis, pyelonephritis, renal abscess, etc;

5) PREVIOUS LABORATORY RESULTS e.g. under the form of interpreted information about hematology, information about the urine such as the estimation of the urine red cells, the urine white cells, the urine glucose, the urine pH and the urine proteins;

6) SAMPLE COLLECTION e.g. general information on the sample taken such as the date when it was taken, the type of examination, the type of sample, the operator, the place where the sample was taken (the emergency room, the operating room); information on the justification of the exam such as monitoring; information on the sample mode used such as direct mode, permanent urine bladder catheter, transitory urine bladder catheter, or renal catheter; information on the status of the catheter, i.e. clean or dirty, whether it is old or new, and so on.

The SCUI application guides the user into collecting specific detailed information, that is, to use the terminology in the CoRe model which is required in order make a microbiological laboratory request. In other words, the CoRe model contains the medical knowledge needed to collect this detailed patient's data (see Section 3.2). The SCUI makes use of the Terminology Server to access the CoRe model. The server provides the application with a set of terminological services such as conceptual and multilingual (see Section 3.3). The detailed clinical information collected from the user is then stored into the SCUI patients' medical records. Once the request has been made, it is available to the microbiology laboratory.

3 Terminology Server

3.1 Terminology Modelling and Representation

The building of a Terminology Server, which is able to conciliate a number of classifications and nomenclatures, requires a powerful knowledge representation schema and accordingly a representation language. GRAIL is such a language, primarily intended to model terminology. It is used in the development of medical domains, and one of the final goals is to design a large medical model in the form of a semantic network in order to accommodate all sensible medical expressions. Two processes are available for this task. First, a sanctioning mechanism allows the specification of what is medically sensible to say, i.e. what is not sanctioned is not accepted. Second, an indefeasible definitional schema is ready to grasp the reality in the model. Any expression in the network is later made available to the outside world in a canonical form [1]. For example, the concepts nephritis and oliguria are defined as follows in the CoRe model, where *which* and *name* are two GRAIL key words: *which* to introduce the definition of a new concept, as a specialization of existing concepts and *name* allows to name that concept.

(InflammatoryProcess which hasLocation Kidney) name Nephritis.

(Micturition which hasFeature (ProcessActivity which hasState Decreased)) name Oliguria.

3.2 Conceptual Models

Given the clinical requirements described in Section 2.3, the SCUI needs to use a variety of concepts. This means, for example, that the TeS needs to know about urinary devices, signs and symptoms (specific and general) as well as geographical locations (e.g. operating room) and actors (e.g. nurse, doctor). In other words, the TeS is to tell the application what can be sensibly and necessarily said about a concept like dysuria, to support the SCUI interface to help the clinician to enter the information. The GRAIL formalism makes a separation between the definition of a concept and its properties. For example, for the concept dysuria, the TeS will obtain from the CoRe model the definition of dysuria, i.e. pain which occurs during micturition (micturition being itself a complex concept), and properties of dysuria like its causal agent and the types of bacterium that causes it. The collection of clinical information for dysuria by the user is then structured and driven by what comes from the model. Furthermore, the data collection can be made more or less detailed according to what the end-user wants to express about some particular clinical information.

Although it is expected that the CoRe model will be able to support the terminological needs of the application, it is acknowledged that the application has terminology requirements which remain "local" to the application. These local requirements are twofold. First, the terminological extension model i.e. extra detailed concepts specific to the application and which are not found in the Core model. Second, the dialogue model which uses the terminology as a "hook" to add extra information like what concepts are interesting in a particular section of the SCUI application. This results in a situation where there is a terminology "linkage" between the CoRe model and the SCUI specific model. For instance, the CoRe model will be able to provide the application with sensible statements of what can be said about dysuria, while the SCUI model will specify that dysuria should be asked for in the "Signs and Symptoms" section of the application.

3.3 Services of the Terminology Server

The Terminology Server [3], [4] can be viewed as a networked interface to a set of terminological services. The TeS contains different modules, each of which provides a specific set of services to clinical applications, for example the SCUI (see Figure 1.).The *Concept Module* (CM) [16] provides GRAIL concept services: it allows the navigation and manipulation of expresssions in the CoRe model to answer questions like, for example, "what can be said about dysuria".

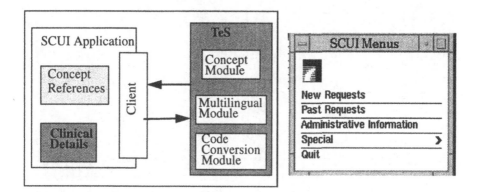

Figure 1. Connection SCUI & TeS   **Figure 2.** SCUI Menus

The *Multilingual Module* (MM) [6] provides naming services: it generates natural language (e.g. mostly in English and in French in the current version) from expressions in the CoRe model. For example, from the expression: "Ulcer which hasLocation Stomach" the generator will produce "stomach ulcer" in English and "ulcère de l'estomac" in French. This module provides the application with a language service, particularly useful when developing a single application to be used in more than one natural language context (see Section 4.3). The *Code Conversion Module* (CCM) [17] provides coding services: it relates concepts of the CoRe model to codes in existing systems such as ICD-9 or Snomed. For instance, the CCM can answer the request "return the code for the concept: Ulcer which hasLocation Stomach" with ICD-9 531.1.

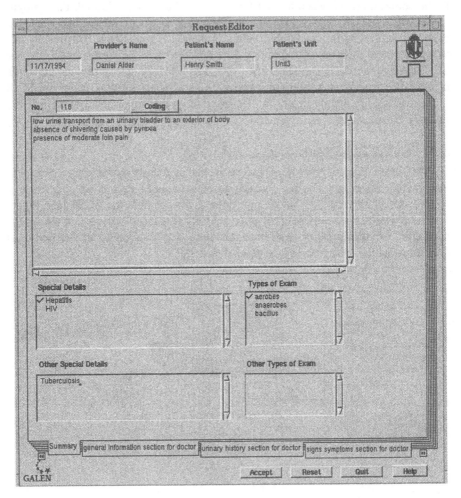

Figure 3. Example of the main page of the request editor
while a request is being made

The main page of the request editor includes: a patient location pane at the top; the request number with the set of data which has been collected so far by the user and generated in natural language; the option to view codes for these data; special details and types of exam to be selected from given lists or to enter as free text into a text editor. The buttons to access the clinical sections and the current summary pane are shown on a slider. By clicking on a section name, the model-based interface for that section appears (see Figure 4). Control buttons to save or dismiss the request are at the bottom of the window.

4 SCUI Application

4.1 Overview

In order to run the SCUI, the application needs to establish a connection with the TeS. A *connection session* between the application and the TeS consists of: connecting to the server, performing a series of requests (e.g. specify a terminological operation like "specialise relationships" to get the attributes that are used to specialise a given concept), and disconnecting from the TeS. By connecting to the server, the application becomes its client (see Figure 1.). The TeS which runs as a network server, communicates with the SCUI client using TCP/IP over sockets [16]. The SCUI client is provided with a library which is used to interact with the server. The SCUI is developed in Smalltalk/VisualWorks^{TM} .

We have separated the overall task of building the clinical application into two: a) a terminological part that the TeS handles and b) an application part which can be built by the application builders without needing to worry about implementing the terminology. This separation constitutes a new approach in the task of developing clinical applications. The clinician or the nurse starts the SCUI application by giving his/her user name and password. From then on, the clinician interacts mostly with the application using the mouse. A set of menus (see Figure 2) is available to the user. For example, the user can enter a new SCUI request (see Figure 3), or visualise a previous request and possibly restart a request which was not complete. The user can also access administrative information such as the patient's personal details and the patient's SCUI medical record, and make use of configurational options (e.g changing font or language) found under the Special menu (see [18] for details).

The SCUI connects to the TeS whenever, depending on the user's action, it needs to access concepts from the CoRe model and in a given language. Figure 3 shows the main page of the request editor where a set of clinical information has already been collected. These data are presented in a natural language form generated from the MM of the server. To get the model of a specific section (see Figure 4), the user clicks on the button with the section name on it e.g. signs symptoms section.

4.2 Model-Based Interface

There are two important aspects to stress here about the interface related to the conceptual models (found in the request editor window). First, the interface construction for each section of the SCUI scenario (see Section 2.3) is not fixed but rather is constructed *dynamically* based upon what comes from the CoRe model via the TeS. This is an advantage of the system over traditional form-based interfaces. The application is much more flexible than traditional systems and much better tailored to the particular clinical situation as it has access to a server of terminology rather than to static terminology. This means that if a new concept which is of interest to the SCUI is entered in the CoRe model, it will appear in the corresponding section.

Second, when making a request, the clinician clicks on the various concepts and associated attributes which are available to him/her and which are relevant for the given patient. By doing so, the clinician constructs simple or complex concept entities to express and detail the patient's clinical information. For example, he/she can express the fact that a loin pain has a moderate severity (LoinPain whichHasSeverity Moderate in GRAIL), and thus builds GRAIL expressions. Figure 4 shows an example of the model-based interface of the request editor for the "Signs Symptoms Section" of the SCUI application. The user clicks on the concepts and their corresponding values e.g. rigor and absence.

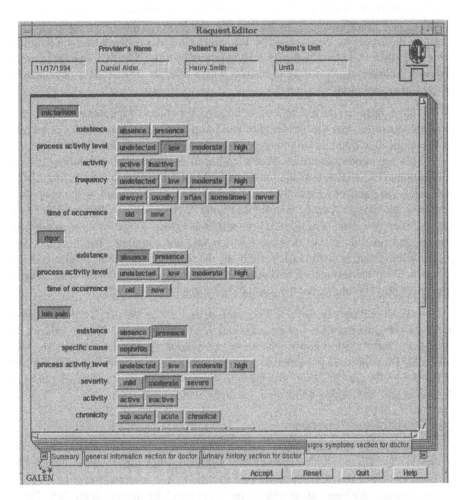

Figure 4. Example of the model-based interface of the request editor

Figure 4 shows elements of the model for the "Signs Symptoms Section" of the SCUI application. Patient's detailed clinical data are collected by selecting elements of the GRAIL model with the mouse. Each concept like rigor is associated with one or more relations like existence and its corresponding values e.g. presence, absence. The data collected and generated in natural language are shown in the summary pane (as illustrated in Figure 3).

4.3 Other Particularities

Multilingual Feature.

The work carried out on the MM module of the TeS has focused on the generation for English and French. Preliminary experiments with other languages (such as German, Dutch, Finnish, Italian and Spanish) have started for the generation of noun phrases. The results are encouraging. Figure 5 shows the screen dumps for the phrase "presence of acute cystitis" as generated by the MM in various languages and which is displayed on the summary pane of the request editor window.

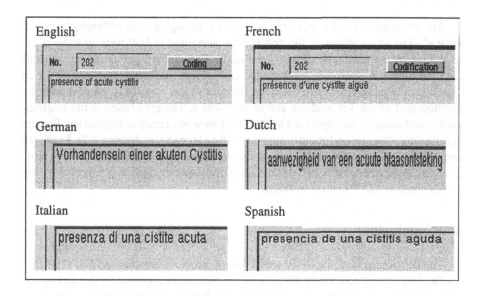

Figure 5. Examples of the use of TeS multilingual services

for the SCUI application

The noun phrase "presence of acute cystitis" generated by the MM of the TeS in various European languages and as it appears to the end-user on the summary pane of the request editor.

Types of End-Users.

As part of testing the Galen technologies, the SCUI application has the option to view different models according to the kinds of clinical users i.e. a nurse or a doctor who is using the application. This switch is important to illustrate the flexibility of the application to use different conceptual models according to the kind of users who is running the application. In the current prototype, the nurse has only four clinical sections against the doctor's six, and the contents of all the clinical sections are different: there are concepts which appear on both forms and also on one but not the other. The model shown in Figure 4 is for a user-doctor, and the section names in Figures 3 and 4 also specified that there are for a user-doctor.

Tools for customizing the clinical sections.

The SCUI includes a tailoring tool which allows the clinician to customize the contents of the clinical sections according to his/her needs [18]. In particular, the aim of this tool is let the user add and remove concepts from a clinical section, and to create a new section if necessary. This tool makes use of a complementary tool called Confuser (CONcept Finder for USERs) which contains a series of functionalities to browse through the CoRe model [23].

5 Prototypes and Evaluation

The development of the SCUI has been going through a cycle of "prototyping and testing". The expectations of the first version described in [19] were to explore the modelling process and make use of some of the Terminology Server services such as the multilingual module. The work for the first version produced requirements for an "intelligent" dialog. The current version (version 2) has evolved in parallel with the development of the Tes modules and of the models. The application is run multilingual: annotations of concepts used by the SCUI have been done in English and French, and that task has started for other languages like German (see Section 4.3). Feedback and comments on the prototypes have been obtained from doctors at the Geneva Hospital and within the Galen consortium. Larger scale user evaluation is expected for the current version SCUI in the near future.

6 Conclusions

The development of the SCUI stems from a previous experience [20], [21]. In a similar line of research, the SCUI presents an innovative approach in the design and development of clinical systems: it not only has features which are not usually found in other traditional systems (in regards to its dynamic interface construction), but also has access to new technologies (through a range of terminological services).

Key aspects relevant to the SCUI include the following: First, the application uses a separate generic server which provides terminological services (e.g. multilingual) and which can be used by other applications simultaneously. This corresponds to a new approach in the development of clinical systems and the sharing of medical knowledge [22]. Second, the conceptual models which the SCUI makes use of are developed with the new formalism GRAIL for representing conceptual knowledge. Third, the SCUI exploits new technology for model-based interface applications, and finally, the application deals with collecting patients' detailed clinical information. Although the SCUI is still at the prototype stage, it provides promising grounds and is an illustration of how the Galen technologies can be used. Moreover, the SCUI offers a framework for the construction of practical and usable medical applications in the future.

Acknowledgements

The current developments, carried out at the Geneva University Hospital, are part of the GALEN project (A2012) and have been made possible thanks to the funding of the AIM programme of the EU, for which the Swiss government (CERS) has given full support. Galen documents are available from the Medical Informatics Group of Manchester.

7 References

[1] AL Rector, WA Nowlan, A Glowinski, G Matthews. *The Master Notation.* Galen Deliverable 6, March 1993.

[2] AL Rector, WA Nowlan, S Kay. *Conceptual Knowledge: The CoRe of Medical Information Systems,* MEDINFO 92, K C Lun et al. editors, North Holland, pp1420-1426.

[3] TW Rush, W Claassen, I Piumarta, P Pomper,WA Nowlan, P Zanstra. *Consolidated Specification of the Terminology Server and PUPPI.* Galen Deliverable 9, June 1993.

[4] AL Rector , WD Solomon, WA Nowlan, TW Rush. *A Terminology Server for Medical Language and Medical Information Systems.* IMIA Working Group 6 on Natural Language and Medical Concept Representation. Vevey, Switzerland, May 94.

[5] LL Alpay, C Lovis, WA Nowlan, WD Solomon, TW Rush, RH Baud, C Juge and J-R Scherrer. *Constructing Clinical Applications: The Galen Approach.* Accepted to MEDINFO95, Vancouver, Canada.

[6] J Wagner, A-M Rassinoux, R Baud, J-R Scherrer. *Generating Nouns Phrases from Medical Knowledge Representation.* MIE 94, Lisbon, Portugal, May 94.

[7] RL Thompson. *Surveillance and reporting of nosocomial infections.* In: Prevention and Control of Nosocomial Infections. Wenzel RP (ed.), Williams & Wilkins, Baltimore, 1987, pp. 70-82.

[8] G Ayliffe, E Lowbury, A Geddes, J Williams. *Control of Hospital Infection.* A Practical Handbook. Chapman & Hall Medical, London, 1992, pp. 1-11.

[9] J Garner. *CDC definitions for nosocomial infections, 1988.* J. Infect. Control 16 : 128-140, 1988.

[10] W Schaffner. *The global impact of hospital-acquired infections.* In: Prevention and Control of Nosocomial Infections. Wenzel RP (ed.). Williams & Wilkins, Baltimore, 1987, pp. 13-18.

[11] R Wenzel, S Streed. *Surveillance and use of computers in hospital infection control.* J. Hosp. Infect. 13 : 217-229, 1989.

[12] D Pittet, L Herwaldt, R Massanari. *The intensive care unit.* In: Hospital Infections. Bennett JV, Brachman PS (eds.). Little, Brown, and Company, Boston, 1992, pp. 405-439.

[13] R Weinstein. *Epidemiology and control of nosocomial infections in adult intensive care units.* Am. J. Med. 91, suppl 3B, 179-184, 1991.

[14] E Safran, L Alpay, C Lovis, R Baud, J-R Scherrer. *User Centred Design and Requirements for and Specification of Structured Clinical User Interface (SCUI).* Galen Deliverable 5, November 1992.

[15] Scherrer J-R, Baud RH, Hochstrasser D, Ratib O. *An Integrated Hospital Information System.* MD Computing, 7, 81-89, 1990.

[16] WD Solomon, TW Rush. *The consolidated Terminology Engine and SET Tools and Terminology Server version 2.* GALEN Deliverable 15, March 1994.

[17] F Lorino, A Rossi Mori, P Agnello, E Galeazzi. *The Consolidated CoRe model and Code Conversion Information Module.* GALEN Deliverable 16, January 1994.

[18] L Alpay. *SCUI User Guide.* Galen Documentation. October 1994.

[19] L Alpay, C Lovis, J Wagner, P-A Michel, R Baud. *SCUI version 1 and Evaluation.* Galen Deliverable 11, November 1993.

[20] WA Nowlan, AL Rector, C Goble, B Horan, T Howkins, A Wilson. *PEN & PAD: a Doctor's Workstation with Intelligent Data Entry and Summaries.* SCAMC 90, RA Miller (ed), IEEE Press, Washington, pp941-942.

[21] WA Nowlan, AL Rector, S Kay, B Horan, A Wilson. *A Patient Care Workstation Based on User Centred Design and A Formal Theory of Medical Terminology: PEN & PAD and the SMK Formalism.* AMIA 1992, pp855-857.

[22] WA Nowlan, AL Rector, TW Rush, WD Solomon. *From Terminology to Terminology Services.* SCAMC 94, J.G. Ozbolt (ed), Hanley & Belfus publishers, Philadelphia, pp150-154.

[23] S Bechhofer. *Confuser User Guide.* Galen Documentation. July 1994.

[24] Lagana M, Berney J-P, Schulthess P, Nawrocki B. *Guide Utilisateur UNILAB.* Internal Report. September 1993.

A Knowledge-Based Modelling
of Hospital Information Systems Components

Henry KANOUI[1], Michel JOUBERT[1,2], René FAVARD[1]

1. IIRIAM. Technopôle de Château-Gombert. Europarc, bat. C. 13453 Marseille Cedex 13. France
2. CERTIM. Faculté de Médecine. Boulevard Pierre Dramard. 13326 Marseille Cedex 15. France.

In this paper, we point out the necessity for large information systems, and especially hospital information systems, to encompass a knowledge-based model of the domain covered. We discuss the characteristics of such a model and present the knowledge representation adopted in previous projects. The XQL formalism, which enables application programs to query the model at run-time, is then introduced. The theoretical model and operational semantics of XQL are presented and discussed.

1. Introduction

The ultimate goal of an Information System (IS), in the hospital as well as in any organisation, is to provide the professional end-users with efficient means to access, structure and retrieve the professional information they need to achieve their tasks. In addition to traditional relational and documentary technologies, new approaches as hypertexts (more generally hypermedia) have recently emerged and come into wide use. However, in front of the variety and the huge amount of information concerned, the problem of overload as well as the problem of information sharing still remains prominent. The study of large information systems shows that the variety and the complexity of information databases make it mandatory the provision of an efficient support to information retrieval by the means of intelligent navigation tools and suitable user interfaces. In other terms, the professional users of information systems expect this latter to behave as an intelligent assistant with which a high-level dialogue can take place. This entails the necessity for the IS to encompass part of the domain-knowledge covered and of the professional know-how of the users.

In addition, and this is particularly true in the case of Hospital Information Systems (HIS), users of such systems are high-level professionals, who have a very large autonomy in performing their tasks. This requires the IS to be very flexible in order to really personalise the organisation, the processing and the display of information it manages. In summary, a HIS has :
- to efficiently support the professional users,
- to be customisable to the particular needs of these latter,
- to manage not only multimedia information databases, but also knowledge bases,

- to offer a dynamic model of the information managed, directly exploitable by users applications.

The considerations above lead to consider a formal representation of the information databases in terms of abstract objects, classifications, semantic relationships and behaviours with respect to some situations, events or operations. Such a representation is nothing else that a knowledge-based model of the information system components, sufficiently flexible and dynamic to be activated and exploited at run-time by the application programs. In summary, such a model acts as a mediator between the users (represented by application programs) and the information databases by achieving the "mapping" between domain objects pertaining to the user's professional paradigm and the actual data stored under the form of relational tables of multimedia documents.

This approach has been investigated and implemented in the past few years in the frame of several projects, in particular RICHE (Esprit n° 2221) and NUCLEUS (AIM n° 2025), both dedicated to the area of hospital information systems [1,2]. It allows:
- to extend the relational model of information to a knowledge model,
- to structure and handle domain objects at a high abstraction level,
- to formulate requests of information at an equally high level of abstraction.

A dedicated query language has been designed and implemented which processes queries to the knowledge bases at run-time level [3]. The queries regard both the consultation of knowledge bases and their exploitation by users applications.

2. The Knowledge Representation Model

The knowledge representation model has been directed from requirements of HIS. There was a need to represent concepts expressed as classes of objects and semantic relationships able to model concepts of the hospital domain. Moreover, arrangements of objects in taxonomies supporting attributes values inheritance were strongly required. Finally, the usage of a model built along these lines appeared to be quite particular. Unlike the "traditional" AI approach as usual in expert systems where the model is "wired" in the application itself (under the form of objects or rules), the knowledge-based model for information systems (especially HIS) has to be dynamically invoked by application programs, that is queried and exploited at run-time. All these considerations led to the adoption of an object-oriented approach which has been implemented by relying on the original notion of Knowledge Servers.

The knowledge representation model is built in successive levels of abstraction from generic types of objects (class-models and binary relationship-models). These elementary objects allow in particular to build semantic nets which are the support of hypertextual-like navigation. This latter is therefore a functionality offered to any application built upon the kernel above mentioned. The generic models of objects are then refined into models depending of the application domain. In general each of these latter is associated with a concept of the domain. The notion of complex objects, compound objects and taxonomies, is then introduced. As usual in the object-oriented approach, the structure of the objects modelled is described in terms of typed attributes, possibly multivalued. When defined for a model, this structure is inherited by the sub-models. The same goes when a binary relationship has been declared between two models.

2.1. Generic types

They are the very basic elements which serve to model a hypernet (hypertextual/hypermedia network). They are :
- *Class-models*: the model of objects which are the nodes of a hypernet. Among class-models attributes, *content* refers to the information actually carried by the node. In general, *content* refers to an application dependant information encapsulated in the node.
- Binary *Relationship-models*: the model of objects which are the arcs of a hypernet. Among binary relationship-model attributes, *source* and *destination* refer to nodes of the hypernet. Each binary relationship-model is associated with the dual relationship. Binary relationship models is subdivided into *Structural-relationship-models* and *Semantic-Relationship-models*. The first sub-category features system built-in relations which serve to model structured objects. The other one features associations having a precise meaning in the considered domain.

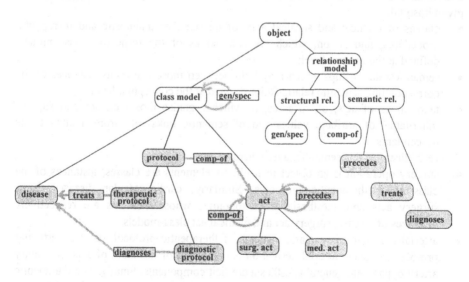

Figure 1. The general knowledge model

2.2. Structured types

Structured types model complex objects built by assembling together other objects. To do so, two types of structural relationships are defined: composition (*composed-of*) and generalisation/specialisation (*gen/spec*). They allow to model the following structured objects:

- *Compound objects*: the model of objects built by associating, via the composition relationship, other objects designated as the components of the compound object. The class which each component depends of is determined in advance. Each component is given a name. The compound objects are a means to group several objects into a single one, to designate and to handle the group under a unique name. A compound object is a heterogeneous structure, since its components are not necessarily of the same class-model.
- *Taxonomies*: the model of objects which feature, via the generalisation/ specialisation relationship, a classification, with respect to a given criteria, of a set of objects depending of a same class-model. A taxonomy is thus a homogeneous structure.

Both generic types and structured types are summarized in figure 1, where white boxes represent generic types and greyed boxes more specific types. For instance, taxonomies are defined through the "gen/spec" structural relationship which applies to "class model" reflexively; in a same way, the "comp-of" relationship, which applies to "act", defines compound acts.

2.3. Knowledge bases

It is then possible to refine the above general model in defining sub-classes. As such, this level is customisable to the particular situation of a hospital, a unit, an individual health care professional, ... It contains domain objects actually found in a given hospital:

- classes of medical and surgical acts, of diseases, of diagnostic and therapeutic procedures, and so on, which are sub-classes of the respective class models defined at the general level;
- semantic relationships which may hold between those classes as instances of the corresponding semantic relationships defined at the conceptual level;
- taxonomies of acts, of, diseases, ... which organise thoses classes and support inheritance of attribute values and of semantic links set from a class to its descendants.

Figure 2 illustrates this semantic layer where:

- surgery operation is an object in itself. Its elements are classes, instances of the class-model therapeutic protocols. Similarly, cardio-vascular diseases, heart surgery acts and anesthesia are taxonomies whose elements are respectively instances of disease, surgery act and medical act class-models.
- arterial derivation is a class, instance of therapeutic protocol, itself a particular protocol. As such, arterial derivation is composed of instances of acts: coronary artery bypass and general anesthesia are licit components. Finally, as a therapeutic protocol, arterial derivation may treat cardio-vascular arteriosclerosis which is an instance of disease.

The Data layer is built by instanciation of above classes and semantic links which form together the informational content of the patients data including:

- actual acts as instances of classes of acts,
- associative links between actual acts as instances of semantic relationships, and,
- other objects as protocols under performance, diagnoses of diseases, ... as instances of related classes of the above layer.

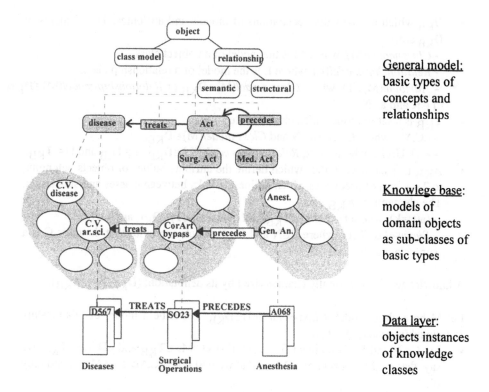

General model:
basic types of
concepts and
relationships

Knowlege base:
models of
domain objects
as sub-classes of
basic types

Data layer:
objects instances
of knowledge
classes

Figure 2. From the general model to actual data

3. The Query Language XQL

An eXtended Query Language, named XQL, has been defined and implemented
to consult and exploit the knowledge bases developed under the general model as an
extension of the so-called SQL. An XQL query has the traditional SQL appearance:

```
SELECT    <select-clause>
FROM      <from-clause>
WHERE     <where-clause>
```

in which the <select-clause> indicates the objects attributes to be selected, but in
which the <from-clause> indicates the involved objects and relationships in the
selection, and the <where-clause> defines conditions to be satisfied by the objects
attributes values. The complete syntax of the XQL select statement is given in annex
A.

3.1. Denotation of a knowledge base

Let **N** be a set of identifiers. Let $\mathbf{K}=\mathbf{K}_{string}\cup\mathbf{K}_{integer}\cup\{\bot\}$ be a set of constants.
We define:

- $\mathbf{D_{KB}}$ which contains the declarations of models and attributes. The elements of $\mathbf{D_{KB}}$ are:
  - *Class-model*(M), where M is the model of an object class,
  - *Relationship-model*(L), where L is the model of a relationship class,
  - *Attribute*(M,a,T), where *Class-model*(M)∈$\mathbf{D_{KB}}$ or *Relationship-model*(M)∈$\mathbf{D_{KB}}$, T∈**K** and a∈**N**.
- $\mathbf{T_{KB}}$ is a set of terms which elements are:
  - <C,M> where C∈**N**, M∈**N** and *Class-model*(M)∈$\mathbf{D_{KB}}$,
  - <<T,U>,L> where L∈**N**, *Relationship-model*(L)∈$\mathbf{D_{KB}}$, T∈$\mathbf{T_{KB}}$ and U∈ $\mathbf{T_{KB}}$.
- $\mathbf{A_{KB}}$ is a set of equalities which define the attribute values of objects belonging to $\mathbf{T_{KB}}$ and of predicates which define the links between classes and subclasses. The elements of $\mathbf{A_{KB}}$ are:
  - (a(t)=k), where t=<C,M>∈$\mathbf{T_{KB}}$, *Attribute*(M,a,T)∈$\mathbf{D_{KB}}$ and k∈**K**,
  - *Super-class*(C,D) where C∈**N** and D∈**N** which expresses that the class C is a superclass of the class D.

A knowledge base is formally characterized by its denotation: $(\mathbf{D_{KB}},\mathbf{T_{KB}},\mathbf{A_{KB}})$.

Let M be a class model, *Class-model*(M)∈$\mathbf{D_{KB}}$. Let L be a model of relationship, *Relationship-model*(L)∈$\mathbf{D_{KB}}$.
- Let C∈**N** and D∈**N** be identifiers such that <C,M>∈$\mathbf{T_{KB}}$ and <D,M>∈$\mathbf{T_{KB}}$, we say that <C,M> subsumes <D,M>, that we will note <C,M>≥<D,M>, if and only if :
  - either C=D,
  - or there exists identifiers $C=C_1,C_2, ..., C_n=D$ so that $\forall i \in \{1,2, ... ,n-1\}$: <$C_i$,M>∈$\mathbf{T_{KB}}$ and *Super-class*(C_i,C_{i+1})∈$\mathbf{A_{KB}}$.
- Let T,U,V,W∈$\mathbf{T_{KB}}$ terms such that <<T,U>,L>∈$\mathbf{T_{KB}}$ and <<V,W>,L>∈$\mathbf{T_{KB}}$, we say that <<T,U>,L>≥<<V,W>,L> if and only if T≥V and U≥W.

The subsumtpion relation defines a pre-order on the set of terms $\mathbf{T_{KB}}$. Let T∈$\mathbf{T_{KB}}$, we will note by $\mathcal{A}$(T) the set of the ancestors of T and by $\mathcal{B}$(T) the set of the descendants of T in $\mathbf{T_{KB}}$:

$$\mathcal{A}(T) = \{U \in \mathbf{T_{KB}} \mid U \geq T, \ U \neq T\}$$
$$\mathcal{B}(T) = \{U \in \mathbf{T_{KB}} \mid T \geq U, \ U \neq T\}$$

For every T∈$\mathbf{T_{KB}}$, the set $\mathcal{A}$(T) is ∅ or has a most specific element noted $Min_\leq$(T), i.e. the father ot T, and the set $\mathcal{B}$(T) is ∅ or has a subset of most generic elements $\mathcal{C}$(T)⊆$\mathcal{B}$(T), i.e. the children of T.

3.2. Interpretation of a XQL query

In what follows, we will define the interpretations of the clauses which constitute an XQL query statement to a given knowledge base KB. After what we will define the interpretation of the query statement. Let $V=\{x_1,x_2,...,x_n\}$ a set of variables. An allocation of V in T_{KB} is a set of equalities $(x_i=T_i)$ where $T_i \in T_{KB}$. Let $\{()\}$ the empty allocation, i.e. the allocation of no variables.

3.2.1. Interpretation of a <from-clause>

A <from-clause> is a sequence $\mathcal{F}$ of expressions, $\mathcal{F}=F_1,F_2,...F_n$, where each F_j is of the form "X AS x_j", where X is a class model reference in most cases, or a class model connected from a class by the means of a relationship (see annex A), and x_j is a variable which does not occur in the previous F_i, $i<j$. The interpretation $I_{KB}(\mathcal{F})$ is built recursively. For every $i \in \{1, ... , n\}$, $I_{KB}(F_1, ... , F_i)$ is a set of allocations to the variables $x_1,x_2,...,x_i$ processed from F_i and $I_{KB}(F_1, ... , F_{i-1})$. The detailed proccess is given in annex B.

3.2.2 Interpretation of a <where-clause>

A <where-clause> is a sequence of expressions $\mathcal{W}=W_1,W_2,...,W_n$, where each W_i has the following appearance: $W_i=x_i.a_i$ *operator* v_i, where $x_i \in V$, a_i is such that $Attribute(M,a_i,T) \in D_{KB}$ and *operator* is a traditional operator on strings or integers, according to the type T. The interpretation $I_{KB}(\mathcal{F})(\mathcal{W}) \subseteq I_{KB}(\mathcal{F})$ is defined as follows:

$$I_{KB}(\mathcal{F})(\mathcal{W})=\{ (x_i=T_i) \in I_{KB}(\mathcal{F}) \mid Value(x_i.a_i)\ operator\ v_i \}$$

where *Value* is defined as follows, for each i: if there exists $v \neq \bot$ such that $(a_i(T_i)=v) \in A_{KB}$, then $Value(T_i.a_i)$ is v, else $Value(T_i.a_i)$ is $Value(Min_{\leq}(T_i),a_i)$.

3.2.3 Interpretation of a <select-clause>

A <select-clause> is a sequence $\delta=S_1,S_2,...,S_n$, where each S_i has the following appearance: $S_i=x_i.a_i$, where $x_i \in V$ and $Attribute(M,a_i,T) \in D_{KB}$. The interpretation $I_{KB}(\mathcal{F})(\mathcal{W})(\delta)$ is defined as follows: let be i_1, i_2, ..., i_q being such that x_{i_1}, x_{i_2},..., x_{i_q} are the only variables which appear in the <from-clause> and have an occurence in δ:

$$I_{KB}(\mathcal{F})(\mathcal{W})(\delta)=\{Value(T_j.a_j) \mid \exists\ T_1,...,T_{j-1},T_{j+1},...T_n:$$
$$(x_1=T_1,...,x_{j-1}=T_{j-1},...,x_n=T_n) \in I_{KB}(\mathcal{F})(\mathcal{W})\}$$

$I_{KB}(\mathcal{F})(\mathcal{W})(\delta)$ is a set of attributes values of objects which are referenced in the <select-clause> of the query.

3.2.4. Semantics of a XQL query

Finally, the operational semantics of a XQL query

SELECT δ FROM $\mathcal{F}$ WHERE $\mathcal{W}$

when applied to a knowledge base KB is:

$$S_{KB}(\delta, \mathcal{I}, \mathcal{W}) = \{false\} \qquad \text{if } I_{KB}(\mathcal{I})(\mathcal{W})(\delta) = \varnothing,$$
$$S_{KB}(\delta, \mathcal{I}, \mathcal{W}) = I_{KB}(\mathcal{I})(\mathcal{W})(\delta) \qquad \text{else.}$$

3.3. Queries to knowledge and data bases

Let consider a knowledge base KB as presented in 2.3. In KB, "Surgical_Act" is a class model. "Coronary_Artery_Bypass" is an sub-class of the above class. The following XQL query allows to retrieve the identifiers of the classes linked to this latter class by the means of relationships whose model is "Precedes":

```
SELECT   Act.Id
FROM     Surgical_Act AS Surg,
         Surg.←Precedes AS Link,
         Link.ORG AS Act
WHERE    Surg.Id="Coronary_Artery_Bypass"
```

The XQL engine processes as follows:
- ⟨from clause⟩: it selects all the classes which are origins of "Precedes" relationships to other classes,
- ⟨where clause⟩: then, it selects in the above classes those which are linked to the class whose identifier is "Coronary_Artery_Bypass",
- ⟨select clause⟩: finally, it gives as result the identifier of the selected classes.

For instance, the class "General_Anesthesia" will be listed in the result.

The following XQL query shows an example which operates the bridge between the model and the data. It allows to retrieve the identifier and its related code of the disease which have been treated by a the surgical "Coronary_Artery_Bypass" #S023 in the patients database.

```
SELECT   ClassDisease.Id, ClassDisease.Code
FROM     Disease.←Treats AS Link,
         Link.ORG AS Surg,
         Disease.SUPN AS ClassDisease
WHERE    Surg.Id='S023'
```

The process operated by the XQL engine is different than previously. The navigation between objects is operated in the database firstly from the surgical act which is the origin of a link "Treats" to the disease diagnosis report. Then, by the means of the key-word SUPN, the bridge is processed between the object "Disease" and the class "ClassDisease" from which it is an instance. In this case, the result will be the identifier "Coronary_Arteriosclerosis" and its associated code which depends on the nomenclature or taxonomy used for coding patients data.

4. Discussion

The need to represent the objects involved in a HIS and their properties in knowledge bases able to be exploited at run-time by application programs was our starting point. Several knowledge representation methods were good candidates. Since such knwoledge bases should be able to represent both concepts and relationships between them, semantic networks were envisaged, such as KL-ONE or KL-ONE-like knowledge representation systems [4,5]. Since automatic inheritance of concepts attributes values was needed to propagate concepts properties all along hierarchies, description of taxonomies of concepts were chosen [6]. The flexibility and the powerness of conceptual graphs were also interresting for our purpose [7,8]. But, even if conceptual graphs have been formalized [9], the lack of effective implementations of systems based on conceptual graphs, and the difficulties encountered to design the needed lattice of concepts [10], were not encouraging.

Facing the above considerations, and taking into account the real powerness of relational databases management systems to store information and not only data, we decided to design and implement an object-oriented system able both to represent knowledge regarding HIS components naturally and to exploit them. The well known capabilities of SQL queries to navigate in databases and to be embedded in application programs have been extended to knowledge bases in the XQL query language. The capabilities of such a representation and exploitation system have been tested on the frame of the projects RICHE and NUCLEUS. In the frame of NUCLEUS, two pilot sites, in Italy and Great-Britain, use it in workfloor applications.

Acknowledgements

This work was is supported by the Commission of the European Community under the Esprit project RICHE n° 2221 and the AIM project NUCLEUS n° 2025. The authors are indebted to all their colleagues of the RICHE and NUCLEUS consortia.

References

[1]   Joubert M, Kanoui H. The knowledge-based management of medical acts in NUCLEUS. In: Andreassen S, Engelbrecht R, Wyatt J, (eds), *Proc. AIME 93*. IOS Press, 1993: 377-380.
[2]   Kanoui H, Joubert M, Riouall D, Favard R. Customisation environment for an act-based hospital information system. In: Reichert A, Sadan BA, Bengtsson S, Bryant J, Piccolo U (eds), *Proc. MIE'93*. Freund Publish. 1993: 241-245.
[3]   Kanoui H, Joubert M, Favard R. Knowledge-based model and query language to medical databases in a hospital information system. In: Barahona P, Veloso M, Bryant J, (eds), *Proc. MIE'94*. 1994: 379-383.

[4]   Brachman RJ, Schmolze JG. An overview of the KL-ONE knowledge representation system. *Cognitive Science* 9. 1985: 171-216.

[5]   Brachman RJ, McGuinness DL, Patel-Schneider PF, Resnick LA, Bordiga A. Living with CLASSIC: when and how to use a KL-ONE-like language. In: Sowa JF (ed), *Principles of semantic networks: exploration in the representation of knowledge*. Morgan Kaufmann Publish. 1991: 401-456.

[6]   Woods WA. Understanding subsumption and taxonomy. In: Sowa JF (ed), *Principles of semantic networks: exploration in the representation of knowledge*. Morgan Kaufmann Publish. 1991: 45-94.

[7]   Sowa JF. *Conceptual Structures: information processing in mind and machine*. Addison Wesley, 1984.

[8]   Sowa JF. *Conceptual analysis as a basis for knowledge representation*. Tutorial Handbook. MEDINFO 92. 1992.

[9]   Chein M, Mugnier ML. Conceptual graphs: fundamental notions. *Revue d'Intelligence Artificielle* 6. 1992: 365-406.

[10]   Volot F, Zweigenbaum P, Bachimont B, Ben Said M, Bouaud J, Fieschi M, Boisvieux JF. Structuration and acquisition of Medical knowledge using UMLS in the conceptual graphs formalism. In: Safran C (ed), *Proc. 17th SCAMC*. McGraw-Hill. 1993: 710-714.

Annex A

The following describes the syntax of a XQL query:

```
<XQL-query>      ::= SELECT <select-clause> FROM <from-clause> WHERE <where-clause>
<select-clause>  ::= {<select-expr>}*
<from-clause>    ::= {<from-expr>}*
<where-clause>   ::= {<where-expr>}*
<select-expr>    ::= <variable.<attribute-name>
<from-expr>      ::= <class-model-ref>AS <class-variable> |
                     <class-model-ref> .SUB AS <class-variable>|
                     <class-model-ref>.SUPER AS <class-variable>|
                     <class-model-ref>.SUBN AS <class-variable>|
                     <class-model-ref>.SUPN AS <class-variable>|
                     <model-ref>.→ <link-model-name>AS <link-variable>|
                     <class-model-ref>.← <link-model-name>AS <link-variable>|
                     <link-model-ref>AS <link-variable>|
                     <link-model-ref>.DEST AS <variable>|
                     <link-model-ref>.ORG AS <variable>|
<where-expr>     ::= <variable>.<attribute-name><operator><value>
<link-model-ref>::= <link-model-name>| <link-variable>
<class-model-ref>::= <class-model-name>| <class-variable>
<variable>       ::= <link-variable>| <class-variable>
<link-variable>  ::= <identifier>
```

<class-variable> ::= <identifier>
<operator>   ::= <string-op>| <integer-op>
<string-op>   ::= = | LIKE
<integer-op>   ::= = | < | >

Annex B

The following describes the process operated to interpret a <from-clause>:

- $F_0=\{()\}$

- F_k is M AS x_k, *Class-model*(M)$\in\mathbf{D}_{KB}$ or *Relationship-model*(M)$\in\mathbf{D}_{KB}$
$\mathbf{I}_{KB}(F_k)=\{(x_1=T_1,...,x_k=<C,M>) \mid (x_1=T_1,..., x_{k-1}=T_{k-1})\in\mathbf{I}_{KB}(F_{k-1})\}_{C\mid<C,M>\in\mathbf{T}_{KB}}$

- F_k is x_j AS x_k, j<k
$\mathbf{I}_{KB}(F_k)=\{(x_1=T_1,...,x_k=T_k) \mid (x_1=t_1,...,x_{k-1}=T_{k-1})\in\mathbf{I}_{KB}(F_{k-1})\}$

- F_k is M.SUB AS x_k , *Class-model*(M)$\in\mathbf{D}_{KB}$
$\mathbf{I}_{KB}(F_k)=\{(x_1=T_1, ... ,x_k=T_k) \mid (x_1=T_1,... ,x_{k-1}=T_{k-1})\in\mathbf{I}_{KB}(F_{k-1})$
$\wedge T_k \in \mathcal{C}(<C,M>)\}_{C\mid<C,M>\in\mathbf{T}_{KB}}$

- F_k is x_j.SUB AS x_k , j<k
$\mathbf{I}_{KB}(F_k)=\{(x_1=T_1, ... ,x_j=T_j,...,x_k=T) \mid (x_1=T_1,...,x_j=T_j,...,x_{k-1}=T_{k-1})\in\mathbf{I}_{KB}(F_{k-1})\}_{T\in\mathcal{C}(T_j)}$

- F_k is M.SUPER AS x_k , *Class-model*(M)$\in\mathbf{D}_{KB}$
$\mathbf{I}_{KB}(F_k)=\{(x_1=T_1, ... ,x_k=T_k) \mid (x_1=T_1,... ,x_{k-1}=T_{k-1})\in\mathbf{I}_{KB}(F_{k-1})$
$\wedge T_k= Min_\leq(<C,M>)\}_{C\mid <C,M>\in\mathbf{T}_{KB}}$

- F_k is x_j.SUPER AS x_k , j<k
$\mathbf{I}_{KB}(F_k)=\{(x_1=T_1, ... ,x_j=T_j,...,x_k=Min_\leq(T_j)) \mid (x_1=T_1,...,x_j=T_j,...,x_{k-1}=T_{k-1})\in\mathbf{I}_{KB}(F_{k-1})\}$

- F_k is M.SUBN AS x_k , *Class-model*(M)$\in\mathbf{D}_{KB}$
$\mathbf{I}_{KB}(F_k)=\{(x_1=T_1, ... ,x_k=T_k) \mid (x_1=T_1,... ,x_{k-1}=T_{k-1})\in\mathbf{I}_{KB}(F_{k-1})$
$\wedge T_k\in\mathcal{B}(<C,M>)\}_{C\mid <C,M>\in\mathbf{T}_{KB}}$

- F_k is x_j.SUBN AS x_k , j<k
$\mathbf{I}_{KB}(F_k)=\{(x_1=T_1, ... ,x_j=T_j,...,x_k=T) \mid (x_1=T_1,...,x_j=T_j,...,x_{k-1}=T_{k-1})\in\mathbf{I}_{KB}(F_{k-1})\}_{T\in\mathcal{B}(T_j)}$

- F_k is M.SUPN AS x_k , *Class-model*(M)$\in\mathbf{D}_{KB}$
$\mathbf{I}_{KB}(F_k)=\{(x_1=T_1, ... ,x_k=T_k) \mid (x_1=T_1,... ,x_{k-1}=T_{k-1})\in\mathbf{I}_{KB}(F_{k-1})$
$\wedge T_k \in\mathcal{A}(<C,M>)\}_{C\mid <C,M>\in\mathbf{T}_{KB}}$

- F_k is $x_j.\text{SUPN AS } x_k$, $j<k$

$$I_{KB}(F_k)=\{(x_1=T_1, \dots ,x_j=T_j,\dots,x_k=T) \mid (x_1=T_1,\dots,x_j=T_j,\dots,x_{k-1}=T_{k-1})\in I_{KB}(F_{k-1})\}_{T\in A(T_j)}$$

- F_k is $M.\rightarrow L \text{ AS } x_k$, $Class\text{-}model(M)\in D_{KB}$ or $Relationship\text{-}model(M)\in D_{KB}$, and $Relationship\text{-}model(L)\in D_{KB}$

$$I_{KB}(F_k)=\{(x_1=T_1, \dots ,x_k=<<<C,M>,D>,L>) \mid$$
$$(x_1=T_1,\dots ,x_{k-1}=T_{k-1})\in I_{KB}(F_{k-1})\}_{C,D \mid <<<C,M>,D>,L>\in T_{KB}}$$

- F_k is $x_j.\rightarrow L \text{ AS } x_k$, $j<k$, $Relationship\text{-}model(L)\in D_{KB}$

$$I_{KB}(F_k)=\{(x_1=T_1, \dots , x_k=<<T_j,D>,L>) \mid$$
$$(x_1=T_1,\dots,x_j=T_j,\dots,x_{k-1}=T_{k-1})\in I_{KB}(F_{k-1})\}_{D \mid <<T_j,D>,L>\in T_{KB}}$$

- F_k is $M.\leftarrow L \text{ AS } x_k$, $Class\text{-}model(M)\in D_{KB}$ or $Relationship\text{-}model(M)\in D_{KB}$, and $Relationship\text{-}model(L)\in D_{KB}$

$$I_{KB}(F_k)=\{(x_1=T_1, \dots , x_k=<<D,<C,M>>,L>) \mid$$
$$(x_1=T_1,\dots ,x_{k-1}=T_{k-1})\in I_{KB}(F_{k-1})\}_{C,D \mid <<D,<C,M>>,L>\in T_{KB}}$$

- F_k is $x_j.\leftarrow L \text{ AS } x_k$, $j<k$, $Relationship\text{-}model(L)\in D_{KB}$

$$I_{KB}(F_k)=\{(x_1=T_1, \dots , x_k=<<D,T_j>,L>) \mid$$
$$(x_1=T_1,\dots,x_j=T_j,\dots,x_{k-1}=T_{k-1})\in I_{KB}(F_{k-1})\}_{D \mid <<D,T_j>,L>\in T_{KB}}$$

- F_k is $L.\text{DEST AS } x_k$, $Relationship\text{-}model(L)\in D_{KB}$

$$I_{KB}(F_k)=\{(x_1=T_1, \dots , x_k=D) \mid (x_1=T_1,\dots ,x_{k-1}=T_{k-1})\in I_{KB}(F_{k-1})\}_{G,D \mid <<G,D>,L>\in T_{KB}}$$

- F_k is $x_j.\text{DEST AS } x_k$, $j<k$

$$I_{KB}(F_k)=\{(x_1=T_1,\dots,x_j= <<G,D>,L>, \dots , x_k= D) \mid$$
$$(x_1=T_1,\dots,x_j= <<G,D>,L>,\dots , x_{k-1}= T_{k-1})\in I_{KB}(F_{k-1})\}$$

- F_k is $L.\text{ORG AS } x_k$, $Relationship\text{-}model(L)\in D_{KB}$

$$I_{KB}(F_k)=\{(x_1=T_1, \dots , x_k=G) \mid (x_1=T_1,\dots ,x_{k-1}=T_{k-1})\in I_{KB}(F_{k-1})\}_{G,D \mid <<G,D>,L>\in T_{KB}}$$

- F_k is $x_j.\text{ORG AS } x_k$, $j<k$

$$I_{KB}(F_k)=\{(x_1=T_1,\dots,x_j= <<G,D>,L>, \dots , x_k= G) \mid$$
$$(x_1=T_1,\dots,x_j= <<G,D>,L>,\dots , x_{k-1}= T_{k-1})\in I_{KB}(F_{k-1})\}$$

Use of a Conceptual Semi-Automatic ICD-9 Encoding System in an Hospital Environment

C. Lovis[1], PA. Michel[2], R. Baud[2], JR. Scherrer[2]
University State Hospital of Geneva
[1]Department of Medicine
[2]Informatic Center

Abstract
The necessity of encoding medical diagnosis has become essential, not only for medical purposes, but also for community-based research, epidemiology and economy, but there is a real lack of tools ensuring good quality and exhaustivity of diagnosis encoding. To achieve this goal whilst avoiding the need of to much computer processing power, we have built a tool using some natural language processing techniques, like a simple semantical representation and partial symbolic queries. We have also tried to build a cost-effectiveness system, which will run on any PC-Windows based system, with a good user-friendliness quality.

1   Introduction

The necessity of encoding medical diagnosis has become essential, not only for medical purposes, but also for community-based research, epidemiology and economy. This necessity exists at least since the 17th century [1], but the problem remains increasingly crucial owing to the rapid evolution of medicine [2]. Although one can nowadays use many different classifications, some of them extremely large, like SNOMED International [3], there is a real lack of tools ensuring good quality and exhaustivity of diagnosis encoding. In this field, Chute et al. report that only up to approximatively 60% of encoding fully matches the original diagnosis[4]. This may have important consequences if one thinks of all the economic decisions that are taken on behalf of these data. There are few stages between a clinical diagnosis and its expression as a code in a knowledge representation or disease classification such as ICD-9. At each stage, aspects of the semantic content can be lost and specific means must be found to minimise this loss of information. We now distinguish three main stages : a) the voicing of the diagnosis in the physician's usual language, b) the understanding of the coding scheme the physician has, and c) the projection of the diagnosis in the classification scheme.

The WHO International Classification of Diseases version 9 (ICD), possibly with the CM modificators, is the most currently used disease classification [5]. Our approach is based on literature [6, 7, 8] as well as on our hospital experience in coding [9] and in natural language processing [10]. We developed an informatic tool that largely simplifies the diagnosis encoding for ICD-9 and significantly increases the encoding quality. This tool has some very important characteristics :

a) it is user friendly, since it uses a graphical interface,

b) it is a low cost solution, as it runs on any Personal Computer with Microsoft Windows,

c) it is reliable, based on a large corpus of controlled expressions (more than 40'000)

d) it is *physician oriented*, our tool builds a set of controlled expressions centred on the diagnosis in French natural language.

2   The Encoding Paradigm

To develop a conceptual representation in a given system based on a pragmatic diagnosis of a given patient implies several stages and many constraints. The first step is that the physician must express the diagnosis. A study done in our hospital shows that more than 30% of all written diagnoses in discharge letters are complex associations of elementary diseases that can be found in the ICD-9 set. Therefore, a physician should probably learn, during his studies, to establish a diagnosis in a more synthetic language and avoid complex associations. The second step relates to the necessity of « translating » the written diagnosis into the adequate ICD-9 scheme. The success of the second operation can be measured by the match between the written diagnosis and the code(s) obtained. As said before, 60% is at best fulfilled. But, in most cases even this score is not obtained because the encoders are physicians, not specialists of classifications. Two factors have influence on this process. First, the classification must at least contain the semantic elements of the written diagnosis, and second, one must be able to find these elements scattered in the classification and combine them.

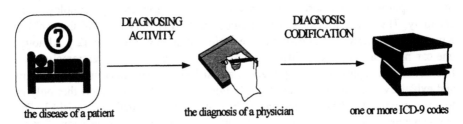

Fig. 1. From patient to code

2.1   Expression of the Diagnosis

The written language used to express diagnosis has some very interesting features. Most sentences are nominal forms, and the syntactic structure is often simple and stereotyped. In French, the length of sentences varies from one to fifteen words on average. In our corpus, the mean number of words is 4.92 per sentence in discharge letters (37.8 characters). Despite this apparently simple linguistic form, the content is highly significant and almost each word is important. This is consistent with the large corpus of specialised words used in the medical sub domain.

In fact, a physician does not express a diagnosis for the purpose of encoding, but he wishes to express as precisely as possible and with the minimum of words, the exact situation of the patient.

2.2 Knowledge Representation

The ICD-9 classification was originally designed to express diseases and not diagnoses. For a physician, this is an important difference. Therefore, a lot of written diagnoses (40% in our study) have no direct equivalence in ICD-9, but can be found either as a partial representation or as a complex association. Moreover, as will be shown further, a lot of medically related concepts are disseminated in the ICD-9 classification, owing to its monoaxial structure. In such a situation, it is necessary to have a deep understanding and a long practice of ICD-9 in order to obtain high quality encoding taking every exception and exclusion into account. Our tool is able to compute a pertinence index for ICD-9 expressions, given a specific written diagnosis, taking partially into consideration the semantic of the expression.

2.3 From Diagnosis To Diagnostic Code

One can evaluate both the quality and the completeness of the encoding of a natural language diagnosis into its representation in a disease classification. The quality, or precision of the code, can be expressed on a scale that measures the proximity of the concepts in the natural language sentence to the one in the coding scheme. To measure exhaustivity, one must make sure that all the concepts that can be found in the sentence in natural language can be found in the codes.

Example 1:

décompensation cardiaque gauche sur fibrillation auriculaire post-opératoire (left heart failure due to atrial fibrillation after surgery)

a) quality:

décompensation cardiaque (*heart failure*)	428.9:	insufficient
décompensation cardiaque gauche (*left heart failure*)	428.1:	good
fibrillation auriculaire (*atrial fibrillation*)	427.3:	insufficient
fibrillation auriculaire post-opératoire (*atrial fibrillation after surgery*)	997.1:	good

Note here the position of 997.1 in the hierarchy and the illustration of *exceptions.*

b) completeness

428.1 or 997.1:   insufficient
428.1 & 997.1:   good

In the previous examples, each code matches with a specific and distinct part of the sentence. In this example, one can also note that expressions that seem near in terms of semantical content may be distant in the hierarchy of ICD-9 (like 427.3 and 997.1, both applying to atrial fibrillation). Moreover, as illustrated in the next example, there may be no direct similarity between a given code and a part of the sentence taken from a linear cut-off parsing (see line *b* and *c*). Sometimes, important information is implied and not explicitly written and belongs to « medical sense » like some information can belong to « common sense » (see line *d* , bone metastases is inferred from pathological fracture in a context of prostatic neoplasm). In such situations, classical approach of surface linguistic string processing is not sufficient and a medical knowledge representation with inference processing is necessary. This has still to be implemented in our tool. Presently, we only use semantical associations instead of a real knowledge representation.

Example 2 :

adénocarcinome de la prostate avec fracture pathologique du col fémoral droit
(prostate adenocarcinoma with a pathological fracture of the right femur hip)

a)	adénocarcinome de la prostate : *(adenocarcinoma of the prostate)*	*tumeur de la prostate*	*185*
b)	fracture du col fémoral droit *(fracture of the right femur hip)*	*fracture du col du fémur*	*820.8*
c)	fracture pathologique *(pathological fracture)*	*fracture pathologique*	*733.1*
d)	(métastase osseuse) *(metastase in the bone)*	*métastase osseuse*	*198.5*

Finally, it is very important to be able to deal with some exclusion groups in ICD-9 or associations of codes apparently justified, but in fact contradictory, due to the hierarchy of ICD-9. In the next example, the correct association *is Chronic obstructive pulmonary disease (496)* and *bronchitis (491.1),* since *bronchitis* under *490* should not be associated with *chronic obstructive pulmonary disease,* whereas *bronchitis* under *491* is correct in a context of acute worsening of a COPD.

Example 3:

Aggravation d'un syndrome obtructif respiratoire chronique (SOC) sur bronchite
(Aggravation of chronic obstructive pulmonary disease (COPD) due to bronchitis)

COPD	*COPD*	*496*
Bronchitis (in case of COPD)	*bronchitis*	*491*
Bronchitis (acute in case of COPD)	*bronchitis*	*491.1*

These examples show that it is difficult to perform adequate diagnosis encoding, and both a good understanding of the disease classification and a good knowledge of medicine is needed. *"The coding activity should be reserved to professional encoders".* The possibility to have professional people dealing with diagnosis encoding represents an economic challenge. In most European countries, diagnosis encoding is performed directly by physicians or by secretaries; the latter leading often to poor quality. The physician knows medicine and his patients well and we provide a tool that knows ICD-9. The association of both has given preliminary results that are encouraging.

3   Methodology

There are several research groups working on natural language processing in Europe and around the world, but only few applications, at the present time, are sufficiently achieved to allow professional use in hospitals [11 , for example]. Nevertheless, the simplified linguistic structure of diagnosis sentences and the compact semantic content allows a very efficient informatic approach to this specific task, using AI technics. In our tool, we have chosen methodological solutions that only use partial AI technologies, but that are fast and immediately usable.

3.1   Morphological Step

The first step of the analysis is linguistics. In this part of the analysis, we ensure that each words in the sentence can be recognised. This analysis uses two distinct techniques. The first one is the word mapping against a medical dictionary. Each lexicon entry contains the basic form and a morphological and syntactic description of a word. This makes possible to recognize almost all the inflected forms (plural, gender, ...) of a word. The second technique is used whenever the word is not in the dictionary. This is the intra-word processing step, and we use the approach of morpho-semantems [12 , 13 , 14]. This analysis is a rule-based decomposition of words into semantical units and uses a concept-oriented affix dictionary. That medical jargon is largely built with combinations of latino-greek roots allow us to considerably reduce the size of the dictionaries whilst keeping a very large corpus of known words (like *ectomy, hyper, appendic*, etc ...). This is also an elegant and powerful method to deal with neologisms.

Example 4:   *Ganglions axillaires de type lymphomatoïde*
                (Axillary lymph nodes with lymphomatoïd characteristics)
                        First step :   *Ganglions axillaires*
                    = *ganglion & axillaire*
                        (axillary lymph node)
                    Second step : *lymphomatoïde*
                    = *lymphom_at_oïde* = *aspect de lymphome*
                        (looks like lymphoma)

3.2   From Words to Sense of Expression

When each word of the written sentence has been recognised, we have access to the set of all concepts pointed to by these words. This allows us to create a meaning set of the sentence. As a result, we always have a given set of known words in our hierarchy, and this represents the first main projection of the sentence in our conceptual world. This reduction is illustrated in figure 2, where we show different ways (in this case, trivial ways) of expressing the same concepts.

words for the concept A : *tumeur maligne, adénocarcinome, cancer, néoplasie*
                            (malignancy, adenocarcinoma, cancer, neoplasm)
words for the concept B : *prostatique, prostate*
                            (prostate, prostatic)

words for the concept C :   *métastatique, plurimétastatique, métastases*
                            (metastatic, metastase, multimetastatic, etc ...)
expression A1xB1xC1:   *tumeur maligne prostatique métastatique*
                            (malignancy of the prostata metastatic)
expression A1xB1xC2:   *tumeur maligne de la prostate multimétastatique*
                            (malignancy of the prostata mulitmetastatic)
expression A4xB2xC4:   *néoplasie de la prostate avec métastases*
                            (neoplasia of the prostata with metastase)

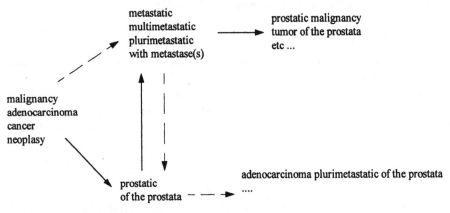

Fig. 2. Combinatorial example with key words AxBxC

3.3   Conceptual Step

In this part of our tool, we project the set of concepts found in the written sentence against the expressions found in ICD-9, regarding their associativity. This is possible through a large corpus of possible associations of concepts from ICD-9.

We obtain a set of ICD-9 expressions that are centred on the set of concepts given in the written sentence. This set allows us to generate the corresponding ICD-9 controlled expressions in order to hand them over to the physicians in charge of encoding. This helps us to give to the physicians a limited set of pertinent expressions scattered in the original ICD-9 hierarchy. It is equivalent to let the physician choose the best expression among a dynamic list of pertinent expressions.

This is illustrated in the following example :

Diagnosis : **Prostatic adenocarcinoma**

The result of the analysis is an association of concepts **PROSTATE** and **MALIGNANT_TUMOR**. This system will parse all expressions to obtain a subset corresponding to the semantic meaning of *prostate* and *malignant_tumor* :

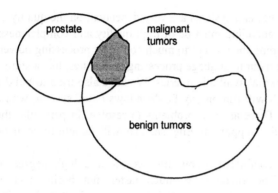

Fig. 3. The subset of target concepts

The system will then project all these concepts into the representation of the hierarchy of ICD-9 to retrieve all controlled expressions that could have a meaning pertinent to the target concepts. It should be noted that these expressions are disseminated in the ICD-9 hierarchy, so that the query is rather complex if it is done *manually*. The same mechanism is repeated for each subset of concepts in a complex expression, for each expression in a complex list of diagnoses. For each query, we obtain a sublist of pertinent expressions.

This is illustrated as follows :

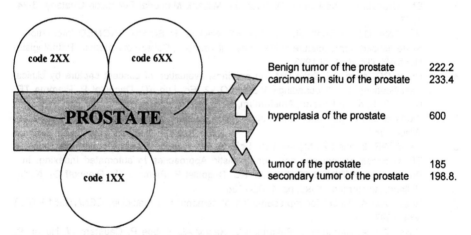

Fig. 4. Projection of a subset of concepts on the ICD-9 hierarchy

4   CONCLUSION

The encoding is very crucial. It is used for research purposes in epidemiology, it is also a community health indicator. More and more countries use it in order to try to understand and control the economic aspects of medicine, like the DRG's in USA. The diagnosis encoding is also used to make comparisons between hospitals and to evaluate cost-effectiveness of medical procedures. All this data is unusable if based on invalid or incomplete encoding, and systems are very dependent on the quality of the inputs [15]. Therefore, the encoding must be of high quality and completeness. To achieve this goal whilst avoiding the need of to much computer processing power, we have built a tool using some natural language processing techniques, like a simple semantical representation and partial symbolic queries. We have also tried to build a cost-effectiveness system, which will run on any PC-Windows based system, with a good user-friendliness quality. There are still problems to resolve, in particular the processing of exclusions and the support of English, that will hopefully soon be available.

The first clinical experiments conducted in our hospital show a high degree of satisfaction from physicians. The encoding is much faster, not boring and of significantly better quality and exhaustivity. Physicians can encode directly using narrative language and they do not need any information on the exact hierarchy of ICD-9. This tool seems to have very exciting and promising possibilities.

5   References

1   Graunt J. Natural and Political Observations made upon the Bills of Mortality. Baltimore MD : The Johns Hopkins Press, 1932

2   EM. Grundberg. The Failure Of Success. Milbank Memorial Fundation Quaterly, 3-24, 1977

3   RA. Côté, DJ. Rothwell, JL. Palotay, RS. Beckett, L: Brochu. SNOMED International, Systematized Nomenclature of Medicine. Introduction, College of American Pathologists, Northield, Illinois, April 1993

4   Chute CG, Atkin GE, Ihrke DM. An empirical evaluation of concept capture by clinical classifications. In : Proceedings MEDINFO 92 (Ed. Lun KC, Degoulet P, Piemme TE, Rienhoff O), North-Holland, Amsterdam, 1992, pp.1469-1474

5   World Health Organisation. International Classification of Diseases, Version 9. Geneva: WHO, 1980

6   Hersh WR, Evans DA, Monarch IA, Lefferts RG, Handerson SK, Gorman PN. Indexing Effectiveness of linguistic and non-linguistic Approaches to automated Indexing. In : Proceedings MEDINFO 92 (Ed. Lun KC, Degoulet P, Piemme TE, Rienhoff O), North-Holland, Amsterdam, 1992, pp. 1402-1408

7   Rossi Mori A. Model for representation of semantics in medicine. CEN/TC251/PT003 N34. 1992

8   Rossi Mori A, Bernauer J, Pakarinen V, Rector AL, Robbè P, Ceusters W, Hurlen P, Ogonowski A, Olesen H. Models for Representation of Terminologies and Coding Systems in Medicine. In : Proceedings of 'Opportunities for European and US cooperation in standardization in Health care Informatics', 1992, Geneva

9   P. Frütiger. Prospective encoding, transcoding and pragmatic representation of medical language. In : Computerized Natural Medical Language Processing for Knowledge Engineering (Ed. JR. Scherrer, RA. Côté, SH. Mandil) Elsevier, Amsterdam, 83-94, 1989

10   Baud RH, Rassinoux AM, Scherrer JR. Natural Language Processing and Semantical Representation of Medical Texts. Methods of Information in Medicine, 1992, 31: 2.

11   B. Birgl, M. Mieth, R. Haux, E. Glück. The LBI-method for automated indexing of diagnoses by using SNOMED. Part 1. Design and Realization. Int J BioMed Computing 37:237-247, 1994

12   LM. Norton, MG. Pacak. Morphosemantic Analysis of Compound Word Forms Denoting
        Surgical Procedures. Meth Inform Med, 22: 29-36, 1983
13   S. Wolff. The Use of Morphosemantic Regularities in the Medical Vocabulary for
        Automatic Lexical Coding. Meth Inform Med, 23: 195-203, 1984
14   C. Lovis, PA. Michel, R. Baud, JR. Scherrer. Word Segmentation Processing : a way to
        exponentially extend medical dictionnaries. To be published, Medinfo 1995
15   Heckerling PS, Elstein PS, Terzian CG, Kushner PS. The effect of incomplete
        knowledge on the diagnoses of a computer consultant system. Med Inf (London), 1991;
        16: 363-370

Neural Networks and Image Interpretation

Quality Assurance and Increased Efficiency in Medical Projects with Neural Networks by Using a Structured Development Method for Feedforward Neural Networks (SENN)

T. Waschulzik[1], W. Brauer[2], M. Förster[3], K. Kirchner[4],
R. Engelbrecht[1], T. Schütz[1], T. Koschinsky[4], G. Entenmann[4]

[1] GSF-MEDIS-Institut, Ingolstädter Landstr.1, Neuherberg, D 85758 Oberschleißheim, Germany
[2] Technische Universität München, Institut für Informatik, Arcisstr.21, D 80290 München 2, Germany
[3] Institut für Medizinische Informatik und. Biometrie, Fetscherstr. 74, D 01307 Dresden, Germany
[4] Diabetes Forschungsinstitut, Auf'm Hennekamp 65, D 40225 Düsseldorf, Germany

Abstract

The growing number of projects using neural networks in medical care makes it necessary to examine how productivity can be increased and how quality can be assured. This examination addresses the problem specification, data preparation as well as the development of appropriate representations, the selection of suitable encodings and the combination of encodings. Network paradigms with fast learning properties and network structures that can be analysed and interpreted after the training process have been successfully applied to medical tasks. The associative recall of examples (ARE) can be used to verify the quality of representations and encodings. Furthermore it is possible to evaluate the competence of a neural network for a specific task by an ARE. Finally, a standard approach and the hereafter presented method applied for a medical project are compared. The comparison of these two approaches and the collection of the medical data is part of the DIADOQ-project[1].

1 Introduction

An increasing number of papers applying neural networks to medical problems [Andreassen 1993, Ifeachor 1994] have been published for the last years. In medical applications, we have often to deal with the problem that sufficient examples for the set-up of the knowledge bases are not available. If the number of available examples is to small for an appropriate statistical analysis, neural networks should not be applied either. Ignoring this may lead to severe quality assurance problems. In order to improve the efficiency and to assure the quality of the projects a method for a structured development of neural networks is necessary. This method has been developed based on experiences in industrial and medical projects [Waschulzik 1990a].

2 Medical Problem

Type II diabetes is characterised by an abnormal insulin secretion and a decreased insulin effect on insulin sensitive tissues (insulin resistance) [Yki-Järvinen 1994]. During the first stage of Type II Diabetes diet alone or in combination with oral antidiabetic drugs can compensate for diabetes related metabolic changes.

---

[1] DIADOQ - Optimized Care Through Knowledge-based Quality Assurance: Diabetes Mellitus. The organisations involved are: Institute for Diabetes Research, Düsseldorf; GSF - MEDIS Institute for Medical Informatics, Neuherberg; Institute for Diabetes, Karlsburg; University Hospital of Erlangen-Nürnberg, Nürnberg; IMIB - Institute for Medical Informatics and Biometry, Dresden; Department of Diabetes, München-Bogenhausen; Boehringer Mannheim GmbH, Mannheim; Institute for Mathematics, Ludwig Maximilians University, München. DIADOQ is part of the German MEDWIS (Medical Knowledge-bases) research programme

Several factors like obesity concomitant disease, the duration of diabetes, poor meta-bolic control and/or treatment, e.g. with sulfonylureas, can finally result in an additional need for long-term insulin treatment to achieve stable, near normal metabolic control. This stage of type II diabetes is referred to as secondary failure (SF) of the oral diabetic drug treatment. Such a persistent SF has to be distinguished from temporary metabolic derangements due to e.g. transient infections or the use of diabetogenic drugs for a limited period of time.

Obviously, there is an urgent need for a reliable and real-time differential diagnose of SF in order to select and start an appropriate treatment assuring the best quality of life with the smallest degree of diabetes related complications [European NIDDM 1990].

The problem "secondary failure in type II diabetes" was approached in two different ways to compare the proposed method with a standard method. Data of 573 diabetics were

Data Set	SF	No SF	Total
Training Set	219	113	332
Test Set	114	53	167
Assessment Set	55	19	74
Total	388	185	573

Table 2.1:*Number of cases in the data sets*

available (table 2.1) for the differential diagnosis of SF which can be considered a classification problem. Additional data are being collected for purposes of validation.

3 Structured Development Method for Feedforward NNs - SENN

The SENN method (Strukturierte Entwick-lungsmethodik für vor-wärtsgerichtete Neuronale Netze) is based on a system life cycle model (see fig 3.1) and refers to techniques commonly used in the literature. The new aspect is the integration and combination of these techniques for quality assurance and efficiency purposes and their support by appropriate tools. In the following we will be discussing how systems (e.g. a knowledge base for this differential diagnoses) are realised in projects involving neural components.

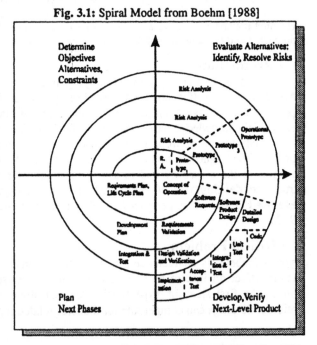

Fig. 3.1: Spiral Model from Boehm [1988]

3.1 System Life Cycle

The lack of knowledge extraction algorithms which does not allow the set-up of an optimized system within a single step leads to a cyclic development process. We have to build a set of prototypes for the problem solution in order to stepwise improve the knowledge extraction. To enable appropriate testing, all prototypes have to be complete systems where only the user interface and the integration into other components may be prototypical. This cyclic development process directly corresponds to the spiral model from [Boehm 1988]. The aim of a development method for neural networks is to assure the quality of a project under these cyclic circumstances as well as to reduce the costs of the project. It is also interesting to enhance the domain knowledge for further projects and system maintenance.

3.2 System Analysis and Design

The set of input and output variables and a set of possible dependencies for each of the output variables from the input variables is one of the results of the system analysis in neural network projects. These dependencies as well as additional information on the appropriate representation and encoding of these variables are documented in a table of dependencies which is the basis for further system development. We distinguish between the representation of the variables (for example, whether a disease is classified in ICD9 or SNOMED) and the encoding of this representation for the neural network.

In some projects such as the set-up of a system for the support of the differential diagnosis of SF, the table of dependencies can degenerate since the only output variable simply depends on a list of input variables. Only unstructured variables are relevant for the diagnosis of SF. Thus, no special techniques for the representation and encoding of the variables had to be used. This is favourable for the conventional approach and supports the comparison of the methods. In the system analysis, it is also determined how many examples will be necessary and how they are available.

3.3 Realisation of the System

The realisation of the system starts with the collection of the data sets and the selection of the examples. Here, the information about SF and the input variables in table 3.1 had to be registered. Afterwards the appropriate representation has to be chosen followed by the encoding of the variables and the network configuration. The neural network can be integrated into the system after the training and testing phase.

num-ber	variable	type
1	sex	C B
2	age	N
3	duration of diabetes	N
4	duration of treatment with sulfony-lureas	N
5	duration of poor metabolic control	N
6	body-mass-index (BMI)	N
7	weight change the last three months	C O
8	fasting blood glucose	N
9	postprandial blood glucose	N
10	HbA1	N
11	acetone	C O
12	leukocytes	N
13	cholesterol	N
14	triglycerides	N
15	sedimentation rate of blood (BSR)	N
16	diabetic factor	C B
17	infections	C
18	heart disease	C
19	thyroid disease	C
20	urinary tract infection	C

Table 3.1: Features used (C=categorical, B=binary, O= ordinal, N= number)

Representation of the Variables

Sometimes it is advantageous to pre-process raw data before applying it to the network. For example, in object recognition, it is not useful to present the raw pixel information to the neural network since cases with similar outputs might have very different representations at the input. It is necessary to enhance the similarities between the input and output of the neural network. This can be done by ignoring useless information because, generally, neural networks are continuous approximators that map similarities from the input to the output with some non-linear distortions. Thus, we have to switch from grey values to contrast pictures. Another step to enhance similarities is to switch to histograms of edges and combinations of edges. This simple representation is good for translation- and scale-invariant object recognition.

The quality of the representation can be scrutinised by using an ARE (Associative Recall of Examples). This is achieved by calculating the distance between the representations of the data set and the current example. The representation is suitable if similar patterns yield small distances in between, otherwise the representation has to be improved. The efforts for the development of representations and the development of networks must be well-balanced. The importance of the representation is also stressed in the literature: "A good representation does most of the work. Neural networks no matter what kind are weak computers, but a proper representation, coupled with the abstracting and co-operative effects of the network, can become extremely fast, powerful, and accurate." A. Rosenfeld in [Anderson 1988] according to [Nelson 1990, page 155]. In further investigations we will check whether the results of the ARE using the representations and later the encodings can be used as early quality indicators in the quality assurance process.

Encoding of the Variables

The encoding has to present the representations to the network in an appropriate way. It can also be used to combine variables in different ways. In the following, a subset of possible encodings and combination techniques for variables is being described.

Categorical Variables:

A metrics over values of categorical variables is not defined not even if an order for possible values exists. This fact forbids to map values from categories to numbers and to use these numbers for the activation of neurons in a neural network. Although this procedure of encoding different values of one category into a single neuron is frequently applied we have to be aware that this mapping implicitly introduces a metrics over the possible values which is used by the neural network. Doing this is like adding a wrong formula to a set of formulas that are used for the solution of a problem. Consequently, the "one of n"-encoding should be preferred. It uses one neuron for each possible value of the category. A value is encoded by activating only the according neuron. In case an order is defined over the possible values, the neighbourhood in the order can be presented to the network by activating those neurons (with lower activation) which encode the categories neighboured to the active category. An example of an encoding for hierarchical categorical variables (e.g. variables encoded with ICD9 or SNOMED) can be found in [Waschulzik 1993].

Numbers:

Over numbers a metrics is always defined. Hence, we can initialize the activity of the encoding neuron with an activity calculated from the value of the variable. We prefer the topological encoding [Geiger 1990]. This distributed encoding for numbers allows to model even complex functions with a simple perceptron [Eldracher 1993] and thus avoids the need of a hidden layer in the neural network.

Combination of Variables by Encoding

In most cases a set of input variables is used to determine the output variable(s). According to the desired properties of the combination, there are different possibilities for the combination of variables.

Parallel Encoding:

Parallel encoding is the most common type of variable combination. Encodings of the variables are simply put side by side without carrying out specific operations. This simple approach is advantageous when many variables of different types have to be combined.

Conjunctive Encoding:

The conjunctive encoding is built upon the basic encodings of the variables that are to be combined. One neuron is used for every possible combination of the neurons from the different basic encodings. The activity of the resulting neuron is calculated by multiplying the activities of the neurons from the basic variable encodings that are combined. The conjunctive encoding in principal allows to model every possible mapping from the input to the output space as it was shown for RBF-Networks in [Park 1991, Benaim 1994]. The disadvantage of conjunctive encoding is the large number of needed neurons. The conjunctive encoding is formally discussed in [Hinton 1986, p. 90].

One result of our investigation is that conjunctive encoding was less important in our practical applications than we had expected. In most of our applications, the conjunctive encoding did not yield better results than the parallel encoding with appropriately represented variables. These results lead to the impression that the information about the conjunction of variables in most practical applications is less important.

Network Structure

The first applications we rendered by neural networks were completely neural solutions involving complex, structured neural networks for pattern recognition tasks. The complexity of the networks was reduced stepwise to improve efficiency and quality assurance. For the same reason we directed our attention towards the development and improvement of the representations and encodings.

For the majority of the applications we prefer the simple perceptron with one input and one output layer where each input neuron is connected with each output neuron [Rosenblatt 1958]. The directed connections are optimized by a gradient descent algorithm, the so called δ-rule [Widrow 1960]. The advantages of the perceptron are its good generalisation properties, the fast convergence of the training algorithm and the easy interpretation of the network structure.

3.4 Evaluation

The evaluation consists of verification, validation and assessment [Engelbrecht 1994].

Verification

The verification is made by analysing the structure of the trained network which is important especially when the results on the assessment data set are better than expected. The competence of a network in a special context can be evaluated by an ARE - retrieving cases similar to a particular example from the training set (which is representing the context) using the same representations and encodings that were used for the initialisation for the input and output neurons. If the ARE finds many cases that are similar to such an example from the applicational view the competence of the network is good in the given context. Furthermore, the training data set can be split up into two equally sized parts and the resulting network structures can be compared after training.

Validation
In addition to the tests with the test set a data set with artificial examples, e.g. proto-typical cases, can be set up and the results can be compared with textbook knowledge and be discussed with an expert of the domain. It is also interesting to analyse the examples that yield poor results.

Assessment
The assessment is entirely done by tests with the assessment data set. Statistical tests are recommended to decide whether the results obtained by the neural network differ significantly from random choice.

4 A Conventional Approach
In our first approach, we used conventional tools and mainly conventional techniques despite some influences from SENN introduced by former discussions about neural network development methods.

4.1 Pre-Processing, Creation and Simulation of the Neural Networks

DataSculptor® V1.53 was used for data pre-processing. It imports different database formats and offers many ways of transforming raw data.

NeuralWorks Professional® V5.0 was used for creating the networks and for performing the simulations. The creation of each network has to be done manually by using dialogues in a graphical environment where architectures and learning paradigms can be chosen from a wide range. Each network has to be connected with two pre-processed files: one containing the training set, the other containing the test set. All network and learning parameters including the number of input and output variables must be set by the user since they can not be derived from the data files. During learning and testing, different protocols are written to disk (for later analysis) and to the screen for instant information. Only a limited batch-processing facility comes with NeuralWorks. Unfortunately, it exclusively serves for training and testing existing networks. There is no easy-to-use net description language that could be used to automatically create networks based on the set of variables in the data sets. In practice, networks are being created manually and then trained and tested in batch-mode where the creation not only represents a bottleneck but also a very error-prone activity because of the manual mapping from data fields to input and output neurons.

4.2 Post-Processing, Analysis and Selection
These tasks are performed by different tools (DataSculptor, spreadsheet, word processor, shell-scripts) each processing the protocol files created during the simulations. The difficulty is to select the right files of the right version and the right information in them to compare different networks. This is not supported by the used

commercial tools and is hence a very time consuming process. The selection of a "best network" is made based on the following criteria:

- mean value of sensitivity and specificity: MSS = (sensitivity + specificity) / 2 on the test set which should be maximal. Note that this is quite different from the "classification rate" which uses the a-priori probability
- difference(classification rate (test set), classification rate (training set)) which should be minimal
- coverage = (number of cases that could be classified at all) / (total number of cases) which should be maximal [Egmont-Peterson, 1994]
- complexity factor c_k=(total number of free model parameter)/(total number of training cases) which should be minimal

Although the used tools are very flexible, their main drawback is that they are not working together in an efficient way. While being useful for research this approach is too time consuming for real-world projects.

4.3 Description of the Simulations

Binary features were coded by one neuron; ordinal features were coded by 0 to n neurons using a thermometer code and all other categorical features were coded as 1-of-n. Different coding schemes were used for numeric features. Missing values were estimated in rare cases: the a-priori probability was used for categorical and the mean for numeric features. In order to obtain a minimal net with best generalisation characteristics the following approaches were undertaken (see also [Foerster 1994]):

A1) permanent decrease in the total number of free parameters of a multi-layer perceptron (MLP) trained by back-propagation [Rumelhart 1986] by training over-dimensioned (large) nets and afterwards eliminating weights of minor importance (weight elimination) and / or pruning of neurons in the hidden layers that had only minor influence on the overall performance of the net

A2) permanent increase in the total number of free parameters of the net by adding new neurons until no further improvement of the classification quality can be reached (Cascade Correlation) [Fahlman 1991]

B) numeric features were coded in three different ways:
  B1) scaled input, unchanged distribution
  B2) scaled input after suitable transformation of the distribution (e.g., log()) in order to achieve a more homogeneous distribution
  B3) like B2, in addition, each numeric input is coded as a fuzzy variable where every neuron is activated according to a membership function (5 neurons per feature, e.g. blood glucose: very low, low, medium, high, very high).

This fuzzy coding was used in order to preserve similarities of the transformation to a maximal extent which would have been partly lost by applying the straight-forward splitting into categorical medical categories.

4.4 Results of the Simulations

The criteria for stopping the learning procedure was the (at a certain point decreasing) classification rate on the test set instead of the mean square error on the training set. Therefore, a cyclic testing based on the test set of the net during learning was performed and the currently best net was saved. The results of the simulations can be summarised as follows:

A1) A bad convergence during learning produced relatively poor initial configurations. Neither weight elimination nor pruning achieved better generalisation using these starting configurations

A2) Nets of different complexity evolved that did not achieve better generalisation than nets of smaller complexity.

After these disappointing results, different coding schemes for numeric input features were applied (B1-B3) the results of which are summarised below. The results show the mean of 10 runs. During training, every class is presented with the same frequency, i.e., the information of the a-priori probability is discarded. To verify the results, logistic regression (C) was performed using BMDP®. Afterwards, another net (C1) was created using only features selected by logistic regression.

Keeping in mind that a subset of features produces similar simulation results, three different subsets of features were tested (C2, C3, C4). The feature selection was based on sensitivity analysis of trained nets as well as on some early results obtained by HEAD. A fuzzy coding similar to B3 was used.

Table 4.1 shows that:
- a suitable representation of the features (B3) not only yields better generalisation but facilitates the network structure (no hidden neurons were needed) as well;
- only a small subset of features is needed to make the decision (C1-C4).
- a better generalisation based on MSS was obtained with a decreasing number of free parameters (B2, B3 vs. B1)

Simulation			Test Set			Training Set
	Input variables / topology (numbers from table 3.1)	number of free parameters	classification rate	sensitivity	specificy	classification rate
B1	all input variables (37-6-1)	235	63%	64%	62%	76%
B2	all input variables (37-0-2)	76	67%	66%	69%	74%
B3	all input variables (80-0-1)	81	70%	70%	69%	75%
C	logistic regression[2]	12	71%	84%	45%	67%
C1	input variables of C and coding of B3 (33-0-1)	34	70%	69%	73%	73%
C2	4, 6, 8, 10 (27-0-2)	56	63%	60%	69%	72%
C3	4, 6, 8, 10, 12 (30-0-2)	62	68%	71%	61%	78%
C4	4, 6, 7, 8, 10, 11, 12, 19 (37-0-2)	76	69%	71%	67%	77%

Table 4.1: Simulation results on the training set (conventional approach)

The focus on the representation and the use of a simple perceptron instead of a MLP resulted in a remarkable speedup in the learning time and in an easy, "glass box" interpretation of the nets. In contrast to the logistic regression (C), sensitivity and specificy are not very different using neural networks. Finally, all networks were tested on the assessment data set as it is shown in table 4.2.

Although sensitivity and specificy are comparable for both the test and the training set the assessment set shows a different behaviour. Only a SF is recognized with some certainty (high sensitivity) whereas the networks are roughly guessing when

---

[2] The variables 4,6,7,9,10,11,13,15,16 from table 3.1 were used by the logistic regression

they are faced with patients without SF (specificy near 50%). Interestingly this is very similar to the result of the logistic regression (C) in table 4.1.

	Assessment Set (network trained with training set)			Assessment Set (network trained with training and test set)			Training Set & Test Set (network trained with training and test set)		
	classification rate	sensitivity	specificy	classification rate	sensitivity	specificy	classification rate	sensitivity	specificy
B1	67%	76%	42%	68%	76%	44%	74%	77%	69%
B2	72%	81%	47%	69%	77%	45%	72%	71%	74%
B3	72%	80%	52%	64%	59%	80%	79%	81%	74%
C1	68%	65%	78%	71%	66%	84%	72%	72%	73%
C2	67%	74%	47%	73%	79%	57%	70%	71%	67%
C3	77%	85%	52%	80%	87%	57%	76%	80%	69%
C4	78%	85%	57%	78%	83%	63%	76%	80%	70%

Table 4.2: Simulation results on the test set (conventional approach)

On the assessment set, five to eight input variables (C3, C4) yield even better results than networks using all possible inputs (B1-B3). The fact that the classification rate on the assessment set was higher than that of the test set deserves further discussions.

5 Approach Using SENN

The same diagnosis of the SF was solved by using the SENN-approach supported by HEAD (Hybrid Expert system for the Analysis of large Data sets) [Nöhmeier 1994]. HEAD is an integrated tool for the data preparation and the development of neural networks using SENN. It is based on experiences developed for the analysis of epidemiological studies using neural networks and on an optimized simulator for perceptrons. The simulations were performed independently from the conventional approach.

5.1 Data Preparation, Creation and Simulation of the Neural Networks

It was agreed upon that the available data set should be completely used for the simulations and that the assessment should be done with examples collected later. The available data set was split up automatically by HEAD into a training and test set. Within HEAD, a network is designed by selecting the variables and their encodings for the simulations. The selection of the encodings is supported by proposals suggested by HEAD after the analysis of the available values for each variable. After these selections, the network is automatically generated, trained, tested and the results are analysed. The results are presented on-line and are also available in a file that can be easily analysed by shell-scripts and programs. This approach reduces the amount of time required for one development cycle and avoids errors as well. Categorical variables are encoded like in the conventional approach. Applying topological encoding, numeric variables were coded by different numbers of neurons.

5.2 Results

On the first day of the project, we prepared the data set for the application of HEAD and made simulations in order to optimize the number of neurons for the representation of the numeric variables. In order to improve the mean value of sensitivity and specificy both classes were trained with the same frequency. At the end of the first day, we tested the first combinations containing all input variables and obtained the results of S1. On the second day, we tried to combine only a subset of the variables. The best result (MSS = 72,38%) was obtained in simulation S2. On the third day, we changed to the routine level of HEAD and computed all combinations with three input-variables using both parallel and conjunctive encoding. Simulation S3 was the best combination and considering only three input variables. The simulations using conjunctive encoding achieved worse results than those using parallel encoding.

Examining the results of S3, the effect of overadaption on the test set can be recognized (the results on the test set are better than on the training set). The reason for this is the small number of free parameters which inhibits the overadaption effect on the training data set. The selection of the best results from a large number of simulations encourages the overadaption on the test set which is stronger than the overadaption effect on the training set. On the third day, we also tested some combinations of up to 6 variables where the simulation S4 achieved the best results.

Simulation			Test Set			Training Set
	Input variables (numbers from table 3.1)	number of free parameters	classification rate	sensitivity	specificy	classification rate
S1	all input variables	418	71%	85%	43%	90%
S2	4, 6, 8, 10, 12	60	71%	69%	75%	73%
S3	4, 6, 12	40	74%	78%	66%	71%
S4	1, 4, 6, 8, 9, 12	76	73%	84%	50%	77%
S5	2 X 4 X 6	128	74%	94%	32%	71%
S6	2 X 4 X 6, 16	132	76%	92%	43%	72%
S7	2 X 4 X 6, 19, 11	150	71%	71%	71%	73%

Table 5.1: Simulation results on the test set (approach with SENN)

	Assessment Set (network trained with training set)			Assessment Set (network trained with training and test set)			Training Set & Test Set (network trained with training and test set)		
	classification rate	sensitivity	specificy	classification rate	sensitivity	specificy	classification rate	sensitivity	specificy
S1	72%	89%	26%	74%	96%	10%	83%	93%	63%
S2	71%	72%	68%	79%	90%	47%	78%	87%	60%
S3	64%	67%	57%	64%	69%	52%	71%	78%	58%
S4	71%	80%	47%	81%	83%	73%	71%	70%	72%
S5	74%	98%	5%	78%	100%	15%	73%	93%	32%
S6	72%	92%	15%	60%	80%	5%	71%	82%	48%
S7	74%	94%	15%	75%	100%	5%	72%	93%	30%

Table 5.2: Simulation results on the assessment set (approach with SENN)

On the fourth and fifth day, we altered the training process in such a way that the classes were trained taking into account their a-priori probabilities. Now simulations using conjunctive encoding achieved better results. For the resulting networks, a big difference between the sensitivity and the specificy can be stated on the test set. On the sixth and seventh day, we ran the last simulations for the moment in order to evaluate the results on the assessment set. The tests with the assessment data set show that S1, S5, S6 and S7 reach a very good sensitivity but such a bad specificy that they will be hardly of any use in the practical application. The overadaption effect in simulation S3 was confirmed. The results on the assessment set of S2 and S4 are astonishing good. We can not expect to obtain these results in a practical application. It has to be critically analysed whether the assessment set is statistically representative. In addition, further assessment cases have to be collected. Before a medical interpretation of these results can be made, a lot of evaluation work, e.g. the reinterpretation of resulting network structure and the analyses of the wrong classified examples have to be carried out. All the achieved results are summarised in table 5.1 and 5.2.

6 Comparison of the two approaches

The conventional and the SENN-approach led to similar results after additional information on a proper encoding of the variables and on good combinations of data fields both obtained by SENN was used in the conventional approach. Until now, the main difference is to be found in the total time spent in each approach (table 6.1).

Further advantages are expected in the upcoming evaluation phase where the networks will be analysed to obtain a medical interpretation of the results. Mistakes during the analysis in the conventional approach due to the complexity and the number of different protocol files created during simulations and tests are another important disadvantage of the conventional approach. Because of the lacking integration of tools in the conventional approach, this approach could not cover as many trials as SENN did.

activity	conventional approach	SENN approach
total spent time	3 month	2 weeks
pre-processing	35%	
creation of networks	15%	40%
simulations	5%	
analysis and selection of networks	35%	40%
evaluation	10%	20%

Table 6.1: Total time spent for the development

7 Summary

Efficiency and quality assurance in medical projects can be enhanced by using the SENN method and neural networks without hidden layers. The higher efficiency mainly depends on the approach of developing proper representations and encodings for the combination of variables instead of testing different network types (like cascade correlation or back propagation) with a time consuming parameter adjustment. The SENN-based development of neural networks can be better supported by tools because network parameters for the training of the perceptron can be adjusted automatically. Thus, the training and the test of networks can be conducted without any further manual interaction of the user. This also improves the quality of the project. The use of the representations and encodings for an associative recall of examples might also enable both an early evaluation of the first project phases as well as a context dependent analysis of the competence of the network.

References

Anderson J.A., Rosenfeld E.: Neurocomputing - Foundations of Research MIT Press (1988)

Andreassen S., Engelbrecht R., Wyatt J. (eds), Technology and Informatics 10, Artificial Intelligence in Medicine, p 466-476. IOS Press, Amsterdam (1993)

Benaim M.: On functional Approximation with Normalized Gaussian Units Neural Computation 6, p 319-333 (1994)

Boehm B.W.:A spiral model of software development and enhancement, Computer, May 1988, pp.61-72

Eldracher M., Waschulzik T.: A Topologically Distributed Encoding to Facilitate Learning. Journal of Systems Engineering 3, p 110-119. Springer, London (1993)

Egmont-Peterson M., Talmon, J. L., Brender J., NcNair P.: On the quality of neural net classifiers. Artificial Intelligence in Medicine, Vol. 6, No.5, pp. 359-381, 1994

Engelbrecht R., Rector A., Moser W: Verification and validation. In: Assessment and Evaluation of Information Technologies in Medicine (Eds.: van Gennip E.M.S.J.,Talmon J.L.) Amsterdam, IOS Press, p 51-66 (1994)

European NIDDM Policy Group: A Desktop Guide for the Management of non-insulin-dependent diabetes mellitus. IDF Bulletin 25, 1990 1 (1990)

Fahlman S. E., Lebiere C.: The Cascade-Correlation Learning Architecture. School of Computer Science, Carnegie Mellon University, Pittsburg, USA (1991)

Förster M, Kirchner K, Waschulzik T, et al.: DIADOQ: "Untersuchungen zum Aufbau von Wissensbasen in der Diabetologie mittels Neuronaler Netze". 39. Jahrestagung der GMDS, Dresden (1994) (in press)

Geiger H.:Storing and Processing Information in Connectionist Systems.- In: R. Eckmiller, editor, Advanced Neural Computers, p. 271-277. Elsevier Science Publishers B.V. (North Holland) (1990)

Hinton G.E., McClelland J. L., Rummelhart D. E.: Distributed Representations p77-109. In: Rummelhart D. E., McClelland J.L. and the PDP Research Group: Parallel Distributed Processing, MIT Press (1986)

Ifeachor E., Rosen K. (eds), Proceedings of the International Conference on Neural Networks and Expert Systems in Medicine and Healthcare, Plymouth (1994)

Minsky M., Papert S.: Perceptrons, Cambridge, MA: MIT-Press (1969)

Nelson M. M., Inllingworth W.T.: A practical Guide to Neural Nets, Addison-Wesley Publishing Company (1990)

Nöhmeier M., Waschulzik T., Brauer W., Grothe F., Engelbrecht R.: A Rule-based System for Realisation of Medical Projects with Neural Networks (HEAD). In: Ifeachor E., Rosen K. (eds), Proceedings of the International Conference on Neural Networks and Expert Systems in Medicine and Healthcare, Plymouth (1994)

Park J., Sandberg I. W.: Universal approximation using radial-basis-function networks, Neural Computation 3, p 246-257 (1991)

Polzer A.: Invariante Mustererkennung durch neuronale Netzwerke mit großen rezeptiven Feldern.- Diploma thesis, München, 1992

Rosenblatt F.: The perceptron: a probabilistic model for information storage and organization in the brain. Physiological Review 65, p 386-408 (1958)

Rumelhart D. E., Hinton G. E., Williams R. J.: Learning representations by back-propagating errors. Nature 323, p 533-536 (1986)

Waschulzik T., Geiger H.: Eine Entwicklungsmethodik für strukturierte konnektionistische Systeme.- In: G. Dorffner (eds), Informatik Fachberichte 252, Konnektionismus in Artificial Intelligence und Kognitionsforschung, p. 202-206. Springer , Berlin (1990a)

Werbos P. J. : Beyond regression: new tools for prediction and analysis in the behavioral sciences. Ph. D. thesis, Harvard University, Cambridge, MA (1974)

Widrow B., Hoff M. E.: Adaptive switching circuits. 1960 IRE WESCON Convention Record, New York: IRE, p 96-104 (1960)

Yki-Järvinnen H.:Pathogenesis of non-Insulin-dependent diabetes mellitus. The Lancet Vol 343, p 91-95 (1994)

A Prototype Neural Network Decision-Support Tool for the Early Diagnosis of Acute Myocardial Infarction

Joseph Downs, Robert F Harrison

Dept. of Automatic Control and Systems Engineering, University of Sheffield
Mappin Street, Sheffield S1 3JD, UK
e-mail: {downs,r.f.harrison}@acse.shef.ac.uk

R Lee Kennedy

Dept. of Medicine, University of Edinburgh
Western General Hospital, Crewe Road South, Edinburgh EH2 4XU, UK
e-mail: LK@srv0.med.edinburgh.ac.uk

Abstract. An application of the ARTMAP neural network model to the early diagnosis of acute myocardial infarction is described. Performance results are given for 10 individual ARTMAP networks, and for combinations of the networks using "pooled" decision making (the so-called *voting strategy*). Category nodes are pruned from the trained networks in different ways so as to improve accuracy, sensitivity and specificity respectively. The differently pruned networks are employed in a novel "cascaded" variation of the voting strategy. This allows a partitioning of the test data into predictions with a high and a lower certainty of being correct, providing the diagnosing clinician with an indication of the reliability of an individual prediction.

1 Introduction

The early identification of patients with acute ischaemic heart disease remains one of the greatest challenges in emergency medicine. The ECG only shows diagnostic changes in about half of acute myocardial infarction (AMI) patients at presentation [2,15]. None of the available biochemical tests becomes positive until at least three hours after symptoms begin, making such measurements of limited use for the early triage of patients with suspected AMI [1]. The early diagnosis of AMI, therefore, relies on an analysis of clinical features along with ECG data. A variety of statistical and computer-based algorithms has been developed to assist with the analysis of these factors (for review see [13]). Although none of these has yet found widespread usage in clinical practice, this remains an important area of research not only because of its clear potential to improve triage practices for the commonest of all medical problems, but also because of the light it may shed on techniques for the development of decision aids for use in other areas of medicine.

This paper describes the application of the ARTMAP neural network model to this task. This powerful, but little-used, model has a number of advantages for medical domains, outlined in section 2 (see also [11]). Section 3 describes the provenance of the patient data used in this study, as well as the training and testing procedures applied to the ARTMAP model with this data. Section 4 gives performance results for different ARTMAP configurations. Section 5 concludes with a discussion of the strengths and weaknesses of the approach, and suggests areas for future work.

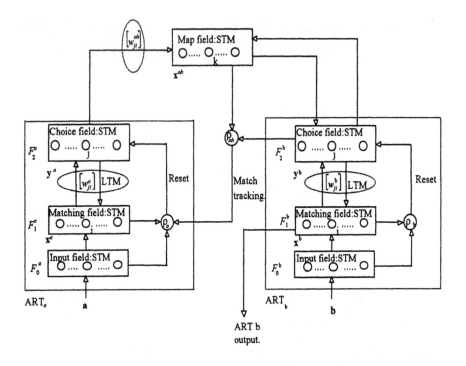

Figure 1: ARTMAP

2 ARTMAP

ARTMAP [6] is a self-organizing, supervised learning, neural network model for the classification of binary patterns. (In actuality, our implementation is most closely akin to Simplified Fuzzy ARTMAP [12] which can process analogue or binary data. However, with the purely binary data of this application, see section 3.1, the implementation coincides with ARTMAP). It is one of a series of models based upon Adaptive Resonance Theory, or ART, [4] an outgrowth of competitive learning which overcomes the stability problems of that paradigm [10]. This is achieved by utilizing feedback between layers of input and category nodes in addition to the standard feedforward connections of competitive learning. Thus, in ART models, an input pattern is not automatically assigned to the category that is initially maximally activated by the input. It should also be noted that most ART models, including ARTMAP, employ a localist representation for category nodes owing to the so-called "winner-take-all" competitive learning dynamics.

ARTMAP itself consists of three modules, two ART 1 systems [3] termed ART_a and ART_b, and a related structure termed the map field (see figure 1). During training, input patterns are presented to ART_a together with their associated teaching stimuli at ART_b. Associations between patterns at ART_a and ART_b are then formed at the map field. During testing, supervisory inputs at ART_b are omitted, and instead the inputs at ART_a are used to recall a previously learned association with an ART_b pattern via the map field. ARTMAP does not directly associate inputs at ART_a and ART_b. Instead, such patterns are first self-organized into prototypical category clusters before being associated at the map field.

This results in the formation of generalized associations. (Although in practice, domains such as AMI diagnosis, which perform many-to-one classification do not usually require generalization of the teaching inputs and a simplified ART_b module can be used which simply codes these patterns directly.)

Training in ARTMAP almost always results in multiple category clusters forming at ART_a for each teaching category present at ART_b. Each such ART_a cluster thus represents a significant sub-region of the overall state space covered by a particular teaching category. It can be seen therefore that ARTMAP instantiates a many-to-one mapping between ART_a input patterns and their actual classification.

ARTMAP has a number of desirable properties for potential use as a decision-support tool in medical domains. First, it has few user-changeable parameters, which allows the model to be tuned to a particular problem without undue effort. The single most important parameter is that controlling the *vigilance* of the ART_a module. This determines how close a match is required between an ART_a input pattern and a category cluster prototype before accepting an input as a member of the cluster. This parameter (indirectly) controls the size of the category clusters that will form, since the higher it is set, the closer acceptable matches must be, and the smaller the coverage of the state space each cluster will have. Generally, higher vigilance provides better classification performance, although this must be balanced against the potential proliferation of category clusters, providing poor data compression and leading the net to become little more than an "look-up table" [14]. Additionally, with small training sets and/or "wide" input vectors with many features, high vigilance can lead to incomplete coverage of the state space by the network.

Second, ARTMAP does not perform optimization of an objective function and is not therefore prone to the problem of local minima as occurs with feedforward networks using backpropagation. Instead, as described before, it self-organizes its own structuring of the data, automatically creating new category nodes for itself as and when they become needed.

Third, the model is able to discriminate rare events from a "sea" of similar cases with different outcomes owing to the feedback mechanism based on top-down matching of learned categories to input patterns. This is again in contrast to feedforward networks using backpropagation, where weights are refined by a process which effectively averages together similar cases and hence fails to acknowledge rare events. ARTMAP is thus suitable for domains where the distribution of data items is highly skewed between different categories. (See [9] for a particularly marked example of this phenomena.)

Fourth, successful learning in ARTMAP can occur with only one pass through the data set (termed single-epoch training). Furthermore, the model is capable of incorporating new data items at a later time without degradation of performance on previous data, or the necessity of retraining on past data. (This solution to the so-called *stability-plasticity dilemma* is claimed to be a feature unique among neural networks to the ART models.)

For the purposes of this paper, two further features of ARTMAP are of particular note, the *voting strategy* and *category pruning*. These are described in detail next.

2.1 Voting Strategy

The formation of category clusters in ARTMAP is affected by the order of presentation of input data items [5]. Thus the same data presented in a different order to separate ARTMAP networks can lead to the formation of quite different clusters within the two nets. This subsequently leads to different categorisations of test data, and thus different performance scores. This effect is particularly marked with small training sets and/or high-dimensional input vectors, where the input items may not be fully representative of the domain, and with single-epoch training.

This effect can be compensated for by the use of the ARTMAP voting strategy [5]. This works as follows: a number of ARTMAP networks are trained on different orderings of the training data. During testing, each individual network makes its prediction for a test item in the normal way. The number of predictions made for each category is then totalled and the one with the highest score (or the most "votes") is the final predicted category outcome. The voting strategy can provide improved ARTMAP performance in comparison with the individual networks. In addition it also provides an indication of the confidence of a particular prediction, since the larger the voting majority, the more certain is the prediction. In particular, this application utilises unanimous verdicts to indicate predictions which have a very high certainty of being correct.

2.2 Category Pruning

An ARTMAP network often becomes "over-specified" on the training set, generating many low-utility ART_a category clusters which represent rare but *unimportant* cases. The problem is particularly acute when a high ART_a baseline vigilance level is used during training. To overcome this difficulty, category pruning can be performed. This involves the deletion of these low utility nodes.

Pruning is guided by the calculation of a *confidence factor* (CF) between nought and one for each category cluster, based upon a node's *usage* and *accuracy*. The usage score for an ART_a node is simply the number of training set exemplars it encodes, normalized through division by the maximum of exemplars encoded by any node with the same category outcome. (Hence, there will be at least one node for each different category class which has a maximal usage score of one.) The accuracy score for a node is calculated as the proportion of predictions that are correct which the node makes on a prediction data set separate to the training data. This score is then normalized, similarly to the usage calculation, through division by the maximum proportion of correct predictions made by any node with the same outcome. (Thus there will be at least one node for every category class which has a maximal accuracy score of one.) The confidence factor for a node is then calculated as the mean of its usage and accuracy scores. All nodes with a confidence factor below a user-set threshold will be pruned. (Full details of the process are given in [7].)

The pruning process can provide significant reductions in the size of a network. In addition, it also has the very useful side-effect that a pruned network's performance is usually superior to the original, unpruned net on both the prediction set and on entirely novel test data. Although category pruning was originally intended largely as a "preprocessing" stage prior to the extraction of symbolic rules from the network, we consider the "side-effects" to have great benefit in themselves. Accordingly, in this paper we address category pruning without any consideration of the subsequent extraction of symbolic rules. (See [8] for a discussion of this latter capability within a medical domain.)

In the original formulation of the pruning process, a uniform CF threshold is used to select nodes for deletion, irrespective of their category class. In this application, we have generalised the pruning process to allow separate CF thresholds for nodes belonging to different category classes. This allows us to vary the proportion of the state-space covered by different categories. This is useful for medical domains since it allows an ARTMAP network to be pruned so as to trade sensitivity for specificity and vice versa.

3 Patients and Methods

3.1 Patients and Clinical Data

The data used in this study were derived from consecutive patients attending the Accident and Emergency Department of the Royal Infirmary, Edinburgh, Scotland, with non-traumatic chest pain as the major symptom. The relevant clinical and ECG data were entered onto a purpose-designed proforma at, or soon after, the patient's presentation. The study included both patients who were admitted and those who were discharged. 970 patients were recruited during the study period (September to December 1993). The final diagnosis for these patients was assigned independently by a Consultant Physician, a Research Nurse and a Cardiology Registrar. This diagnosis made use of follow-up ECGs, cardiac enzyme studies and other investigations as well as clinical history obtained from review of the patient's notes. Patients discharged from Accident and Emergency were contacted directly regarding further symptoms and, where necessary, their General Practitioners were also contacted and the notes of any further hospital follow-up reviewed. The final diagnosis in the 970 patients was Q wave AMI in 146 cases, non-Q wave AMI in 45, unstable angina in 69, stable angina in 271 and other diagnoses in 439 cases. The patients were 583 men and 387 women with a mean age of 58.2 years (range 14–92). Unstable angina was defined as either more than two episodes of pain lasting more than 10 minutes in a 24 hour period or more than three episodes in a 48 hour period, or as angina which was associated with the development of new ECG changes of ischaemia (either at diagnosis or in the subsequent three days).

The input data items for the ARTMAP model were all derived from clinical or ECG data available at the time of the patient's presentation. In all, 35 items were used, coded as 37 binary inputs. For the purposes of this application, the final diagnoses were collapsed into two classes termed "AMI" (Q wave AMI and non-Q wave AMI) and "not-AMI" (all other diagnoses). AMI cases were assigned as positive diagnoses, not-AMI cases as negative diagnoses. Informed consent was obtained from all patients participating in the study which was approved by the local Medical Ethics Committee.

3.2 Method

The 970 patient records were divided into three data sets; 150 randomly selected records formed the *prediction set*, a further 150 randomly chosen records formed the *test set*, and the remaining 670 comprised the *training data*. The prediction set consisted of 28 cases of AMI and 122 not-AMI; the test set of 30 AMI and 120 not-AMI.

The training data was randomly ordered in ten different ways, and each ordering applied to a different ARTMAP network using single-epoch training. The ART_a base-line vigilance was set to a medium level (0.6) for training, all other parameters were set to their standard values [12]. The performance of the ten trained ARTMAP networks was then measured on both the prediction and test sets. During this testing phase the ART_a baseline vigilance was relaxed slightly (to 0.5) to ensure that all test items were matched to an existing category cluster (i.e. forced choice prediction).

The performance of the trained networks on the prediction set alone was then used to calculate accuracy scores for the category nodes in each network, as a prerequisite of the category pruning process described in section 2.2.

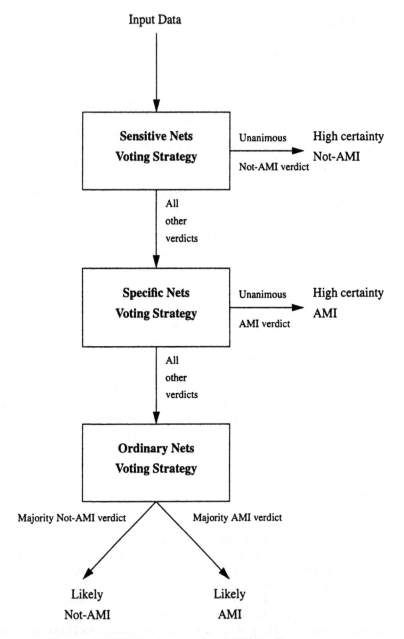

Figure 2: ARTMAP Configuration for AMI Diagnosis

The "standard" form of category pruning [7] was performed on the original networks, such that all nodes with a CF below 0.5 were deleted from the networks to improve predictive accuracy. Performance of the resultant pruned networks was then measured on the prediction and test sets. Vigilance was further relaxed to 0.4 for testing these (and all other) pruned networks, again to ensure forced choice prediction.

The original networks were then pruned using different CF thresholds for the AMI and not-AMI nodes to produce pruned networks which maximized *sensitivity*. CF thresholds of 0.2 for AMI nodes and 0.95 for not-AMI nodes were employed, the criterion for setting the CF thresholds being to produce a mean sensitivity greater than 95% on the prediction set for the 10 pruned networks. Performance of the resultant nets was recorded for both the prediction and test sets. A similar procedure was then conducted to produce 10 networks which maximized *specificity*. CF thresholds of 0.7 AMI and 0.5 not-AMI were sufficient to yield a mean specificity greater than 95% on the prediction set.

The final pruning procedure was to produce 10 networks with approximately equal sensitivity and specificity (ESAS), the criterion for setting the CF thresholds being a performance on the prediction set where sensitivity and specificity were within 5% of each other. The performance of the pruned networks was again recorded on both the prediction and test sets.

Performance results using the voting strategy were then obtained for the unpruned networks and all classes of pruned network. Three voters were used with all networks types, except the ESAS class, where five voters were used. Voters for the unpruned, uniformly pruned, and ESAS network classes were selected on the basis of the networks with the highest accuracy on the prediction set. Selection criteria for the set of sensitive networks was maximum specificity, while maintaining a minimum sensitivity of 95% on the prediction set. The converse criteria were used for the specific networks.

Lastly, a novel "cascaded" variant of the voting strategy was employed utilizing 3 sensitive nets, 2 specific nets and 5 ESAS nets (see figure 2). This operated as follows: data items were first applied to the sensitive voting nets. If these yielded a unanimous (3-0) verdict that the category prediction was not-AMI, this was taken as the final category prediction. If not, the input was presented to the specific voting nets. If these yielded a unanimous (2-0) verdict of AMI, this was taken as the category prediction. Otherwise the final prediction of the category class of the test item was obtained by majority verdict from the ESAS nets.

4 Results

The mean performance on the prediction and test sets for all classes of ARTMAP networks is shown in table 1. As a baseline for comparisons, the Casualty Doctors showed an accuracy, sensitivity and specificity of 83.0%, 81.3% and 83.5% respectively over the entire data set.

Average accuracy for the unpruned networks can be seen to be only slightly below this baseline. However this is largely an artefact of the unequal prior probabilities of the category distributions—specificity accounts for the majority of accuracy, and although the networks' sensitivity is much poorer than the humans', this is compensated for by the slightly superior specificity.

As expected, the uniformly pruned networks show an across-the-board increase in accuracy over the unpruned nets, with a 2.7% increase on the test set, and a 7.3% increase on the prediction set. (The greater increase in performance on the prediction set is explained by the fact that pruning utilized the accuracy scores for this data, and the networks are consequently optimized for this data.) However, the increase in accuracy is largely because of an overall improvement in specificity rather than sensitivity, which actually drops on the test set.

	Prediction Set			Test Set		
Network Type	Accuracy (%)	Sensitivity (%)	Specificity (%)	Accuracy (%)	Sensitivity (%)	Specificity (%)
Unpruned	80.9	51.8	87.5	80.9	59.0	86.3
Uniform Pruning	88.2	60.7	94.5	83.6	52.0	91.5
Pruning for Sensitivity	50.0	96.4	39.3	47.3	94.3	35.5
Pruning for Specificity	86.9	41.8	97.2	84.7	39.7	96.0
Pruning for Equal Sensitivity and Specificity	76.6	76.1	76.7	75.6	80.0	74.5

Table 1: Mean Performance of 10 Differently Pruned Networks

Figures for the sensitive nets show that almost all AMI cases can be diagnosed by the network, while approximately 36% of the not-AMI cases are detected. Conversely, with the specific nets, almost all not-AMI cases are covered while approximately 40% of the AMI cases are also detected.

The performance of the ESAS class networks is most directly comparable with that of the Casualty Doctors, since they are not unduly biased towards specificity or sensitivity. It can be seen that the mean individual accuracy of such networks is approximately 7% worse than the human diagnoses.

When the voting strategy is employed the accuracy of all network types except the specific nets is improved, as shown in table 2. Furthermore, unlike pruning, performance improvements owing to the voting strategy are almost always because of both increased sensitivity and specificity. Accuracy for the ESAS nets is now much closer to that of the Casualty Doctors and sensitivity is slightly better. Accuracy for the unpruned and uniformly pruned networks is now higher than the human diagnoses, particularly with the latter network class. However, this is again because of the networks' very high specificity while their sensitivity remains relatively poor.

Use of the voting strategy with the sensitive networks on the test set results in increased coverage of the AMI cases while trapping even more not-AMI cases than previously. However, the converse is not true for the specific nets, where a gain in not-AMI coverage is offset by poorer coverage of the AMI cases in comparison to the individual network means.

	Prediction Set			Test Set		
Network Type	Accuracy (%)	Sensitivity (%)	Specificity (%)	Accuracy (%)	Sensitivity (%)	Specificity (%)
Unpruned	86.0	64.3	91.0	83.3	56.7	90.0
Uniform Pruning	92.0	78.6	95.1	88.0	56.7	95.8
Pruning for Sensitivity	55.3	96.4	45.9	51.3	96.7	40.0
Pruning for Specificity	88.7	46.4	98.4	84.7	33.3	97.5
Pruning for Equal Sensitivity and Specificity	82.0	82.1	82.0	81.3	83.3	80.8

Table 2: Voting Strategy Performance of Differently Pruned Networks

The best overall network performance was achieved by the cascaded voting strategy, shown in table 3.

	Prediction Set			Test Set		
	Accuracy (%)	Sensitivity (%)	Specificity (%)	Accuracy (%)	Sensitivity (%)	Specificity (%)
High Certainty Voters	100.0	100.0	100.0	96.3	88.9	97.8
Lower Certainty Voters	71.0	73.7	70.3	72.9	81.0	70.7
Overall Performance	82.0	82.1	82.0	82.7	86.7	81.7

Table 3: Performance of the Cascaded Voting Strategy

The cascade's overall performance can be seen to be almost identical to that of the Casualty Doctors. Moreover, the cascade provides a partitioning of input items into those with a high and a lower certainty of a correct diagnosis. Unanimous not-AMI decisions by the highly sensitive networks (i.e. the first stage of the cascade) are almost certain to be correct, similarly unanimous AMI decisions by the highly specific networks (the second stage of the cascade) are also almost certain to be correct. The ESAS class voters then provide lower certainty predictions for the remaining data items at the bottom of the cascade. High-certainty predictions accounted for 38% of items in the prediction set and **36% of items in the test set.**

Perfect performance by the high-certainty voters on the test set was prevented by the occurrence of one false positive case and one false negative case. At least one of these data items is highly atypical, and will be discussed further in section 5.

5 Discussion

We consider the prototype decision-support tool that has been described here to be potentially valuable in assisting the early diagnosis of AMI. Furthermore, the general architecture should be of utility in other medical domains.

The ARTMAP application employs two novel techniques. First, the generalization of the category pruning method to allow for different threshold confidence factors for nodes of differing category classes. Second, the employment of a cascaded voting strategy which uses differently pruned networks.

The use of different CF thresholds for category pruning allows networks to be created which trade either sensitivity for specificity or vice versa. This should be particularly useful in domains where the costs of misdiagnosis of one class are much greater than for another, since it allows biasing of network performance so as to avoid the high-cost misdiagnoses. It also allows for the correction of any biases in classification arising with the initial training of ARTMAP. Such an approach may be compared with weighting "risk" in a Bayesian classification system.

The benefits of the cascaded voting strategy are twofold. First, overall classification performance is improved. Second, it allows those input data items to be identified which ARTMAP has a very high likelihood of diagnosing correctly. We shall now consider these claims in the context of the system implemented for AMI diagnosis.

It is certainly the case that the variable CF thresholds allowed the construction of both highly sensitive and highly specific nets. Moreover, the construction of the ESAS nets demonstrates the use of the technique to correct an initial bias in the network performance—in this case high specificity at the cost of poor sensitivity owing to disparate prior category probabilities.

The drawback of the technique is that it can be difficult to select the best CF thresholds that are needed to cause the desired changes in network performance (cf the use of Receiver Operating Characteristic, ROC, curves in setting appropriate thresholds in Bayesian systems). This problem was particularly acute for the construction of the ESAS class nets, where each individual network required a different CF threshold setting to achieve the desired performance. In the present implementation the CF thresholds were "hand-set" by the system's designer using a rather laborious trial-and-error process. A useful area for future work therefore would be to devise a procedure to automate this process.

Considering the cascaded voting strategy, the overall performance results for this method were the best achieved by any of the ARTMAP configurations tested (see tables 1, 2 and 3). (However, performance of the ESAS voting nets was identical on the prediction set and only slightly weaker on the test set). This performance was very similar to that of the Casualty Doctors and we have some confidence that further improvements could be achieved with an enlarged data set. In particular, we believe the prediction and test sets were probably too small, particularly given the unequal distribution of category classes with relatively few AMI cases. The small number of AMI cases in the prediction set is the cause of most concern, since optimum benefit from category pruning is achieved only if the prediction set is truly representative of the overall domain. Otherwise, pruning will optimize a net's performance on the prediction set, but not will not generalize well to novel test data. It is intended to test this using more data that will shortly become available from the same site.

The cascaded voting strategy was also intended to partition input data so to identify those cases on which the ARTMAP system makes near-perfect predictions. In the AMI domain such cases accounted for over one-third of the test items although perfect performance was prevented by the occurrence of one false positive and one false negative prediction. Examination of the input features for these cases is revealing however.

The false positive case had the following features: age=45-65, smoker, family history of ischaemic heart disease, central chest pain radiating to the jaw, short of breath, nausea, new ST segment elevation, new pathological Q waves and ST segment or T wave changes suggestive of ischaemia. This exhibits almost all of the "classic" features of AMI, the latter three features being regarded as particularly strong AMI indicators.

The false negative case had the following features: age<45, smoker, pain in left side of chest radiating to the left arm, pain described as sharp or stabbing, old ECG features of MI and ECG signs of ischaemia known to be old. This displays none of the classic features of AMI, although the existence of the latter two features should mean a human clinician probably would not entirely discount the possibility of AMI. We conclude therefore that these cases are highly idiosyncratic, particularly the false positive, and would cause most human experts to make the wrong diagnosis. Thus the general ability of the cascaded voting strategy to identify cases with high-certainty of a correct diagnosis is not greatly undermined.

In summary therefore, we have described the application of the ARTMAP neural network model to the diagnosis of acute myocardial infarction, and introduced two new enhancements to this model—the use of variable threshold category pruning, and a cascaded voting strategy. The strengths and weaknesses of these new techniques, and the ARTMAP model in general, have been discussed. We conclude that the model is of potential value in both the diagnosis of AMI and in medical domains generally.

Acknowledgements

Thanks to Shaun Marriott for providing the diagram of ARTMAP.

This research was supported by the Science and Engineering Research Council (SERC) of the UK, grant number GR/J/43233.

References

[1] J.E. Adams, D.R. Abendschein and A.S. Jaffe (1993) Biochemical Markers of Myocardial Injury. Is MB Creatine Kinase the Choice for the 1990s?, *Circulation*, 88, 750–763.

[2] J. Adams, R. Trent and J. Rawles (1993) Earliest Electrocardiographic Evidence of Myocardial Infarction: Implications for Thrombolytic Treatment, *British Medical Journal*, 307, 409–413.

[3] G.A. Carpenter and S. Grossberg (1987) A Massively Parallel Architecture for a Self-Organizing Neural Pattern Recognition Machine, *Computer Vision, Graphics and Image Processing*, 37, 54–115. Reprinted in [4], 316–382.

[4]   G.A. Carpenter and S. Grossberg (Eds.) (1991) *Pattern Recognition by Self-Organizing Neural Networks*. Cambridge, MA: MIT Press.

[5]   G.A. Carpenter, S. Grossberg, N. Markuzon, J.H. Reynolds and D.B. Rosen (1992) Fuzzy ARTMAP: A Neural Network Architecture for Incremental Supervised Learning of Analog Multidimensional Maps, *IEEE Transactions on Neural Networks*, 3(5), 698–712.

[6]   G.A. Carpenter, S. Grossberg and J.H. Reynolds (1991) ARTMAP: Supervised Real-Time Learning and Classification of Nonstationary Data by a Self-Organizing Neural Network, *Neural Networks*, 4(5), 565–588.

[7]   G.A. Carpenter and A.H. Tan (1993) Rule Extraction, Fuzzy ARTMAP, and Medical Databases, *Proceedings of the World Congress on Neural Networks*, Volume I, 501–506.

[8]   J. Downs, R.F. Harrison and S.S. Cross (1995) Evaluating a Neural Network Decision Support Tool for the Diagnosis of Breast Cancer, this volume.

[9]   J. Downs, R.F. Harrison, R.L. Kennedy and K. Woods (In Press) The Use of Fuzzy ARTMAP to Identify Low Risk Coronary Care Patients, to appear in *Proceedings of the 1995 International Conference on Artificial Neural Networks and Genetic Algorithms* (ICANNGA–95). Vienna: Springer-Verlag.

[10] S. Grossberg (1987) Competitive Learning: From Interactive Activation to Adaptive Resonance, *Cognitive Science*, 11(1), 23–63.

[11] R.F. Harrison, C.P. Lim and R.L. Kennedy (1994) Autonomously Learning Neural Networks for Clinical Decision Support. In: E.C. Ifeachor and K.G. Rosen (Eds.) *Proceedings of the International Conference on Neural Networks and Expert Systems in Medicine and Healthcare* (NNESMED–94), Plymouth, UK, 15–22.

[12] T. Kasuba (1993) Simplified Fuzzy ARTMAP, *AI Expert*, 8(11), 18–25.

[13] R.L. Kennedy, R.F. Harrison and S.J. Marshall (1993) Do We Need Computer-Based Decision Support for the Diagnosis of Acute Chest Pain?, *Journal of the Royal Society of Medicine*, 86, 31–34.

[14] S. Marriott and R.F. Harrison (In Press) A Modified Fuzzy ARTMAP Architecture for the Approximation of Noisy Mappings, to appear in *Neural Networks*.

[15] M.E. Stark and J.L. Vacek (1987) The Initial Electrocardiogram During Admission for Myocardial Infarction. Use as a Predictor of Clinical Course and Facility Utilization, *Archives of Internal Medicine*, 147, 843–846.

Integration of Neural Networks and Rule Based Systems in the Interpretation of Liver Biopsy Images

Nadia Bianchi
Università degli Studi di Milano
Dipartimento di Fisica
via Viotti 5 - 20133 Milano
ITALY

Claudia Diamantini
Università degli Studi di Ancona
Istituto di Informatica
via Brecce Bianche - 60131 Ancona
ITALY

Abstract

Treatment of natural images requires, due to their complexity, to exploit high level knowledge, such as domain knowledge and heuristics, which are typically well formalized by rule based systems. However, the intrinsic variability and irregularity of objects in the image makes their characterization in terms of rules often unfeasible. Such variability and irregularity are, on the other hand, the ultimate reason for the existence of statistical methods. For these reasons, a hybrid system, exploiting characteristics of both approaches, may show better performances than purely syntactical or statistical systems in the interpretation of natural images. In this paper we present a hybrid system for image interpretation that integrates a rule based system with a Labeled Learning Vector Quantizer. The rule based system controls the interpretation process, by dynamically determining the interpretation strategy, and the Labeled Learning Vector Quantizer is exploited as classification kernel. The system has been tested on images of liver biopsies. Results on nuclei classification are here discussed.

Index Terms: Hybrid Systems, Rewriting Systems, Pattern Recognition, Adaptive Labeled Vector Quantization.

1 Introduction

This paper demonstrates the feasibility of a hybrid approach to the interpretation of digital images discussing the integration of a neural net classifier in a rule-based, data driven prototype for feature based interpretation, indexing and management of digital images of liver biopsies.

An important problem in image interpretation is to endow an automatic interpreter with the capability of flexibly determining the interpretation strategy, i.e. the sequence of computational actions which leads to the association of sets of pixels (structures) in an image with the unknown objects present in the scene from which the image was drawn. The difficulty in determining such a strategy increases when several objects of different kinds and related by a set of spatial-temporal relations are present in the observed scene. This problem can be effectively handled by exploiting syntactical methods not only to describe structures in the image, but also to dynamically establish the actions and the strategy [2, 3, 14].

However, there is an intrinsic complexity in trying to exactly characterize objects to be recognized. This is due to the fact that such objects have not regular characteristics, and descriptions can vary greatly from one object to another. Human experts deal with this problem by means of their experience, following criteria that have no simple syntactical counterpart.

Supervised learning algorithms for non parametric classification [7], [10, 11] share some similarities with this behaviour; they can construct a classification criterion on the basis of the "experience" done on a set of pre-classified observations, and they do not require the explicit formalization of knowledge. This produces higher classification performance on those domains characterized by strong irregularity and variability. For these reasons, a hybrid system, exploiting characteristics of both statistical and structural approaches, may show better performances than purely syntactical or statistical systems.

In this paper we present a hybrid system for image interpretation that integrates a rule based system with a Labeled Learning Vector Quantizer (LLVQ). The rule based system controls the interpretation process, by dynamically determining the interpretation strategy, and the Labeled Learning Vector Quantizer is exploited as classification kernel. The system has been tested on images of liver biopsies.

The paper is organized as follows. In section 2 the rule based system is briefly introduced highlighting its process control role. Section 3 presents the LLVQ, discussing its advantages and its integration in the structural system. Sections 4 and 5 are devoted to the description of experiments and the discussion of results.

2 A Rule Based System for Liver Biopsy Image Interpretation

A liver biopsy image (see e.g. the top left image in figure 4) shows a variety of differently stained biological structures: hepatocytes and their nuclei, sinusoidal spaces, Kupffer, Ito and lining cells together with nuclei and stained artefacts [6]. The goal of the interpretation activity is to recognize the structures in the image, to count them and to classify them on the basis of geometrical and topological attributes.

The traditional visual interpretation of liver tissue images is based on the contextual evaluation of several visual clues, such as the presence or absence of nuclei inside a nearly closed colour path, adjacency with sinusoidal spaces, different textures and shapes. Biologists therefore rely on the recognition of several structures, which appear clearly in the image, to build the context (of the tissue) necessary to recognize ambiguous structures even if they are heavily affected by noise.

The Assistant Biopsy Interpretation system (ABI), presented in [13] and here refined, follows a similar strategy. It exploits the recognition of well emerging structures to build the context for the recognition of the more ambiguous ones, by adopting contextual rules. The fundamental phases of the automatic interpretation process are shown in figure 1.

The RGB image is acquired and multisegmented to obtain different binary images, each containing traces of the interesting structures. Each segmented image is then described. The description is made by associating with each connected set of pixels (CON in table 1) a name and a set of attributes which characterize it. Next, each connected set of pixel is classified by reasoning on its attributes (Clue classification); in case of uncertainty the classification can be revised, on the basis of its relations with other structures classified so far (Data fusion and Contextual evaluation).

The formalization both of the structures of interest and of the criteria leading to the choice of a (sequence of) action(s) is achieved by using a common notation: *conditional attributed rewriting systems* (CARWs) [2].

A CARW is characterized by an alphabet N of attributed symbols and by contextual conditional production rules. A production rule in a CARW is defined by a syntactic part (a pair <Antecedent A; Consequent δ>) and a semantic part (a pair <Condition Co, Functions F>); the notation $<\alpha,\beta,\gamma>$ is used for A, α and γ being possibly empty strings constituting the context and β a non empty string called the rewriting part of the antecedent. A and δ are used in the usual way for rewriting if Co is satisfied, where the satisfaction of Co depends on attribute values of α, β and γ. In this case the functions in F are used to evaluate attributes of δ.

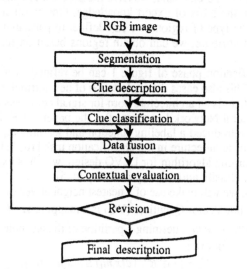

Fig. 1. The scheme of the automatic interpretation process in ABI.

The sets of rules in a CARW are named so that they can be coded into attributed symbols, and strategies can be described as strings of such symbols. Thus CARWs are used in two ways. First attributed symbols are used to denote structures, images, descriptions and actions. CARWs specify the transformations between strings of such symbols. Second, CARWs are used to define a mechanism to progressively determine the patterns of actions to be performed and specialize them to the data at hand, before executing them. In this way, at each step, the process state is described by a string of attributed symbols. The string is composed of symbols describing the data to process and the result produced so far, the already performed actions and the actions to perform next. In this way the language of strategies is used to reason on the language of descriptions, i.e. it is a metalanguage with respect to it [18]. Examples of the use of CARWs in ABI are discussed in section 3.2.

ABI is endowed with the ability to perform what Maes [12] defines as computational reflection, and it exploits metareasoning mechanisms to provide a flexible image interpretation control structure, as advocated by [14, 16].

3 Refinement of the System by the Integration of a Neural Network

In the pre-existent prototype of ABI, object classification is performed by formalizing the expert knowledge in a multivalued logic tree [8, 13], which is a sort of logic or decision table [1, 15]. This kind of structures are widely used for classification tasks, both in purely syntactic [8] and hybrid systems [17].

A logic table is a set of rules applied in a hierarchic fashion, which contain criteria to discriminate between objects. In particular, when objects are described by attribute vectors, as is the case, for example, in Pattern Recognition, rules contain a set of thresholds on each attribute. In this way an observation vector is ascribed to a certain region of the attribute space, and then to a certain class. In practice the logic table makes a partition of the attribute space, in order to define a classification rule.

The main limit of this approach is related to the possible complexity of the shape of regions defining the optimal partition, that is, the partition yielding the minimum classification error. On the one hand, we have the problem to construct a description of such regions, on the basis of expert knowledge. On the other hand, we have a structural limit in the type of regions we can define. In particular, defining a set of thresholds for each attribute, we can obtain regions based on rectangular units, as shown in figure 2a.

The clue classification phase of figure 1 can be refined over this approach, by adopting a LLVQ. This also eases the exploitation of new attributes.

A Vector Quantizer is an architecture born for signal compression purposes [9]. It is known in the Neural Network literature after the papers of Kohonen and his co-workers, who first introduced a labeling function and learning algorithms to exploit characteristics of this architecture in the classification task [10, 11]. We will adopt in this work a new learning algorithm for LLVQ design, which is explicitly devoted to the minimization of classification error probability [4, 5].

In what follows, we will make use of a nearest neighbor vector quantizer.

A nearest neighbor Vector Quantizer (VQ) is a mapping $\Omega : R^n \rightarrow M$, $M = \{m_1, m_2,...,m_M\}$, $m_i \in R^n$, $m_i \neq m_j$, defining a partition of the attribute space R^n. Partition regions are defined as follows:

$$V_i = \{x \in R^n \mid d(x,m_i) \leq d(x,m_j) \, j \neq i\},$$

where d is some distance measure. In what follows we will consider the familiar Euclidean distance. In this case borders of V_i are piecewise linear surfaces. Elements of M are usually called *code vectors*; region V_i is called the *Voronoi region* of m_i. We can adopt a VQ for classification tasks, equipping it with a labeling function $\Lambda: M \rightarrow C$, where $C = \{c_1, c_2,..., c_K\}$ is the set of class labels. In this way, classification of an observed vector x is performed by finding the code vector m_i nearest to x and then by declaring its label:

$$c_j = \Lambda(\Omega(x)).$$

The potential advantage of an LLVQ over decision tables is in the wide generality of regions we can define (see also [9], p.345). Let us consider, for example, the classification problem in figure 2.

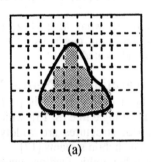 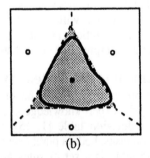
(a)   (b)

Fig 2. Comparison of decision table (a) and LVQ (b) classification performances

The thick line represents the optimal border (also said Bayesian border) which separates observation vectors that are classified in two different classes of objects, say class A inside the border and class B outside. The shaded region in figure 2a is the type of approximation of the optimal region for class A that we can obtain by decision tables (region for class B is its complement). Figures 2b shows the type of solution we can obtain by using an LLVQ to partition the attribute space. Dots represents code vectors, while different colours distinguish their class labels. Note that the Voronoi region of the black dot (the shaded region) is a good approximation of the optimal region for class A. Besides the quality of the approximation, it is also appreciable the difference in the number of regions that have to be defined in the two cases.

3.1 The Learning Algorithm for Classification

In order to obtain the best performance with an LLVQ, partition regions must approximate in the best possible way the optimal partition, as it is the case in figure 2b. To this aim, we can adapt the position of code vector in the attribute space, by a learning algorithm.

The learning algorithm we adopt was presented in [4] and generalized in [5]. It is a supervised learning algorithm for the minimization of error probability (MEP), based on a training set $T = \{(u_1,c_1), (u_2,c_2), ..., (u_T,c_T)\}$. Let the pair $(u, c) \in R^n$ x C denote an element of T. u represents a set of attribute values of an observed object, while c is the class the object belongs to. At each step k, a pair $(u^{(k)}, c^{(k)})$ is drawn from the set T, and the two code vectors $m_i^{(k)}$ and $m_j^{(k)}$ nearest to $u^{(k)}$ are selected. The algorithm is as follows:

$$m_p^{(k+1)} = m_p^{(k)}, \quad p \neq i, j,$$

if $\left\| u^{(k)} - u_{i,j}^{(k)} \right\| \leq \dfrac{\Delta}{2}$ then

$$m_i^{(k+1)} = m_i^{(k)} - \alpha \; \frac{\delta(\Lambda(m_j), c^{(k)}) - \delta(\Lambda(m_i), c^{(k)})}{\left\| m_i^{(k)} - m_j^{(k)} \right\|} (m_i^{(k)} - u_{i,j}^{(k)}),$$

$$m_j^{(k+1)} = m_j^{(k)} - \alpha \; \frac{\delta(\Lambda(m_i), c^{(k)}) - \delta(\Lambda(m_j), c^{(k)})}{\left\| m_i^{(k)} - m_j^{(k)} \right\|} (m_j^{(k)} - u_{i,j}^{(k)})$$

else

$$m_p^{(k+1)} = m_p^{(k)}, \quad p = i, j,$$

where $\delta(x,y) = 1$ if $x = y$; $\delta(x,y) = 0$ if $x \neq y$, $u_{i,j}^{(k)}$ is the projection of $u^{(k)}$ on the border between V_i e V_j and Δ is a parameter which allows to consider only those training vectors falling *near* borders.

In the practice the algorithm performance can be improved by adding to the vector $u^{(k)}$ the realization of a random vector (noise) with a suitable continuous probability density function and operating the training on this new vector [5].

3.2 Integration of the two Approaches

The integration of the network into the system is realized by embedding the network in the semantic part of a CARW rule. This rule transforms the current description of a segmented image into a description in which each connected set of pixels is classified (e.g. as nuclei of hepatocytes, nuclei of not hepatocytes, double nuclei). The network receives in input the attributes of the connected set and determines if it belongs to a meaningful class or it is to be rejected. The result of the network is stored in the data structure describing the connected sets. This gives to the user a visual representation of recognized objects in the image (see figure 4) to validate the interpretation process.

In particular, tables 1 and 2 show the set N of symbols identifying the type of entities (images, descriptions, structures, and actions) and for each $n \in N$ the set of attributes adopted in the strategy for liver biopsy interpretation.

The attributes have the following meanings: ID is an identifier denoting the single entity; ORIGIN indicates the identifier(s) of the entity(ies) from which the entity at hand has been generated; MATRIX contains the matrix of values of an image; MTXR, MTXG, MTXB contain respectively the matrices of the Red, Green and Blue components of a vector-valued image; COMPONENT indicates from which components of a vector-valued image the entity at hand has been derived; STRING contains the strings of attributed symbols (see table 2) describing the recognized structures in the binary image ORIGIN; XC, YC the structure centre coordinates .

For symbols denoting CARWs, <NAME> indicates the CARW $CARW_{<name>}$, TRIED describes to which entities the corresponding rules have been applied; SUCC on which entities the application has been successful; TABLE is used to store in a suitable form the set of production rules in the denoted CARW (see table 4).

Tab. 1 Symbols and attributes of entities in liver biopsies interpretation process		
Entity	$n \in N_1$	Attributes
RGB Image	RGB	ID, MTXR, MTXG, MTXB, ORIGIN
Binary Image	BIN	ID, ORIGIN, MATRIX, COMPONENT
Structural Description	STDES	ID, ORIGIN, STRING
$CARW_{<name>}$	<NAME>	TRIED, SUCC, TABLE

Tab. 2. Symbols and attributes of entities in liver biopsy images		
Entity	$n \in N_2$	Attributes
Connected Set	CON	ID, XC, YC, AREA, ...
Hep. Nucleus	H_NUC	ID, XC, YC, AREA, ELLIPTICITY, ...
Non Hep. Nucleus	NH_NUC	ID, XC, YC, AREA, ELLIPTICITY, ...
Double Nucleus	DH_NUC	ID, XC, YC, AREA, ELLIPTICITY, ...
....		

Let us now describe the steps of the interpretation process that specialize, schedule and execute the CARW integrating the neural network.

In table 3 four states in the process of non contextual classification of nuclei are described as strings of attributed symbols. For each state, the substring before the meta-symbol "§" denotes the data obtained so far (features (k)), the substring between "§" and "‡" denotes the performed actions (p_actions (k)), and the substring after "‡" denotes the action to be performed (f_actions(k)).

Tab. 3 The process state at different steps of the execution of the strategy
......
k)   RGB • BIN$_1$ • BIN$_2$ • ... • § • MStrBiopsy • Acquire • Transform • Multisegment • ... • ‡ • MStr_nuclei • MStrCell • MContex • Revision
k+1)   RGB • BIN$_1$ • BIN$_2$ • ... • § • MStrBiopsy • Acquire • Transform • Multisegment • ... • MStr_nuclei • ‡ • ConDesc • Network • MStrCell • MContex • Revision
k+2)   RGB • BIN$_1$ • BIN$_2$ • ... • STDES$_1$ • § • MStrBiopsy • Acquire • Transform • Multisegment • ... • MStr_nuclei • ConDesc • ‡ • Network • MStrCell • MContex • Revision
k+3)   RGB • BIN$_1$ • BIN$_2$ • ... • STDES$_1$ • § • MStrBiopsy • Acquire • Transform • Multisegment • ... • MStr_nuclei • ConDesc • Network • ‡ • MStrEp • MStrSin • MRevisionContext
......
where k indicates the step of the interpretation process

At step k the acquisition and multisegmentation of the RGB image have already been performed and represented in the feature(k) substring by the symbols "bin$_1$", "bin$_2$",... . The first action to be performed is identified by the symbol "MStrNuclei", whose TABLE attribute is shown in table 4.

At step k+1 CARW$_{MStrNuclei}$ rewrites the symbol "‡" with the string "‡ • ConDesc • Network". The attributed symbol "ConDesc" describes the CARW for the description of the connected sets in a binary image, and the attributed symbol "Network" describes the CARW for the classification of these connected sets. The semantic part of the CARW$_{MStrnuclei}$ evaluates the TABLE attributes for CARW$_{ConDesc}$ and for CARW$_{Network}$.

Tab. 4.   The TABLE attribute for CARW$_{MStrNuclei}$
Syntactic part
<<BIN, ‡ , >; ‡ • ConDesc • Network>
Semantic part
Co: COMPONENTS(BIN) = MTXR (RGB)
F: TABLE (ConDesc):=<u>Specialize</u>(TABLE(ConDesc), ID(BIN), (AREA, ELLIPT., ...)}
TABLE(Network):=<u>Specialize</u>(TABLE(Network),ID(BIN),nuc_vector,(AREA, ELLIPT, ...)}

In order to limit the number of CARWs to be written, the TABLE attribute can be a template to be specialized at run-time to the different cases. In this case the evaluation of the TABLE attribute consists in its specialization (<u>Specialize</u> function).

Table 5 shows the template for CARW$_{Network}$ in which the italic printed terms denote the parameters to be specialized: *orig* indicates the identifier of the binary image to be classified, *code_vector* the trained network to be used for the classification and *set_of_attr* the set of attributes exploited by the network.

Tab. 5. The TABLE attribute for CARW$_{Network}$ before the specialization

Template CARW$_{Network}$ (*orig, code_vector, set_of_attr*)
Syntactic part
<<, STRDES, >; STRDES>
Semantic part
Co: (ORIG(STRDES)= *orig* AND <u>computedattr</u> (STRDES) = *set_of_attr*)
F: STRING(STRDES) = <u>network</u> (STRING(STRDES),*code_vector, set_of_attr*)

Table 6 shows the result of the specialization of this template to the case of nuclei classification.

Tab. 6. The TABLE attribute for CARW$_{Network}$ after the specialization

Syntactic part
<<, STRDES, >; STRDES>
Semantic part
Co: (ORIGIN(STRDES)= ID(BIN) AND <u>computedattr</u> (STRDES)=(AREA, ELLIPT., ...))
F: STRING(STRDES) = <u>network</u> (STRING(STRDES), nuc_vector, (AREA, ELLIPT., ...))

At the step k+2 the execution of the CARW$_{ConDesc}$ has been performed and the obtained description "DesStr$_1$" is added to the substring feature(k+2). This descriptions is updated at the step k+3 by the application of the network that classifies the connected sets.

4 Experimental setup

Images under examination are obtained from a 1.5-2 μ liver section immunostained by ematoxilin and eosine. Images are acquired in three RGB colour bands, of 256 values each, through a DXC-M2PH Sony Colour Video Camera mounted on a Zeiss Photomicroscope III with a dry 40X objective. The dimensions of the digital images are 512x384 pixels.

The following experiments show the classification performance of the proposed system. The basic steps can be summarized as follows.

64 liver biopsy images have been acquired. In order to generate data set to train the net and to test system performances, the image is segmented and described, in terms of connected set. For each connected set attributes are calculated. This produced about 2000 connected sets, part of which corresponds to nuclei in the image, and part to artefacts generated by the segmentation process (i.e. sets of pixels that do not correspond to nuclei in the image). Furthermore, the segmentation process caused the loss of some of the objects present in the original image, and the fusion of several objects in single sets of pixels.

Connected sets so extracted are labeled independently by two expert biologists, on the basis of original images. In particular labels are assigned as follows:
- structures which correspond to nuclei (in the judgement of the expert) are labeled accordingly to the decided class. Considered classes are: hepatocytes nuclei (H_nuclei), non hepatocytes nuclei (NH_nuclei), double nuclei (DH_nuclei);
- structures that do not have such a correspondence are assigned to the class rejected (Rejected);

- structures that are the fusion of two or more nuclei are assigned randomly to one of the fused nuclei classes.

Biologists are used to classify uncertain objects in a "class" called "maybe_nuclei". This allows the reclassification of such objects when more experimental evidence is available. Obviously, this "class" gives a level of confidence for the proposed classification, and does not define a type of object present in the scene. However, at this stage, the architecture of an LLVQ is not able to handle uncertainty, so we decided to assign objects whose classification is uncertain to the most probable class.

In this way, we obtained two distinct labeled data sets, that represent our empirical knowledge. The first one (B1) has been split in a training set and a testing set, in four different, random ways. For each splitting we tested performances of Parzen's method [7]. This allowed us to assess the statistical representativity of the data set and to set the performance we could reach for various choices of attributes.

In the pre-existent prototype, in order to encode the criteria exploited by a biologist in the recognition of different types of nuclei, only the following morphological attributes were considered: area (expressed in number of pixels), perimeter (expressed in number of pixels), ellipticity (squared root of the ratio between minor and major axis), inscribed square side. The introduction of the adaptive system makes easy the experimentation of new attributes to enhance class separability. In particular colour attributes (i.e. average measures of lightness, saturation and hue) were chosen, since different types of nuclei show different reactions to the staining treatment of the biopsy.

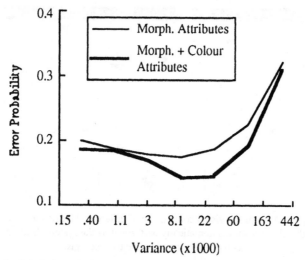

Fig. 3. Gain in introducing colour attributes evaluated by Parzen's method.

In figure 3 the error probability for Parzen's method is shown as a function of the kernel of estimate variance (logarithmic scale). Results refer to the splitting of data that produces the intermediate performance among the four splittings considered. The optimal value of the variance (10.000) is used in order to set the parameter Δ, which results equal to $\Delta = 245$. Note that the introduction of colour attributes improves performances of about 20%for the optimal value of the variance.

5 Discussion of the results

Figure 4 shows, in a page of ABI, the classifications performed by the two biologists and the system on the biopsy shown in the top left. The class of each object is indicated by different grey levels, as shown in the legend. In particular, hepatocites nuclei are in grey, non hepatocites nuclei in black, double nuclei in grey with black border and rejected in white with blak border. We can observe that the set of objects differs in the three classified images. This is due to the fact that the biologists classifications have been performed directly on the real image, classifying only meaningful objects, while the automatic classification has been performed on the segmented image, where some objects are lost and others are generated by the segmentation process.

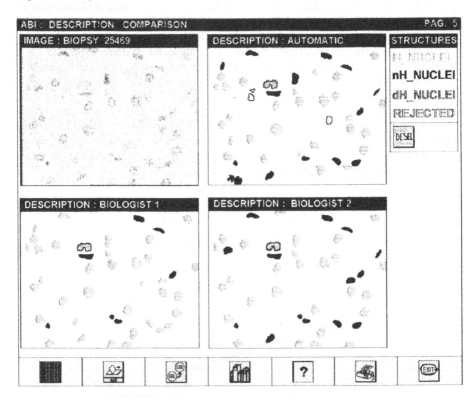

Fig. 4. The page of ABI for the visual comparison of different classifications. Top left: original image. Top right: classification automatically performed by the system. Bottom: classifications manually performed by the two biologists.

Table 6 shows the comparison between the classification of the testing set performed by the first biologist and by a LLVQ with 16 code vectors. The splitting considered is the same as in figure 3.

Net performance computed from table 6 is about 84,6% of correct classification.

The comparison between classifications performed by biologists has resulted in an agreement of 89% on objects derived from segmentation. This percentage reduces to 84% on real images. This is due to the fact that the structures on which the biologists generally disagree have characteristics (e.g. faint colour) such that they are easily lost in the segmentation process.

The performance of the system, considering the objects lost in the segmentation process, has resulted in 71%. This fact suggests that better results can be obtained by refining the segmentation step. This refinement can be also partially performed by the capability of ABI to recover lost structures by exploiting the context [13]. Note however, that the disagreement between the system and the biologist is mainly on those structures on which also the two biologists disagree, as it is evident by figure 4. This means that those structures are actually ambiguous. Biologists have judged performance of the system satisfactory.

LLVQ

		H	DH	NH	RJ	
B	H	561	8	50	0	619
I	DH	5	24	0	0	29
O	NH	51	2	206	1	260
1	RJ	16	0	11	0	27
		633	34	267	1	935

Table 6. Comparison between the classifications performed on the testing set by the LLVQ and the biologist 1

Summarizing, the vector quantizer has shown robustness with respect to the tuning of the parameters and to the initialization of the code vectors. It requires only an amount of 16x7x8 bytes to store the location of code vectors and 1 byte for each class label. The required time for training, using a IBM Risc 6000 workstation, is about 10 minutes. The classification of an image with 20 structures takes about 2 seconds for the classification phase and 4 minutes for the segmentation and attributes computation.

6 Conclusion and future developments

ABI is characterized by its unique capability of translating a colour image into a set of user-defined structures. A formal, efficient definition of the set of structures is obtained by meta reasoning tools.

The use of a CARW formalism makes extremely easy to integrate various methodologies, in particular attributed symbols can be directly used in statistical methods.

The introduction in ABI of the LLVQ trained by the MEP algorithm as classification kernel has resulted in a powerful system for natural image management and process, that also can show high classification performance with low costs in terms of memory and computational time.

This system has been tested for the classification of hepatocytes nuclei in digital images from liver biopsies, obtaining results near to those of biologists.

Other refinements of the system are in progress. They consist in the refinement of the segmentation process, the extension of the use of LLVQ to the recognition of other cellular structures, the exploitation of metareasoning to select such different LLVQs and for the contextual revision of classification.

378

7 Acknowledgements

Thanks are due to Prof. P. Mussio, Ing. A. Spalvieri, Dr. P. Gurzi for their important contribution in system developing , the Centro Studi di Medicina Teoretica of the University of Milan for providing set of images and the biologists Dr. E. Arosio, and Dr. F. Grizzi for providing classified data to train and to test the system.

8 References

[1]   C.G. deBessonet, *A many-valued approach to deduction and reasoning for artificial intelligence*, Kluwer, Boston, 1991.
[2]   P. Bottoni, P. Mussio, and M. Protti, "Metareasoning in the determination of image interpretation strategies", *Patt. Rec. Lett.* vol. 15, pp.177-190, Feb.1994.
[3]   U. Cugini, P. Ferri, P. Mussio, M. Protti, "Pattern-directed restoration and vectorization of digitized engineering drawings", *Comp. & Graph.*, vol. 8, pp.337-350, 1984.
[4]   C. Diamantini and A. Spalvieri, "Vector quantization for minimum error probability," *International Conference on Artificial Neural Networks*, vol. 2, pp. 1091-1094, Sorrento, IT, May 1994.
[5]   C. Diamantini and A. Spalvieri, "Quantizing for Minimum Bayes Risk," *Proc. International Symposium on Information Theory and its Applications*, Sidney, Australia, Nov. 1994.
[6]   N.Dioguardi, "The liver as a self-organizing system", *Res. Clin. Lab.*, vol. 19, pp. 281-326, 1989
[7]   K. Fukunaga, *"Introduction to Statistical Pattern Recognition,"* New York, Acc. Press 1972.
[8]   S. Garibba, E. Guagnini and P. Mussio, "Multiple-valued Logic Tree: Meaning and Prime Implicants", *IEEE Trans. on Reliability,* R vol. 34, 5, Dec. 1985.
[9]   A.Gersho and R. M. Gray *"Vector Quantization and Signal Compression,"* Kluwer Academic Publishers, 1992.
[10] T.Kohonen, G.Barna and R.Chrisley, "Statistical pattern recognition with neural networks: benchmarking studies," *Proc. of the IEEE International Conference on Neural Networks, San Diego, CA,* vol.1, pp.61-68, July1988.
[11] T.Kohonen, "The self organizing map," *Proc. of the IEEE*, vol. 78, n. 9, pp. 1464--1480, Sept. 1990.
[12] P. Maes, "Computational reflection", *Know. Eng. Rev.*, vol.3, 1, pp.1-19, 1988.
[13] P. Mussio, M. Pietrogrande, P. Bottoni, M. Dell'Oca, E. Arosio, E. Sartirana, M.R. Finanzon, N. Dioguardi, "Automatic cell count in digital images of liver tissue sections", *Proc. 4th IEEE Symposium on Computer-based Medical Systems*, pp.153-160, IEEE Computer Society Press, 1991
[14] M.Nagao, "Control strategies in pattern analysis", *Pattern Recognition*, vol. 17, n. 1, pp. 45-56, 1984
[15] N. Rescher, *Many-valued logic*, McGraw-Hill, 1969.
[16] A.Rosenfeld, "Computer Vision: Basic Principles" *Proc. of the IEEE*, vol. 76, n. 8, pp. 863-868, August 1988
[17] P. Smyth, R.M.Goodman and C.Higgins, "A Hybrid Rule-based/Bayesian Classifier," *Proc. of the 9th European Conference on Artificial Intelligence*, pp. 610-615, Stokholm, Sverige, August 1990.
[18] L. Tondl, "Problems of semantics", Reitel, 1981.

A Cooperative and Adaptive Approach to Medical Image Segmentation

C. Spinu, C. Garbay & J.M. Chassery

Lab. TIMC / IMAG - Institut Albert Bonniot
Faculté de médecine - Domaine de la Merci
38706 La Tronche Cedex - FRANCE -
Tel.: (33) 76 54 94 90 ; Fax: (33) 76 54 95 49
email: Corneliu.Spinu@imag.fr

Abstract

The purpose of the paper is to discuss the potential of a multi-agent approach for low-level image analysis. The work is based on the assumption that low-level processings should be adjusted locally, on zones of given characteristics, and be applied in an iterative framework with careful evaluation and control. A multi-agent architecture has been designed to this end, which allows to develop a dedicated low-level analysis strategy for each detected zone in the image. This analysis is based on an initial segmentation result, obtained as the fusion of two maps : a noise map and a texture map, representing regions with similar noise or texture characteristics. For each zone, a filtering / edge detection strategy is carefully selected and adjusted, based on evaluating the resulting edge map. The results are finally combined in a global segmentation image. The potential of the approach is illustrated on a natural MRI image.

1 Introduction

Two main approaches may be distinguished for image segmentation, namely the data driven and model-based approaches. Model-based segmentation is used when the objects in the scene are known exactly or subject only to simple variation [Chin 86, Grimson 90]. Such approach is tuned to operate on a given class of images, and appears to be very sensitive to the input parameters. Data-driven approaches on the contrary are more general, they are used when no a priori knowledge of the object shape is available, and exploit as much as possible the image content to obtain an unbiased result [Dellepiane 89].

Whatever the approach used, it must be conceived as sufficiently adaptive to cope with potential variability in the scene characteristics and object appearance [Venturi 92]. Such variability is known to be one of the major difficulty of medical image analysis. Adaptation may be obtained by the use of flexible models or deformable templates in the model-based approaches [Cootes 94]. It is merely introduced in terms of knowledge about the image processing operator applicability in the data driven approaches. Anyhow, performance evaluation is necessary to decide when the selected model or processing is adequate. Moreover, since there is no prior and precise knowledge to decide what is the best combination between an image and a model or processing strategy, complex optimisation or trial and error strategy are most often developed. The problem is even more difficult in the case of data driven approaches, where no reference model is available to support performance evaluation. Two solutions may be considered in this case : either a reference image is used, which is segmented manually by an expert, or the segmented image is evaluated with respect to some criteria. These criteria usually involve some common sense knowledge about the segmented image quality (presence of large and closed contours, for example) ; it may also involve some statistics about the class of images under interest.

We are concerned in this paper with data-driven edge-based segmentation. The most widely used operators for edge detection may be classified into derivative operators like [Deriche 87, Shen 86], operators based on the surface model [Hueckel 71, Haralick 84] or operators based on Markov fields model [Blake 87, Geiger 89].

These operators are of limited applicability when considered individually, since they involve assumptions about the type of discontinuity present in the image. Moreover, reliable edge detection is known to be conditioned by appropriate image properties, since the presence of noise or texture may blur the edge and region information. Therefore, a variety of pre-processing techniques is often used, in order to reduce the noise and smooth the texture that may be present in the image, while preserving or even enhancing the edge content. Filtering is one of the most known pre-processing method. Filters can be further classified into punctual transformations, noise reduction filters (like linear filters [Pratt 78], order filters [Pitas 90, Zamperoni 92], homomorphic filters [Pitas 90], morphological filters [Serra 88]) or contrast enhancement operators.

Performance of filtering and edge detection operators can drastically vary with the image under interest. From this standpoint stems the idea of adapting the filter and edge detector along with their parameters to image characteristics. This approach can be further refined, by considering the fact that most natural images, and particularly the medical ones, display heterogeneous characteristics that may vary from one zone to the other. This has led us to the concept of local estimation and local adaptation, which is the major guidelining principle of our approach.

Based on these considerations, we propose in this paper a general control model, in which to cope with the selection, adaptation, and cooperative application of adequate image filtering and edge detection techniques. This model is shown to support the derivation of flexible data-driven strategies for medical image segmentation. It involves knowledge of three different kinds, that is computed dynamically, in accordance with the data-driven philosophy :

(i)   knowledge about the image characteristics, based on the local estimation of the noise and texture characteristics ;
(ii)   knowledge of the processing strategy adequacy, based on characterizing the dynamics of the segmentation process (emergence of stable and sufficiently large contours, or increasing apparition of small edge elements, for example)
(iii) knowledge of the processing efficiency, based on estimating the quality of the segmentation result (for example by analyzing the contour length distribution in the zone under interest).

A priori knowledge about the image processing operators applicability is given to reduce the exploration cost, in terms of rough noise, texture and edge characteristics.

The interest of such approach is that it gives the possibility to reason in a reflexive way about the current processing strategy. As a matter of fact, the criteria to adjust an operator does not simply involve describing the situation at hand, but also considering its past adequacy to this situation. As a consequence, the approach provides the further possibility to feedback, and select a different processing strategy.

A multi-agent knowledge-based system is used to formalize and control the above process, whose design and implementation are described in sections 2 and 3. The contribution of the multi-agent structure is discussed in section 3, in terms of system flexibility and modularity, and in terms of task parallelisation and cooperation. Some preliminary experiments are described in section 4, to support the discussion.

2 System Design and Behaviour

As previously explained, the system roles are twofold : to estimate image characteristics and partition the image into corresponding zones, and to proceed to adapted filtering and edge detection. Two main phases, so-called analysis and processing phases, are thus distinguished, and performed sequentially.

2.1 Analysis Phase

The analysis phase consists in dividing the initial image into zones of different characteristics. Noise and texture information is currently considered in this respect. It is estimated separately, thus giving rise to distinct segmentation maps that are fused afterwards.

Based on a local estimator, the noise affecting the image is characterized as additive, multiplicative or impulsive [Chehdi 92]. Its standard deviation is computed in the case of additive or multiplicative noise. Zones of homogeneous noise characteristics are then detected in the image, to constitute the noise map (figure 4). The texture is analyzed in a separate way, based on coarseness estimation, to build the texture map (figure 4). More details on the noise and texture estimators are provided in section 3.

The two maps are then combined in order to obtain a global feature map (figure 4). Each zone in this resulting map corresponds to a given combination of noise and texture. A zone might be defined for example as involving some additive noise, characterized by a given standard deviation and no texture, while another zone would involve a combination of impulsive noise and a texture of a certain coarseness.

2.2 Processing Phase

The processing phase consists in the independent processing of each zone in the global feature map. Such processing implies three main steps, namely filtering, edge detection and evaluation.

An adapted filtering strategy is proposed for each zone in the image, which may imply the successive application of several filters, along with appropriate parameter values. Adapted edge detection operators are then selected, adjusted and applied, according to a similar approach. A partial edge map is thus obtained for each zone. These partial edge maps are then fused, to obtain a global edge map. Each contour element in the global edge map is then described, for example in terms of its length.

Each partial edge map is then submitted to quality evaluation (analysis of the contour length distribution and of the partial edge map evolution for example). It should be noticed that the evaluation is consistent, even if performed independently for each zone, since it is based on information extracted at a global level (case of a contour crossing several zones). This allows to palliate the bounded reasoning capacity of the individual agents.

Filtering and edge detection strategies are then adjusted, or modified, according to the evaluation results, and the process resumed. A processing strategy is defined by the choice of an operator, parameter range and adjustment strategy. At the end of the iteration process, the final results for each zone are merged into a final edge map.

3 Implementation

The system is implemented as a multi-agent system, consisting of an agent network and procedure servers. Inferences are performed in agents, while pure procedural tasks are performed in the procedure servers.

3.1 The Procedure Servers

The external procedures may be invoked by any agent rule ; they are grouped into dedicated procedure servers. These servers are concurrent and can be distributed on several workstations, thus bringing full parallelism (figure 2).

The Noise Estimation Server

For the estimation of noise, its nature and characteristics are supposed to be locally constant. This means that in the neighbourhood of each pixel in the image, the noise can be either additive, multiplicative or impulsive. In the case of additive or multiplicative noise, it is characterized by a certain variance.

Noise estimation is based on the hypothesis that homogeneous regions in the image display only noise statistics [Chehdi 92]. The algorithm involves searching for homogeneous regions in windows centered on each pixel in the image. For this purpose, the Nagao [Nagao 79] regions are computed in each window, and the one with minimal variance is finally retained as local homogeneous region. Noise type and variance can then be estimated by examining the dispersion of the local regions variances [Chehdi 92].

Impulsive noise is simply detected by counting the percentage of impulsive pixels in the window. An impulsive pixel is detected by comparing its value with the values of its neighbours. The noise is estimated as impulsive if this percentage goes beyond a certain threshold.

The Texture Estimation Server

The role of the texture estimator in this case is firstly to discriminate between textured and non-textured zones and secondly to characterize the texture, for example in terms of its coarseness, when it is present. The classical co-occurrence matrix approach is used in this respect. The co-occurrence matrix is a second order statistical measure of image variation, that provides a basis for a number of derived textural features [Haralick 79]. The mere contrast operator is used here, as it can yield an indication of the texture coarseness. The coarseness is estimated locally, as for the noise estimator, based on a window centered on each point.

The Filtering Server

The filtering server contains a rather large set of linear and non-linear operators, such as the linear mean filter, the general convolution filter, various homomorphic filters, some morphological filters like the opening and closure, several order filters (like the median filter), as well as two contrast enhancement operators (laplacian and comparison-selection operators). All these filters are not used at present, but it is intended to use them in future developments of the system.

The Edge Detection Server

The edge detection server currently contains a variety of commonly used derivative operators such as the Roberts, Prewitt, laplacian, Canny, Deriche, or Shen-Castan detectors. The optimal edge detection operators of Deriche and Shen-Castan are the only used at present, because of their well known performances.

The Performance Evaluation Server

The performance evaluation server merely involves two different evaluators, respectively dedicated to evaluating the segmentation quality and the processing efficiency. Segmentation quality is estimated by the amount of small contours which are present in the edge map. Generally speaking, a consistent edge map should contain only a small amount of such contours. Processing efficiency is evaluated by the match between two successive edge maps obtained with different parameters on the same zone in the image. This kind of evaluation is currently used to control the iterative parameter adjustment processes : this adjustment stops when a convergence is detected. No alternative strategy is selected.

Although conceptually very simple, these evaluators prove useful to control the filter / edge adaptation process.

3.2 The Multi-Agent System

As already mentioned, the system is designed according to a multi-agent knowledge-based architecture. This architecture has been implemented under COALA, a generic programming environment which is briefly described below [Baujard 94]. The proposed multi-agent architecture is afterwards presented into some depth.

COALA

COALA is based on the distinction between two basic agent types : Knowledge Servers, or KS agents, and Knowledge Processors, or KP agents. KS agents have been designed to handle the problem elements, while KP agents have been designed to handle the processing and reasoning methods. Any COALA agent is further defined as a full expert system, possessing proprietary production rules and inference engine, and communicating with other agents by means of message sending. The communication is performed according to either the asynchronous or synchronous modes, depending on the message type (transmission or request of information).

Control messages are used in addition to information messages. Any agent is given by them the possibility to modify the system knowledge by creating or deleting new instances, and to modify its configuration as well, by dynamically creating new agents to which pending sub-problems are distributed, according to the current problem solving state. Thus, an agent can duplicate itself (and in this case the new agent inherits its resources from the original agent) or create a new agent based on a predefined structure (pre-compiled agent).

The control cycle in each agent merely reduces to the one of a rule-based system.

Any application problem is thus modelled under COALA as a networked set of KS and KP agents placed in alternation. The calling sequence finally defines the problem solving strategy that is developed by the system. Furthermore, it should be noticed that solving a problem may lead to a dual dynamics, operating at the object level on one hand (creation of new object instances constituting new problem solving elements) and at the agent level on the other hand (evolution of the system configuration by the creation of new agents). This kind of system thus display an actual adaptation capacity, by adapting the agent organisation to the physical and/or logical structure of the problem under interest, and by selecting the proper treatments, based on the situation under interest.

Multi-Agent Architecture

The multi-agent architecture that has been designed for the present study is presented in figure 1. As can be seen, the system operates according to two main steps : a preliminary analysis step, followed by an adapted processing step, as described in section 2. Several agent groups may furthermore be distinguished, which operate under the analysis or processing phase. The analysis phase involves one agent group, while the processing phase involves several groups, each group being associated to a given zone in the global feature map.

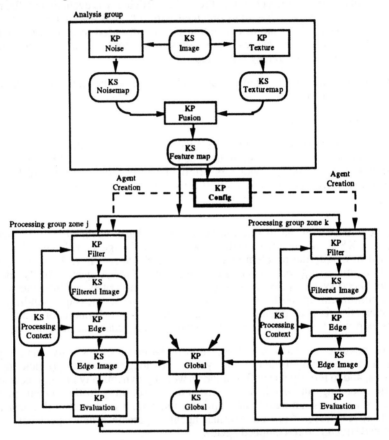

Fig. 1. The multi-agent architecture.

A brief description of each agent as well as some details about communication and cooperation between agents are provided in what follows.

The analysis group : The role of KS Image is to represent the initial image along with its properties (like size, format, etc). It possesses rules for requesting lacking information to the user and for proposing the initial image to KP Noise and KP Texture agents.

The role of KP Noise is to generate the noise map of the input image which it has received from KS Image, using the noise estimator stored in the procedure server. The noise map is transmitted to KS Noisemap when computed. The role of KP Texture is to generate the texture map independently of KP Noise. It makes use of the texture detector stored in the procedure server and proposes the texture map to KS Texturemap.

The role of KS Noisemap and KS Texturemap is merely to store and transmit the corresponding information to KP Fusion. KP Fusion merges the obtained noise and texture maps, using simple rules. Both texture and noise information are preserved in general, except when texture is present together with additive or multiplicative noise : the sole texture information is preserved in that case, as sufficient to guide the processing phase. The resulting feature map is then transmitted to KS Featuremap.

The role of KS Featuremap is to store and transmit the zones represented in the global feature map to KP Config agent. KP Config plays a central role with respect to the system configuration, since it is responsible for dynamically creating and initializing specific processing tracks, one per zone, in terms of a dedicated network of agents. The network is currently initialized according to a fixed and a priori known processing model. Knowledge about the current processing status, e.g. the image, edge map, and zone characteristics, are moreover transmitted to each of the KS Processing Context agents. These agents play a central role with regards to the system dynamics, by keeping track of the processing context and history, in a decentralized (zone-dependent) way.

The processing group : The role of KP Filter is to reason about the situation displayed in the zone at hand, and select accordingly the most adapted filter strategy, as discussed in section 2.2 and illustrated below :

```
UNSIGNED Premise()
        {String type, variance;
        Get(KS_Processing_Context, Noise,Type,type);
        Get(KS_Processing_Context, Noise,Variance,variance);
        return EQ(type,"multiplicative_noise") && EQ(variance,"low");
        }
UNSIGNED Conclusion()
        {String image, filtered_image, zone_mask;
        String size="small";
        Get(KS_Processing_Context, Image,Name,image);
        Get(KS_Processing_Context, Zone_Mask,Name,zone_mask);
        Create_New_Image(filtered_image);
        Call_Homomorphic_Filter(image, zone_mask, filtered_image, size);
        Set(KS_Filtered_Image,Filtered_Image, Name, filtered_image);
        return 1;
        }
```

As shown in this example, a small size homomorphic filter is selected, based on the type and characteristics of the noise present in the zone under interest (a multiplicative noise with a low variance). The resulting filtered image is then transmitted to KS Filtered image.

The role of KS Filtered image is merely to store and transmit information about the filtered image.

The role of KP Edge is again to reason about the situation depicted in the zone at hand, and adapt accordingly the parameters of the edge detector at hand (Deriche operator at present). Moreover, some of these parameters may be adjusted at each iteration step, depending on the evaluation result. For example, for a zone with additive noise, the scale parameter is adapted according to the noise variance. For a textured zone, the scale is adapted according to texture coarseness, and hysteresis thresholds are adjusted at each iteration. The resulting edge map is then transmitted to KS Edge image. A rule example for this latter case is given below :

```
UNSIGNED Premise()
        {String presence, coarseness, type, scale, low_threshold, high_threshold;
        Get(KS_Processing_Context, Deriche_Param,Scale,scale);
        Get(KS_Processing_Context, Deriche_Param, Low_Threshold, low_threshold);
        Get(KS_Processing_Context, Deriche_Param,High_Threshold,high_threshold);
        Get(KS_Processing_Context, Texture,Coarseness,coarseness);
        Get(KS_Processing_Context, Texture,Presence,presence);
        Get(KS_Processing_Context, Evolution,Type,type);
        return EQ(presence,"present") && EQ(type,"unstable");
        }
UNSIGNED Conclusion()
        {String image, edge_image, zone_mask;
        scale=f(scale,coarseness);
        low_threshold=Increment(low_threshold);
        high_threshold=Increment(high_threshold);
        Get(KS_Processing_Context, Image,Name,image);
        Get(KS_Processing_Context, Zone_Mask,Name,zone_mask);
        Create_New_Image(edge_image);
        Call_Deriche_Detector(image, zone_mask,
                edge_image, scale, low_threshold, high_threshold);
        Set(KS_Processing_Context, Deriche_Param,Scale,scale);
        Set(KS_Processing_Context, Deriche_Param, Low_Threshold, low_threshold);
        Set(KS_Processing_Context, Deriche_Param, High_Threshold, high_threshold);
        Set(KS_Edge_Image,Edge_Image, Name, edge_image);
        return 1;
        }
```

In this example, a Deriche operator is selected with given scale and hysteresis thresholds, based on the coarseness of the texture present in that zone. Moreover, the segmentation result evolution is used to control the convergence of the parameter adjustment process.

The role of KS Edge image is merely to store and transmit information about the edge image. This information is transmitted to the KS Processing Context agent, which keeps track of the agent group processing history.

The local edge maps obtained by the individual KS Edge Image agents are then transmitted to the KP Global agent, whose role is to fuse the local maps into a global one, and transmit the resulting edge map to the KS Global agent. This agent is moreover in charge of labelling the connected components and compute some basic features such as the contour element lengths.

The individual KP Evaluation agents are then in charge of computing local statistics describing for example the distribution of the contour element lengths in their zone of interest and transmit the results to KS Processing Context. They also evaluate the evolution that can be observed between the successive processings, either in terms of the match between two successive edge maps, or in terms of the evolution of some feature (shift of the contour length distribution towards either small or large contour elements for example). The evolution features are also transmitted to KS Processing Context agent. The role of KS Processing Context is to store and transmit information about the edge map evaluation.

The iteration process stops in case of a convergence, independently for each processing group. A completion flag is then transmitted to KS Global. It should be noticed that processings of different complexity may occur, in terms of number of iterations and type of processings, depending on the situation given initially to the processing agent group. The program terminates when convergence has been reached for each group, and the results in KS Global displayed.

3.3 Programming Issues

COALA may be seen as a specialization of the C++ language, by addition of a predefined class library. The structure of each agent can be specified separately, using these predefined classes. The specification is done according to a hybrid approach using mixed object, logic and procedural representations.

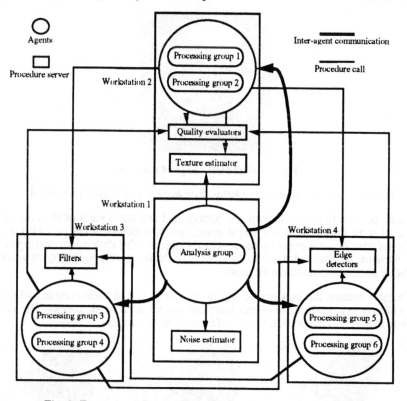

Fig. 2. Example of physical distribution of agents and servers.

The compilation of each C++ source module will generate an executable file for each agent. The agents will then run as independent UNIX processes on one or several networked workstations and communicate by the TCP/IP network protocol. The distribution of agents on physical sites may be static, based on some a priori configuration file, or dynamic, by using dedicated system primitives for real-time agent creation, duplication or deletion, which can be called from the rules of an agent.

The external procedures are grouped into several procedure servers and can be invoked by agent rules. These servers are designed as concurrent and can be distributed (like the agents) on several workstations. Procedure calls from agents are done using a remote procedure call (RPC) protocol. Procedures belonging to different servers (running on different workstations) can thus be activated in parallel, while procedures of the same server can be executed concurrently.

An example of the physical distribution of the agents and procedure servers on a workstation network is illustrated in figure 2. This example concerns a particular case involving 6 processing groups corresponding to 6 different zones in the image, uniformly distributed on 3 workstations. The analysis group runs on workstation 1.. The procedure servers are also distributed on the 4 available workstations.

4 Experiments

We will discuss here some preliminary results obtained for a noisy MRI image of the heart (figure 3).

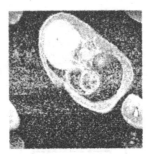

Fig. 3. Initial MRI image of the heart.

The computed noise, texture and global feature maps are displayed in figure 4.

As may be seen in the noise map, the background and some parts of the objects are covered with impulsive noise (light zone), while a combination of additive and multiplicative noise appears on the rest of the objects. The dark zones correspond to low variance additive noise, while different grey shades correspond to higher variance additive noise and multiplicative noise.

As regards texture, the background appears as textured (with various estimated values for coarseness), while most object parts are not textured (dark zones). The global feature map shows the presence of 16 different zones to process.

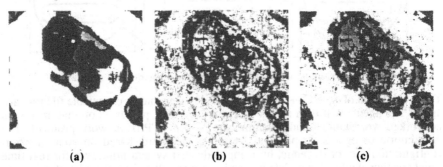

(a)   (b)   (c)

Fig. 4. MRI image of the heart : (a) noise map, (b) texture map, (c) global feature map.

Results obtained during the processing phase are presented in figure 5. As already mentioned, this image is characterized by the presence of a large textured zone (background) surrounding objects with no texture but additive noise. Figure 5 illustrates 3 iterations performed on the sole background regions, by adjusting the hysteresis thresholds of the edge detection operator ; the number of small edges will then decrease, and the process will converge.

Fig. 5. MRI image of the heart : three edge maps obtained at successive iteration steps.

The final result is presented in figure 6 together with two results obtained with two different applications of a global Deriche operator ; in edge map 1, the parameters have been manually adjusted to obtain the most robust image contour, while in edge map 2, the parameters have been manually adjusted to eliminate as much contours as possible on the background (textured zone).

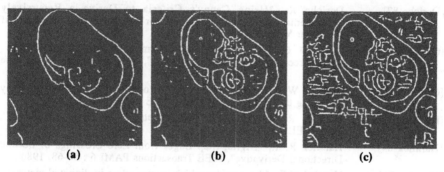

(a)   (b)   (c)

Fig. 6 : MRI image of the heart : (a) final result (b) edge map 1 (c) edge map 2
(for edge map 1 a = 0.5, lowthres = 10, highthres = 20 ;
for edge map 2 a = 0.7, lowthres = 50, highthres = 50).

As may be seen in edge map 1 and edge map 2, the difficulty is to preserve contour information while smoothing textured areas. Such difficulty demonstrates the potential interest of the approach, that is to adapt locally the filtering / edge detection strategy.

5 Conclusion

This paper has shown the potential of a multi-agent approach for medical image analysis. The basic principle is to apply carefully selected and adapted processings, in local zones of given characteristics. These processings are applied iteratively, in the framework of a general control model involving knowledge about the image characteristics, and the processing strategy adequacy and efficiency. A multi-agent architecture has been designed to this end, which allows to develop a dedicated analysis strategy for each detected zone in the image. The system organization is in fact dynamically modified, to cope with the progressively elaborated model of the problem (image, processing adequacy and segmentation quality). Such dynamic configuration operates at the logical and physical levels, thus solving the compromise between flexibility and efficacy. Preliminary results have been discussed, which demonstrate the approach potential, when applied to medical image analysis. Experimental analysis should be pursued on a larger scale, to derive more robust estimation and analysis rules, and to enrich the expertise in medical image analysis.

References

[Baujard 94]   Baujard O., Pesty S., Garbay C., "MAPS : a language for multi-agent system design", Expert Systems, 11(2) : 89-98, 1994.

[Blake 87]   Blake A., Zisserman A., "Visual Reconstruction", MIT Press, Cambridge-MA, 1987.

[Chehdi 92]   Chehdi K., "A new approach to identify the nature of the noise affecting the image", Proc. IEEE ICASSP'92, pp III.285-III.288, 1992.

[Chin 86]   Chin R., Dyer C. R., "Model-Based Recognition in Robot Vision", Computing Surveys 18 : 67-108, 1986.

[Cootes 94]   Cootes T. F., Taylor C. J., "Using Grey-Level Models to Improve Active Shape Model Search", 12th IAPR International Conference on Pattern Recognition, pp 63-67, oct. 1994.

[Dellepiane 89]   Dellepiane S., Ghilino G., Vernazza G., "Biomedical structures recognition by an opportunistic sequence of different segmentation methods", Medical Imaging, 1989.

[Deriche 87]   Deriche R., "Using Canny's Criteria to Derive a Recursively Implemented Optimal Edge Detector", International Journal of Computer Vision, 1(2) : 167-187, 1987.

[Geiger 89]   Geiger D., Girosi F., "Parallel and Deterministic algorithms for MRFs : surface reconstruction and integration", MIT AI Memo 1114, June 1989.

[Grimson 90]   Grimson W. E. L., Lozano-Perez T., "Localising Overlapping parts by Searching the Interpretation Tree", IEEE PAMI 9, 1987.

[Haralick 79]   Haralick R. M., "Statistical and structural approaches to texture", Proc. IEEE, 67 : 786-804, 1979.

[Haralick 84]   Haralick R.M., "Digital Step Edges from Zero-Crossings of Second Directional Derivative", IEEE Transactions PAMI 6 : 58-68, 1984.

[Hueckel 71]   Hueckel M.F., "An operator which locates edges in digitized pictures", J. Ass. Comput. Mach., 18(1) : 113-125, 1971.

[Kirsch 71]   Kirsch R., "Computer Determination of the Constituent Structure of Biological Images", Computer Biomedical Research, 4 : 315-328, 1971.

[Nagao 79]   Nagao M., Matsuyama T., "Edge preserving smoothing", Computer Graphics and Image Processing No. 9 : 394-407, 1979.

[Pitas 90]   Pitas I., Venetsanopoulos A.N., "Non linear digital filters. Principles and applications", Kluwer Academic Press, 1990.

[Pratt 78]   Pratt W.K., "Digital Image Processing", John Wiley, 1978.

[Serra 88]   Serra J., "Image analysis and Mathematical Morphology : Theoretical advances", Academic Press, 1988.

[Shen 86]   Shen J., Castan S., "An Optimal Linear Operator for Edge Detection", Proceedings CVPR'86, pp 109-114, 1986.

[Venturi 92]   Venturi G., Capitani P., Carboni M., "A target oriented adaptive segmentation method", Proc. IEEE 14th Int. Conf. of Engineering in Medicine and Biology Society, pp 1441-1444, 1992.

[Zamperoni 92]   Zamperoni P., "Adaptive rank-order filters for image processing based on local anisotropy measures", Digital Signal Processing, 2 : 174-182, july 1992.

Posters

COBRA: Integration of Knowledge-Bases with Case-Databases in the Domain of Congenital Malformation

Shusaku Tsumoto, Hiroshi Tanaka, *1
Hiromi Amano, Kimie Ohyama, and Takayuki Koroda *2

*1 Department of Information Medicine
Medical Research Institute,Tokyo Medical and Dental University
1-5-45 Yushima, Bunkyo-ku Tokyo 113 Japan
TEL: +81-3-3813-6111 (6159) FAX: +81-3-5684-3618
E-mail:tsumoto@quasar.tmd.ac.jp, tanaka@cim.tmd.ac.jp
*2 Department of Orthodontics(II),
Faculty of Dentistry, Tokyo Medical and Dental University
1-5-45 Yushima, Bunkyo-ku Tokyo 113 Japan

Extended Abstract

There have been developed many medical decision support systems, such as MYCIN since the end of 1970's [2]. One of the most important problems of these decision support systems is that their input interface is based on verbal information and that they cannot handle multi-data, which causes the following two problems. First, users have to know correctly the meaning of medical technical terms on the input screen. However, medical experts who majors in respiratory diseases do not know technical terms in neurology usually. Therefore it is hard even for medical experts to use a medical expert system, which makes them difficult to apply to real-world situations. Second, we often need non-verbal information, such as visual images in order to understand medical technical terms, since those technical terms are closely related to those kinds of data. For example, a neurological symptom "ptosis" can be explained in words as shown in [1]. However, even medical experts cannot easily understand several technical terms in the explanation. Thus, if we can use visual images then it may be easier to understand those terms.

Interestingly, those problems are also closely related not only to practical use, but also medical education. Medical education also needs non-verbal information, and the way how to describe non-verbal information by using technical terms. For example, medical students do not exactly know about ptosis. In order to understand this symptom, we need a picture or a photograph which shows a typical case of ptosis. In other words, medical education can be viewed as a process which learns the relations between verbal information and non-verbal information.

Therefore, these problems suggest that medical intelligent systems should deal with many different kinds of data for practical use. Actually, medical data consist of many kinds of data from different resources, such as natural language data, sound data from physical examinations, numerical data from laboratory examinations, time-series data from monitoring systems, and medical images (for example, X-ray, Computer Tomography, and Magnetic Resonance Image). Furthermore, since medical technical terms are closely related to the characteristics of those kinds of data, we need to represent the semantic relations between medical terms and non-verbal information in order to understand medical terms and their meaning.

Therefore it has been pointed out that medical databases should be implemented as multidatabases.

However, there have been few systems which integrates these data into multidatabases. In this paper, we introduce a object-oriented system, called CO-BRA (Computer-Operated Birth-defect Recognition Aid), which supports diagnosis and information retrieval of congenital malformation diseases and which integrates natural language data, sound data, numerical data, and medical images into multidatabases on syndrome of congenital malformation [3, 4]. Furthermore, since it also has an expert-system module and a module of case-based reasoning [5], COBRA can diagnose future clinical cases.

This system is implemented on several kinds of object-oriented databases [6] or programming language, such as ONTOS system in Sun SPARC Station, SuperCard in Macintosh and Visual Basic in PC. It consists of the following four knowledge-bases, called *ontology*, which are implemented as object-oriented databases, and three modules, which are implemented by object-oriented programming language.

These knowledge-databases in COBRA are easily implemented in the object-oriented scheme, which suggests that these clinical databases should be implemented as object-oriented databases.

References

1. Adams, R. D. and Victor, M. *Principles of Neurology*, 5th edition, McGraw-Hill, NY, 1993.
2. Buchnan, B. G. and Shortliffe, E. H.(eds.) *Rule-Based Expert Systems*, Addison-Wesley, MA, 1984.
3. Goodman, R. M., Golin, R. J. *Atlas of the face in genetic disorders 2nd edition*, C.V. Mosby. Saint Louis, 1977.
4. Jones, K. L. *Smith's Recognizable Patterns of Human Malformation 4th edition*, W. B. Saunders Philadelphia, 1988.
5. Koldner, J. *Case-based Reasoning*, Palo Alto, CA, 1994.
6. Taylor, D. A. *Object-Oriented Technology: A Manager's Guide*, Addison-Wesley, 1990.

Case-Based Medical Multi-Expertise : an Example in Psychiatry

Isabelle Bichindaritz

Université René Descartes, LIAP-5, UFR de Math-Info,
45 rue des Saints-Pères, 75270 Paris Cedex 6, France
CMME, Hôpital Sainte-Anne, 100 rue de la Santé, 75014 Paris, France

Abstract. *Case-based reasoning* particularly studies how expertise takes advantage from experience. The case-based reasoner presented in this paper is applied to the clinical expertise of *eating disorders in psychiatry*. Its peculiarity is to be able to reason from different points of view. Each point of view is associated with one of the various cognitive tasks it can perform, among which are diagnosis, treatment and clinical research.

1 Introduction

Case-based reasoning proposes an artificial intelligence methodology for the processing of *empirical* knowledge. By definition, a *case* is a set of empirical data. *Case-based reasoning* fundamental principle is that it is preferable, in order to process a new case, to use one or several already met, and memorized cases. While traditional case-based reasoning systems are devoted to the realization of a single task, the case-based reasoner presented here is capable of realizing *several cognitive tasks*, and of *adapting* to the task it performs.

2 Domain of Application

Psychiatry is a complex real-world domain, and a weak-theory one. The motivation for this work is that clinical expertise is not uniform. On the contrary, it assumes, for the same clinician, many skills, such as *diagnosis*, *therapy* and *research*. The aim of this system is to give a case-based assistance to clinical expertise in a service specialized in eating disorders.

3 The Memory

The memory of the system gathers all the previously met, and processed, cases. It is a network of *cases and concepts* organized in hierarchies dependent upon the points of view. Each *point of view* is associated with a single cognitive task, and serves as a filter for the description elements retained as significant for the task to realize.

Memory Composition : The cases are both patients cases (restrictive and bulimic anorexics) and control cases.

Memory Organization : The nodes of the hierarchies in memory are *concepts*. Each node is linked to the more general nodes above it in the hierarchy, and to the more specialized nodes under it, as well as cases directly indexed under it.

4 The Reasoning Process

The reasoning process passes through *several steps*, which are the same for analysis and synthesis tasks, and which are represented on Figure 1. For *analysis tasks*, such as diagnosis and treatment planning, the best analogue case is used to propose a diagnosis for the new case (with 96% successful results after a training of 30 cases) ; the treatment planning it followed is adapted to the new case. For *synthesis tasks*, such as clinical research, the concepts learnt by the system during analysis tasks are used to construct an *interpretation* of the data.

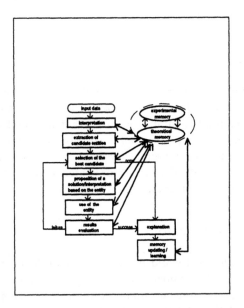

Fig. 1. Functional architecture of the system

5 Conclusion

The content and the organization of the memory give this system its multi-expertise ability. More research needs to be done to better take advantage from the *theoretical* knowledge available. It will then be applied to *other domains* of psychiatry first, and later of medicine.

TIME-NESIS: A Data Model
in Managing Time Granularity
of Natural-Language Clinical Information

Carlo Combi (°), Francesco Pinciroli (°*) and Giuseppe Pozzi (^)

° Dipartimento di Bioingegneria del Politecnico di Milano
* Centro di Teoria dei Sistemi del CNR, Milano
^ Dipartimento di Elettronica e Informazione, Politecnico di Milano

Extended Abstract

In the database field, the need of time management at different levels of granularity has been on for some years [Tansel93]. It has been recently emphasized. Such a need is widespread when we have to manage medical clinical information. The temporal dimension is normally given at different granularities. In database systems based on the calendar time, granularity is the unit of measure used for the time scaling [Montanari93]. Temporal information related to the events the patient narrates, for instance, frequently has a different accuracy depending on whether the information is related to remote events or to closer ones. Furthermore, when a physician makes a diagnosis, he frequently considers information at a increasing granularity with the temporal distance between the particular described event and the actual time.
Let us consider, for example, the following sentences:
1) "In 1990 the patient took a calcium-antagonist for three months"
2) "The patient suffered from abdominal pain from 5 p.m. to 7 p.m., March 27, 1989"
3) "At 4:45 p.m., October 15, 1990, the patient was hit by myocardial infarction"
4) "On November 30, 1991, in the afternoon the physician measured a blood pressure of 120/80 mmHg on the patient"
5) "At 4:30 p.m., October 26, 1991, the patient finished suffering from renal colic, it lasted five days"
6) "The patient suffered from a tachycardia episode for 150 seconds on October, 26 1991, at 3:22 p.m."
Sentences 3) and 4), at a first analysis, describe events by assertions predicated for temporal instants: in the first case, the instant is defined at level of minutes, while in the second case the punctual event of a measurement is temporally located with an accuracy coarser than in the previous case. The remaining sentences, instead, describe events by asserting that some statements are true in temporal intervals, characterized in many different ways: by starting instant and duration, by starting and ending instant, by duration and ending instant, and by different and mixed time units.
The need to store in an appropriate way the contents of any of the above sentences is present in the medical clinical field, e.g. during the collection of anamnestic data. Furthermore, this heterogeneity in expressing temporal clinical information does not weaken the need of establishing temporal relationships between the stored temporal information. This is the only way to suitably reconstruct and manage the whole clinical history of a patient.

To manage the temporal dimension of anamnestic data given at various and mixed levels of granularity, we defined the temporal data model TIME-NESIS (TIME in AnaNESIS). The concept of *temporal assertion* shapes the entire temporal information: it is composed by a *propositional part* and by an *interval*. An *interval* is based for its computational properties on the concepts of *instant* and *duration*, in their turn characterized in respect with the concept of *elementary instant*, i.e. the point on the time axis. The model provides temporal dimension to the data by using intervals that can be specified by different granularities. TIME-NESIS supports a three-valued logic, where *True, False* and *Undefined* are the truth values. So, TIME-NESIS allows to manage some degrees of uncertainty in establishing temporal relationships between intervals or between temporal assertions, expressed at different granularities. The logical connectives and quantifiers manage any of the three truth-values.

We defined the following temporal relationships between intervals (or temporal assertions): temporally expressed as, contemporary to ... at granularity ..., temporally specifies, before, overlaps, during, starts as... at granularity ..., finishes as ... at granularity ..., meets ... at granularity ..., has duration expressed as, lasts as ... at granularity ..., specifies the duration of , lasts minus than.

In order to manage in a sound and global way the temporal dimension of medical data we applied TIME-NESIS to a Time-Oriented Medical Record. The implemented Medical Record deals with data from patients who undergo a coronary-artery angioplasty [Pinciroli92, Combi94]. These patients after a 2-3 day-long hospitalization go through some periodical follow-up visits, aiming at verifying the efficacy of angioplasty. Management of temporal aspect of these data is therefore relevant; the clinical history of these patients has to be managed, in order to relate follow-up parameters with clinical events, such as pathologies or therapies. The temporal data model has been implemented by means of an Object-Oriented Data Base Management System (OODBMS). Many typical clinical temporal assertions have been defined: the temporal assertion *diagnosis*, in order to modelling anamnestic data related to previous pathologies the patient suffered; the temporal assertion *therapy*, in order to modelling anamnestic data related to the previous therapies the patient assumed; the temporal assertions related to the parameters measured at any periodical visit, such as systolic and diastolic blood pressures or heart rate.

References

[Combi94] Combi C, Pinciroli F, Pozzi G. Temporal Clinical Data Modeling and Implementation for PTCA Patients in an OODBMS Environment, Proceedings of Computers in Cardiology, IEEE Computer Society Press, Los Alamitos, 1994: 505-508.

[Montanari93] Montanari A, Pernici B. Temporal Reasoning, in [Tansel93]: 534 - 562.

[Pinciroli92] Pinciroli F, Combi C, Pozzi G. Object-orientated DBMS Techniques for Time-orientated Medical Record, Medical Informatics 1992; 17(4): 231-241.

[Tansel93] Tansel AU, Clifford J, Gadia S, Jajodia S, Segev A, Snodgrass R. Temporal Databases: Theory, Design, and Implementation, The Benjamin/Cummings Publishing Company, Redwood City, CA, 1993.

Induction of Expert System Rules from Databases based on Rough Set Theory and Resampling Methods

Shusaku Tsumoto and Hiroshi Tanaka

Department of Information Medicine
Medical Research Institute,Tokyo Medical and Dental University
1-5-45 Yushima, Bunkyo-ku Tokyo 113 Japan
TEL: +81-3-3813-6111 (6159) FAX: +81-3-5684-3618
E-mail:tsumoto@quasar.tmd.ac.jp, tanaka@cim.tmd.ac.jp

Extended Abstract

One of the most important problems in developing expert systems is knowledge acquisition from experts [6]. While there have been developed a lot of knowledge acquistion tools to simplify this process, it is still difficult to automate this process. In order to resolve this problem, many methods of inductive learning, such as induction of decision trees [1], rule induction methods [2] and rough set theory [5, 7], are introduced and applied to discover knowledge from large database, and their usefulness is in some part ensured.

However, most of the approaches focus on inducing some rules, which classifies cases correctly, and extract only information on classification. On the contrary, medical experts learn not only such knowledge but also information which is important for medical diagnostic procedures. Focusing on these learning procedures, Matsumura et al. [4] analyzes the not-classification information of medical experts, and proposes a diagnosing model, which is composed of the following three kinds of reasoning processes: exclusive reasoning, inclusive reasoning, and reasoning about complications.

Applying this model to headache and facial pain, they have developed a system, RHINOS (Rule-based Headache and facial pain INformation Orgranizing System), where the above three processes are supported by exclusive rules, inclusive rules, and disease image, respectively. This system makes a diagnostic procedure in the following way. First, exclusive reasoning is the one that when a patient does not have a symptom which always appears in a disease in the candidates, such a disease can be excluded from them. Second, inclusive reasoning is the one that when a patient has symptoms specific to a disease, the disease can be suspected(inclusive rules have two kinds of certainty factors: accuracy and coverage(true positive rate)). Finally, reasoning about complications is that when some symptoms which cannot be explained by that disease, complications of other diseases can be suspected.

Since these processes are found to be based on the concepts of set theory, as shown in [4], it is expected that set-theoretic approaches can describe this model and the procedures of knowledge acquisition. In order to characterize these procedures, we introduce and extend the concepts of rough set theory, which is developed to describe how to classify a certain set (denoted as a "class") by intersection or union of some sets which satisfy one equivalence relation. Based on this theory, we develop a program, called PRIMEROSE-REX (Probabilistic Rule Induction Method based on Rough Sets and Resampling methods for Expert systems), which is based on this method and which extracts rules for an expert system from clinical databases.

PRIMEROSE-REX induces three kinds of rules in the following way. First, the system exhaustively searches for exclusive rules, candidates of inclusive rules, and disease images through all the attribute-value pairs. Then, it selects inclusive rules from the candidates in a heuristic way. Finally, PRIMEROSE-REX applies two kinds of resampling methods [3], cross-validation method and the bootstrap method in order to estimate two certainty factors of this model.

In order to evaluate PRIMEROSE-REX, we apply this system to the domain of diagnosis of headache and facial pain, headache(RHINOS domain), whose training samples consist of 1477 samples, 10 classes, and 20 attributes. The experiments are performed by the following three procedures. First, original samples are randomly splits into training samples(S_1) and test samples(T_1). Second, the system induces rules and the statistical measures from S_1. Third, the induced results are tested by T_1. These procedures are repeated for 100 times.

The experimental results can be summarized to the following three points. First, the induced rules perform a little worse than those of medical experts. Second, our method performs a little better than classical empirical learning methods, CART and AQ15. Finally, third, cross-validation estimators and bootstrap estimators can be regarded as the lower boundary and the upper boundary of each rule accuracy. Hence the interval of these two estimators can be used as the estimator of performance of each rule.

References

1. Breiman, L., Freidman, J., Olshen, R., and Stone, C. 1984. *Classification And Regression Trees*. Belmont, CA: Wadsworth International Group.
2. Clark, P., Niblett, T. 1989. The CN2 Induction Algorithm. *Machine Learning*, 3, 261-283.
3. Efron, B. 1982. *The Jackknife, the Bootstrap and Other Resampling Plans*. Society for Industrial and Applied Mathematics, Pennsylvania.
4. Matsumura, Y., et al. 1986. Consultation system for diagnoses of headache and facial pain: RHINOS. *Medical Informatics*, 11, 145-157.
5. Pawlak, Z. 1991. *Rough Sets*. Kluwer Academic Publishers, Dordrecht.
6. Buchnan, B. G. and Shortliffe, E. H.(eds.) *Rule-Based Expert Systems*, Addison-Wesley, MA, 1984.
7. Ziarko, W. 1991. The Discovery, Analysis, and Representation of Data Dependencies in Databases, in: Shapiro, G. P. and Frawley, W. J.(eds), *Knowledge Discovery in Databases*, AAAI press, Palo Alto, CA, pp.195-209.

Sequential Knowledge Acquisition: Combining Models and Cases

B. BRIGL[1], A. GRAU[2], P. RINGLEB[2], TH. STEINER[2], W. HACKE[2], R. HAUX[1]

University of Heidelberg,
[1]Institute for Medical Biometry and Informatics, Department of Medical Informatics,
[2]Department o Neurology,
Im Neuenheimer Feld 400, D-69120 Heidelberg, Germany

1 Introduction

The global aim of medical decision support systems (MDSS) is to improve the quality of patient care. Therefore, the integration of MDSS into clinical workstations ([1]) is one prerequisite ([2]). Another important aspect is the quality of the knowledge base and the knowledge acquisition process ([3]). The integration of decision support functions into clinical workstations and the use of a standardized medical vocabulary ([4]) offers new possibilities for the acquisition of domain knowledge. First, objects of the knowledge base are defined using that vocabulary. Secondly, medical cases - which are documented taking the vocabulary as a basis - serve to fill the knowledge base with case-related knowledge. Thus, the aim of our research is to define a formal method that supports the knowledge acquisition process using a controlled medical vocabulary and documented medical case descriptions. In the following this method is called 'Sequential Knowledge Acquisition' (S-KA). This paper gives a brief description of the requirements and of the basic approach. Finally it discusses the expected advantages.

2 Methodological Approach

The Sequential Knowledge Acquisition (S-KA) approach concentrates on the filling of a domain model with domain knowledge. Therefore, the following requirements must be fulfilled:

- The knowledge-based function is integral part of a clinical workstation.
- There exists a controlled medical vocabulary that at least consists of relevant medical properties and their value sets. This vocabulary is core of a medical workstation, by which, among other things, medical findings are documented.
- There exists a domain model that is described by objects, properties, relations among objects, and relations among property expressions. This model may be developed using for example the concepts of model-based knowledge acquisition methods like KADS ([5]).

For the filling of a knowledge base, S-KA uses case descriptions documented in a patient database, which database relations are generated from a controlled vocabulary. S-KA is based on

- a model of a knowledge base that meets distinct requirements. The knowledge base should for example be able to keep case information and contradictions, which may appear among different cases.
- a general method for the model-based knowledge acquisition based on a controlled vocabulary. A logical term consisting of combinations of property-value-pairs relates each object of the knowledge base to the controlled vocabulary. Dependent on the structure of the controlled vocabulary we have to define

integrity constraints, which describe admissible combinations of property-value-pairs.

- a formalism to process a case description in a way that existing objects of the knowledge base are adapted and, if necessary, new objects are created. The adaption of existing objects includes the definition of new relations between objects as well as the weighting of relations.

3   Discussion

The presented approach of 'Sequential Knowledge Acquisition' offers a number of advantages: (1) A controlled vocabulary is the basis of knowledge acquisition. Thus, problems concerning terminology will not appear any longer. (2) Although the approach is model-based the knowledge of already solved cases is available for future problems. (3) Taking standardized case descriptions as a knowledge source makes sure to actually acquire the relevant knowledge. (4) The close integration of documentation and knowledge acquisition will motivate clinicians to use knowledge-based functions. Nevertheless we have to consider possible side effects concerning the knowledge base, like convergence, variability, etc.

Altogether, we believe that our approach will considerably support the knowledge acquisition process of the knowledge engineer and the clinician. We hope that the acceptance of decision support systems will profit. For that, our approach is viewed as a step towards the improvement of patient care.

Acknowledgements

This research is supported by the German Ministry of Research and Technology as part of the MEDWIS project (Medical Knowledge Bases).

References

[1]   HAUG PJ, GARDNER RM, TATE KE, EVANS RS, EAST TD, KUPERMAN G, PRYOR TA, HUFF SM, WARNER HR (1994): Desicion Support in Medicine: Examples from the HELP System. *Comp. Biomed. Res.* **27**, 396-418.

[2]   HEATHFIELD HA, WYATT J (1993): Philosophies for the Design and Development of Clinical Decision-Support Systems. *Meth. Inform. Med.* **32**, 1-8.

[3]   DAVID JM, KRIVINE JP, SIMMONS R (1993): *Second Generation Expert Systems.* Springer: Berlin., 69-92.

[4]   LINNARSSON R, WIGERZ O (1989): The data dictionary - A controlled vocabulary for integrating clinical databases and medical knowledge bases. *Meth. Inform. Med.* **28**, 78 - 85.

[5]   WIELINGA BJ, SCHREIBER ATH, BREUKER JA (1992): KADS: A Modelling Approach to Knowledge Engineering. *Knowledge Acquisition* **4 (1)**, Special Issue on KADS, 1992, 5-53.

Medical Fuzzy Expert Systems and Reasoning about Beliefs

Petr Hájek, Dagmar Harmancová

Institute of Computer Science, Academy of Sciences
Pod vodárenskou věží 2, 182 07 Prague, Czech Republic

Reasoning on belief using fuzzy logic (as present in some prominent medical fuzzy expert systems like CADIAG [8], MILORD [3]) is examined from the point of view of formal logic.

Our aim is

- to stress the difference between vagueness (fuzziness) on the one hand and uncertainty as degree of belief on the other hand (this difference seems to be disregarded in some medical expert systems)

- to call the reader's attention to a very elegant and simple formal logical system of fuzzy logic

- to present some observations and results on possibilities of handling uncertainty (in particular, probability) in fuzzy logic.

Formally, fuzzy logic can be understood as a many-valued logic with reals from the unit interval [0,1] as truth degrees. Most systems of many-valued logic are *truth-functional:* the truth degree of a compound formula is determined by the truth-degrees of its component. In contradiction with this, degrees of belief are *not* truth functional; e.g. it is clear that we cannot compute the probability of $A \& B$ (A, B being crisp propositions) from probabilities of A and of B; since $P(A \& B)$ is *not a function* of $P(A)$, $P(B)$ (if no additional assumptions of independence etc. are made). This distinction has been observed by several authors, see e.g. [1, 2]. Recall the attempts of probabilistic justification of MYCIN-like systems that turned out to be pseudoprobabilistic, (cf. eg. [7, 6]).

We call the reader's attention to a simplified version of Pavelka's variant of Lukasiewicz's logic as presented in [4]. It deals with *graded formulas*, i.e. pairs (φ, r) where φ is a formula (meaning e.g. presence of a symptom or a diagnosis or a rule relating some symptoms with a diagnosis) and r is a rational number from [0,1]; (φ, r) is understood as saying that the truth degree of φ is at least r. A *fuzzy theory* is a mapping associating to each formula a rational number - its degree of being an axiom (fuzzy set of formulas, rational-valued, or a certain set of graded formulas). In particular, we have the fuzzy theory of logical axioms. There is a (natural) notion of graded proof; $T \vdash (\varphi, r)$ means that (φ, r) is provable in the fuzzy theory T. This logic is called RPL - Rational Pavelka's Logic. We describe two ways of relating RPL to probability.

First, let P be a probability. A fuzzy theory T *respects* P if for each φ, $P(\varphi) \geq T(\varphi)$. We isolate a proper sublogic WPL (Weak Pavelka's Logic) of RPL such that if T respects P and T proves (φ, r) over WPL then $P(\varphi) \geq r$. Thus WPL is a fuzzy logic sound for proving lower bounds of (unconditional) probabilities.

Second, we associate with each propositional formula φ a new propositional variable f_φ (read: φ is $PROBABLE$, or $PROBABILITY\_OF\_\varphi$ is $HIGH$). This is a fuzzy proposition; given a joint probability P, we are free to assign $P(\varphi)$ as the truth-value of f_φ. We have $f_{\neg\varphi} \equiv \neg f_\varphi$. On the other hand, the formulas $[\varphi$ is $PROBABLE$ & ψ is $PROBABLE]$ and $[(\varphi$ & $\psi)$ is $PROBABLE]$ may well have *different* truth values.

One easy way how to read a "knowledge base" consisting of pairs (φ_i, α_i) where φ_i is a crisp formula and $\alpha_i \in [0,1]$ is a fuzzy theory consisting of pairs (φ_i^*, α_i) where φ_i^* results from φ_i by replacing each atom p by f_p (p is $PROBABLE$). Thus for example, a "rule" $(p \rightarrow q, \alpha)$ will be read (p is $PROBABLE \rightarrow q$ is $PROBABLE, \alpha$) and express the fact that $P(p)$ & $\alpha \leq P(q)$.

Here & is Lukasiewicz's conjunction: $x\&y = \max(0, x + y - 1)$. This approach is compatible with full RPL. We formulate the problem of a complete axiomatization of this logic (this has been solved in the meantime in [5]).

A concluding remark discusses probabilities of use of our formalism in fuzzy expert systems to achieve a sound use of probabilities (of some formulas) as truth-degrees (of possibly different formulas) in the frame of fuzzy logic.

The full paper is available as
ftp:// sun.uivt.cas.cz/pub/reports/ICSTR632.ps .

The work on this paper was partially suported by the COPERNICUS grant 10053 - MUM (Managing uncertainty in medicine).

References

1. DUBOIS, D., AND PRADE, H. Non-standard theories of uncertainty in knowledge representation and reasoning. *Knowledge Engeneering Review 9*, 4 (1994), 399–416.
2. GODO, L., LÓPEZ DE MÁNTARAS, R. Fuzzy logic. In *Encyclopaedia of Computer Science*, 1993.
3. GODO, L., LÓPEZ DE MÁNTARAS, R., SIERRA, C., AND VERDAGUER, A. MILORD: The architecture and the management of linguistically expressed uncertainty. *International Journal of Intelligent Systems 4*, 4 (1989), 471–501.
4. HÁJEK, P. Fuzzy logic and arithmetical hierarchy. *Fuzzy Sets and Systems* (to appear).
5. HÁJEK, P., GODO, L., AND ESTEVA, F. Fuzzy logic and probability. submitted, 1995.
6. HÁJEK, P., HAVRÁNEK, T., AND JIROUŠEK, R. *Uncertain Information Processing in Expert Systems.* CRC Press, Boca Raton, Florida, 1992.
7. JOHNSON, R. W. Independence as Bayesian updating methods. In *Uncertainty in Artificial Intelligence* (1986), Kanal and Lemmer, Eds., Elsevier Science Publishers B.V. North Holland, pp. 197–200.
8. KOLARZ, G., AND ADLASSNIG, K.-P. Problems in establishing the medical expert systems CADIAG-1 and CADIAG-2 in rheumatology. *Journal of Medical Systems 10*, 4 (1986), 395–405.

Diagnosis of Human Acid-Base Balance States via Combined Pattern Recognition for Markov Chains[*]

Marek Kurzynski[**], Michal Wozniak[**], Alexandra Blinowska[***]

[*]This work was supported by the Commision of the European Communities under the Copernicus project no CP 10053.
[**]Technical Univ. of Wroclaw, Inst. of Control & Systems Engn, Wyb. Wyspianskiego 27, 50-370 Wroclaw, POLAND (E-mail: kumar@i17unixb.ists.pwr.wroc.pl)
[***]Service d'Informatique Med., Hopital Broussais, Rue Didot 96, 75674 Paris, FRANCE

1. Introduction

A frequent task connected with medical diagnosis consists in reiterated recognition of a patient's state in successive moments on the basis of the same set of examinations, the currently existing state being dependent on the previous course of the disease and the applied therapeutic treatment. The present paper which is a sequel to the author's earlier publications [1,2,5,6] is devoted to this diagnostic (pattern recognition) problem, for the original case when two kinds of information are given: the expert rules and the learning set.

2. Rule Based Pattern Recognition Algorithm with Learning

Let apply to the considered diagnostic problem a probabilistic approach and let assume a first order Markov dependence as a model of interdependence between successive states of the diagnosed patient and the applied treatment. The formalization of the diagnosis implies the setting of a recognition algorithm (let say ψ), which at the n-th moment will give the number of recognized state i_n using all available measurement information, i.e. sequence of examination results (features) $\underline{x}_n=(x_1,x_2,...,x_n)$ and a sequence of previously applied treatment procedures $\underline{u}_{n-1}=(u_1,u_2,...,u_{n-1})$. For the case of complete probabilistic information the optimal recognition algorithm ψ^* indicates this state i_n for which a posteriori probability (app) $p(i_n/\underline{x}_n,\underline{u}_{n-1})$ is the greatest one.

Let us now consider the original and new concept of recognition in which we assume that appropriate probability distributions are not known, whereas the only information is contained in the two kinds of data: learning set (let say S) and set of expert rules (R).

When the only set S is given one obvious and conceptually simple method is to approximate app from set S by estimators $p^{(S)}(i_n/\underline{x}_n,\underline{u}_{n-1})$ and next to use them in the algorithm ψ^* (algorithm k-nearest neighbours (k-NN) is an example of this con-cept [7]). Similarly we can estimate app from the set R (by value $p^{(R)}(i_n/\underline{x}_n,\underline{u}_{n-1})$) which leads to the so-called GAP algorithm [1]. For the considered case, i.e. when both sets S and R are given we propose *the combined* (or *rule based with learning*) *recognition algorithm* $\psi^{(SR)}$, indicating this state i_n for which the classifying function $p^{(SR)} = \alpha\, p^{(R)}(i_n/\underline{x}_n,\underline{u}_{n-1}) + (1-\alpha)p^{(S)}(i_n/\underline{x}_n,\underline{u}_{n-1})$ $(0\leq\alpha\leq1)$ is the greatest one. Technical details can be found in [1].

3. Diagnosis of the Human Acid-Base Balance (ABB) States

In the case of disorders in ABB we distinguish acidoses and alcaloses and each of them can be methabolic and respiratory origin. It leads to the following classification of ABB states:1.respiratory acidosis, 2.metabolic acidosis, 3.respiratory alcalosis, 4. metabolic alcalosis, 5. normal state. For quick diagnosis of ABB states the above presented method of pattern recognition was applied, using results of gasometric examinations (pH of blood, carbon dioxide pressure, actual bicarbo-nate concentration) and taking into account three categories of therapy (respiratory treatment, pharmacological treatment, no therapy).

In the Neurosurgery Clinic of Wroclaw Medical Academy the set of data has been gathered, which contains 88 learning sequences and the set of 55 rules. Results of computer experiments for 3-NN, GAP and combined (3-NN-GAP) algorithms are presented in Fig.1. The superiority of the presented empirical results for the combined algorithm over GAP and k-NN algorithms demonstrates the effectiveness of the proposed concept in such computer-aided medical diagnosis problems in which both the learning set and expert rules are available.

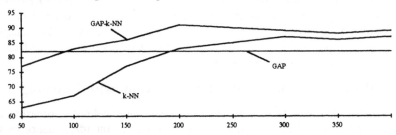

Fig.1. Results of classification accuracy [%] versus the learning set size for GAP, k-NN and GAP-k-NN (combined) algorithms.

References

1. Huzar Z., Kurzynski M., Sas J., *Rule-Based Pattern Recognition with Learning*, Technical University Press, Wroclaw 1984
2. Kurzynski M., Wozniak M., Rule-Based Algorithms with Learning for Recognition of Controlled Markov Chains, *Proc 2nd Nat. Conf. on Expert Systems*, vol 1, pp.173-180, Wroclaw 1993 (in Polish)
3. Kurzynski M., Zolnierek A., A Recursive Classifying Decision Rule for Second-Order Markov Chains, *Control and Cybernetics*, vol.9, no 3, pp.141-147, 1980.
4. Kurzynski M., Compound Pattern Recognition Methods [in:] *Problems of Biocybernetics and Biomedical Engineering*, Nalecz M. [ed.], vol V, pp.159-180, WKiL, Warsaw 1990.
5. Kurzynski M., Sas J., Rule-Based Medical Diagnosis with Learning: Application to the Diagnosis of Acute Renal Failure, *Proc. 14th Annual Int. Conf. of the IEEE Engn. in Medicine and Biology*, vol.3, pp.1258-1260, Paris 1992.
6. Kurzynski M., Sas J., Wikiera I., Rule-Based Medical Decision-Making with Learning, *Proc. 12th World IFAC Congress*, vol.4, pp. 319-322, Sydney 1993.
7. Devijver P., Kittler J., *Pattern Recognition- A Statistical Approach*, Prentice Hall, London 1982

Intelligence Formation Problems in Children at an Early Age Applying New Computer Technologies under Conditions of Rehabilitation Center

I.F. Olkhovsky and S.I. Blokhina

BONUM Center, Ekaterinburg, Russia

Nowadays the problem of computer technologies in formation of intelligence in children at an early age becomes more and more urgent. Among the advertised softwares one can find child's development computer games and test programmes made in the USA, France, Italy, Russia and other countries [1].

Under conditions of rehabilitation center *the focus* of the task of intelligence formation is shifted in the direction of applying medical methods and technologies. It is known that a congenital pathology in children an an early age is accompanied by poor intelligence. In children with maxillofacial pathology this is manifested in speech pathology, weak development of small motility, poor memory, reaction, etc. Moreover in a number of cases the newborns with a congenital pathology have strong defects in development and a number of sensory qualities and intelligence are absent. In children of preschool age psychical deviations like moderate delay in development are observed in 64.6-97.8% of cases [2].

In these cases the task of abilitation practically comes down to the artificial intelligence formation (except the problems associated with medical operations).

A complex programme of rehabilitation and abilitation of children at an early age,worked out and introduced by the rehabilitation center *BONUM* is based on the combination of medical and nonmedical technologies.

In this complex process a great attention is given to computer technologies.

The aim of the work carried out is to form the child's intelligence according to the age standard, applying computer technologies.

The main trends of working out and applying the programme software are as follows:automate testing of a child; development computer games; specialised computer games; databases for storing the results of tests and games; programmes of mathematical processing the database; working out of nonreplicated database, connecting the results of child's examination by the specialists; working out of computer tests and learning programmes for parents.

On the basis of data presented by doctors, psychologists and other specialists specialised tests and development programmes are made which allow both to run the database and apply electronic table for mathematical processing the results of tests and games.

An individual development programme is worked out for each child and includes the studies with computer.

Mathematical processing of a database allows to evaluate the dynamic changes in the quality of a child's intelligence at different stages of rehabilitation process. Special researches are performed with the aim to create mathematical model of rehabilitation process and methods of artificial intelligence simulation [3].

Moreover, the database processing allows to select the control groups of children suffering with different deseases and to perform statistic examination.

One of the trends in the work of the *BONUM* center is working out of specialised programmes with the game functions (for a child) and scientific-research functions (for a teacher and a researcher).

As an example it should be mentioned a learning-research game programme for 3-4 years old children named *Telephone*. This programme allows to obtain graphic information about the process of involvement the functions of small motility and some other functions of child's intelligence during the game.

When the child plays it is measured the time between pushing the buttons of shown on the screen telephone with help of manipulator *mouse* and the maximal and minimal time of attention delay during the game is fixed. In the game and after its end it is possible to interpret graphically the results of the game.

After each game the graphics are stored in the computer and their comparison in different periods of rehabilitation gives the specialists full information on rehabilitation efficiency of small motility functions.

The important moments for successful operation of *BONUM* center are the parents' skills and knowledge in the field of rehabilitation (preparation of a child for surgical operation, postoperative care and other common matters of child's intelligence development at different ages). This problem is a well known one [2, 3, 4] and to help parents the specialized learning computer programmes are worked out. Studies with the parents are conducted according to the system *lesson-hometask*. It allows to make the process more effective as compared to the study on the base of the center alone.

New computer technologies such as CD-ROOM and MULTY MEDIA in rehabilitation center allow to use new learning programmes (programmes of study foreign language, arithmetics, etc.,for example *Little Monster at School*).

Connection of all the trends mentioned in one rehabilitation complex allows to form child's intelligence depending on his age in accordance with the definite scientifically substantiated base standards.

References

1. Olkhovsky, I.F. and Blokhina, S.I. *On application of development computer programmes for teaching and rehabilitation of preschool aged children.* Theses of scientific and practical conference *Family in the system of rehabilitation centers*, Ekaterinburg, 1994, p.6.
2. Blokhina, S.I. *Medico-social rehabilitation of patients with congenital face and palate clefts under conditions of specialized center.* Abstract of thesis for a medical doctor's degree, Moscow, 1992, p.49. twoside
3. Verbuk, V.M. *Criterion for evaluating the results of computer games and tests.* Thesis of scientific and practical conference *Family-94*, September 27-29, 1994.
4. Hohmann, C. *Young Children and Computers.* Ypsilanti: High/Scope Press (313-485-2000), 1994.
5. Neill, S. and Neill, G. *Annual Guide to Highest Rated Educational Software.* ASCD, 1993.
6. Salpeter, J. *Kids and Computers:A Parent's Handbook.* Carmel: Sams (in bookstores), 1992.

Telecardiology

Ian McClelland, Ken Adamson and Norman Black

University of Ulster, Shore Road, Newtownabbey
Co. Antrim, N. Ireland BT38 0QB

Abstract. A prototype system to display diagnostic ECG data across an ISDN connection has been developed in association with a cardiology consultant and general practitioner. It is anticipated that this will offer full 12 lead diagnostic display and will undergo field trials after some further development. A Heterogeneous integrated Multi-media Medical System (HIMMS) is being designed to provide a common patient database along with knowledge based diagnostic and multimedia training tools.

Introduction

It has been shown by Anderson at al [1] that telemedicine can be applied to ameliorate conditions for treatment where specific medium to high risk patients can be identified. They developed and tested a defibrillator and ECG monitoring device which can be remotely controlled by a clinician by telephone line or cellular means. Anderson [1] has shown this device to be successful with some 76% of calls to a mobile coronary unit being backed up by remote monitoring of ECG while a clinician is in transit to the patients house.

Within the area of cardiology, telemedicine has the potential to be a life saver - for example McNeilly and Pemberton [2] have shown that some 40% of deaths due to myocardial infarction take place during the first hours of symptom onset while Guerci [3] pointed out that it normally takes on average around three hours until treatment is administered in this acute phase.

Telemedicine systems are often developed in a disparate manner. HIMMS has been developed as a generic medical environment which will cater for all modalities as and when they are developed. Kruit and Cooper [4] are developing a strategic open systems framework for regional healthcare. HIMMS in itself will provide a complete telemedicine framework allowing many telemedical modalities to be launched from a single 'front end' user interface based around knowledge based systems.

Prototype Analysis & Design

It is proposed that HIMMS will provide a framework to enable disparate medical modalities to be united in an object oriented paradigm with a view to building a completely integrated knowledge based medical diagnostic tool thus allowing each module to share a common patient database. Using this approach each modality would be developed as a complete system to include its own integrated knowledge base. All modalities would have access to a shared patient database which would contain data collected from any source. Patient data would contain information in a structured format indicated as text, video and audio. The ECG cardiology prototype represents one subsystem of this framework which will encompass many modalities including dermatology, orthopaedics, psychiatry and psychology.

The prototype analysis and design is considered in two stages. Firstly a hardware specification - this ensured adequate hardware (both from a computing and telecommunications point of view) for the current project and future enhancements which will result in an integrated hardware / software solution which can be used by general practitioners in the field. Secondly, a requirements analysis was performed to assess how data should be displayed on screen in a manner suitable for diagnosis by a cardiologist and / or GP. In order to achieve reasonable performance 80486DX based PCs with 8Mb RAM were specified using internal ISDN PC Cards.

An ISDN connection was implemented using some commercially available ISDN adapter cards for an IBM compatible PC. Software was tailored for the purpose and allowed either the two 'B' channels to be aggregated or a single 'B' channel to be used for data while the other is used for dialup voice transmission. The latter mode was found to be the most successful.

Prototype screens have been developed developed in Visual BASIC with working buttons and screen links. An evaluation was carried out by a consultant cardiologist and GP and modifications made to meet recommendations made.

Future Work

The prototype will be developed into an integrated multimedia tool for the clinician and will include a knowledge based training module which will be able to offer different levels of training dependent upon the existing users level of knowledge about the system. Patient data will be shared by other modalities in the HIMMS environment.. This will form part of an integrated system which includes a multimedia knowledge based training module and a knowledge based system to aid clinical diagnosis. Evaluation is currently taking place to establish better ways of integrating voice, video and data together in an efficient way whilst taking into account current research in the areas of human-computer interface design and computer assisted training.

References

[1]Anderson J.McC., Dalzell G.W.N. and Adgey A.A.J. Trans-Telephonic Control of Defibrillation; *Automedica*, 13, 1991, p219-226.

[2] McNeilly R.H. and Pemberton J. Duration of last attack in 998 fatal cases of coronary disease and its possible cardiac resuscitation; *British Medical Journal*, 3, 139, 1968.

[3] Guerci A. Sudden death - Medical Staff Conference, *University of California San Francisco, West Journal of Medicine*,133, 1980, p313.

[4] Kruit D., Cooper PA., SHINE: Strategic Health Informatics Networks for Europe. *Computer Methods and Programs in Biomedicine* 45, p155-158, 1994.

Modelling a Sharable Medical Concept System:
Ontological Foundation in GALEN

G Steve, A Gangemi, A Rossi Mori

ITBM-CNR, Roma. E-mail: aldo@color.irmkant.rm.cnr.it

Within the GALEN project, the Core Model design requires an ontological basis. ONIONS, an original methodology for expliciting ontological criteria from terminological knowledge sources, is presented. ONIONS allows explicit ontological analysis and merging of sources at different levels. Assumptions and research background are introduced.

1. Knowledge integration in medicine and formal ontology

Knowledge integration concerns the understandability of information *contained* and *designed* in different sources. Information as symbolic strings can be communicated by common data formatting, but the conceptual design of the same information cannot, because this is a non-trivial operation, which would imply the understanding of the specific conceptualisation used by the knowledge engineer(s) who designed that source.

Although medical community in some way has a shared understanding of its conceptual principles [11], medical languages and medical systems mirror different parts of them. Moreover, they *partially* mirror that understanding. For example, the SnoMed distinction between *Body System* and *Region* does not necessarily mirror all the distinctions a physician can figure out on those items.

If a source of knowledge is theory-laden by its peculiar conceptual principles, knowledge integration requires a process to build a common ground for comparing those principles.

Conceptual principles are introduced here as *ontological criteria* in a process of explicitation and gathering of formal models *(formal ontology)* for medical terminological knowledge integration.

[2] proposes a valuable classification of conventional ontological areas which are expected to underlie a *knowledge organisation system* (KOS). A KOS reveals firstly its *domain ontology*. A KOS should have been developed for a special problem-solving method, which has a *task ontology*. That KOS could have been modelled by means of a knowledge representation notation, which has its *representation ontology*. The particular application which implements the KOS has an *application ontology*. Some current trends in formal ontology [6] take back the research for a common foundation of KOS ontologies to a *general ontology*, in which sometimes a distinction is made between meta-domain categories and inferential ontology.

2. GALEN and ontology: ONIONS methodology

The *AIM-GALEN* (Generalised Architecture for Languages Encyclopaedias and Nomenclatures in medicine) Project [4] is pursuing a line of research which aims at developing an environment in which a core model of medical concepts can be expressed. Such a CORE (Coding Reference Model) should integrate different domain ontologies and should be linked to independently developed concept systems and terminologies. CORE is expressed in the terminological language GRAIL (GALEN Representation and Integration Language, [10]). A main goal of CORE is to allow information exchange among different applications or KOSes in medicine in an easy and efficient way. This task would be easy if the conceptual framework of medicine was explicitly well defined: in this case, CORE would be the finest partition of all concepts expressed in the coding systems to be shared. But since theoretical framework of medicine does not yield explicit and complete definitions, CORE requires a unique framework of reference, which can be designed through ONIONS (ONtological Integration On Natural Systems), a methodology for analysing the domain ontologies underlying different KOSes and for guiding the formalization of a CORE.

Conceptual integration of different KOSes implies a sufficiently high granularity of concepts; as the right granularity is available as compositional definitions in GRAIL language, the actual problem lies in choosing and organising the concepts to represent.

Our assumption is that any natural language construction (mostly noun phrases) used in a KOS can be related to a concept C, whose semantic structure can be defined by two models M and N.

• M locates C in the context of a KOS, providing (new) concepts, used as criteria which allow C:

•) to be included in B, •) to include $D_1...D_n$, •) to be connected to K;

(*include* is used in a loose set-theory sense; *connected* is also used very loosely. B, $D_1...D_n$, and K are concepts in the same system as C).

• N locates criterial concepts from M in the context of meta-domain theories (general ontology without inferential ontology).

3. ONIONS: ontological analysis and modelling

ONIONS is a four-step procedure, described in [5]. The goals of the first three *analytical* steps are:

— Relevant *source concepts* selected from various KOSes:

• SNOMED [1], UMLS [7], Gabrieli Nomenclature [3], ICD10 [13];

• previous experimental GRAIL Model 5 [10]; and some more sources (mostly experts' documents).

— *Surface ontology concepts* (defining source concepts), extracted as differential criteria within a group of concepts from the same KOS. Sometimes surface concepts are already partially explicit in the source.

— *Shallow ontology concepts*, defined as referring to theories which underlie surface concepts.

— In the fourth synthetic step, heterogeneous shallow concepts are merged, allowing the building of a new surface architecture; whereas the integrated model is implemented according to the application ontology of the computational tool employed.

Using integrated ontologies to design the CORE provides the additional feature of getting some *cognitive ergonomy*, say motivated, natural bases for modelling choices [see 5].
The methodology for this analysis/modelling is based on evidence from recent cognitive science research [cf. 6,9,12] which states that a comprehensive framework of ontology could be looked for only in the dispositions humans use in their interaction with environment. The assumption here should be that coding systems share the same basic dispositions with medical natural language, and that medicine shares the same basic dispositions with common sense knowledge [cf. 5]. In the current version of the model, some cognitive adequacy issues have already entered the design, thus enriching the sharable ontological framework. These issues concern the subjectification [8] of the data entering the model through the notions of *referential role* and *interpretation*, the objectification of the data themselves through the notion of *(con)text* of interpretation, and the resulting notion of an informational object, the *Sign*. For example, the inclusion of spatial knowledge of anatomy in the *ReferentialRole* set allows a huge number of definitions of anatomical concepts or abstracted-from-spatial concepts; the inclusion of granularity layers allows to handle the same structural component both in the molecular and the organic frameworks; etc.

4. Outline of top-level concepts in the current model

The current version of our CORE seems quite stable when compared to other general ontology top-levels [cf. 6]. That's a flat list of the concepts in our top-level (a square bracket means 'multihierarchical').

- Object vs. ReferentialRole vs. Text
  Object = ••{Events, Processes, Structures, |Results|, |Signs|, |Roles|, [AetiologicalObjects]}
    Processes = •••{Functions, |Phases|, Traumas, BodyDevelopment, ...}
      Functions = ••••{BiologicFunctions, Activities, ...}
        BiologicFunctions = •••••{BodyFunctions, PhysiologicFunctions, [AbnormalFunctions], ...}
          PhysiologicFunctions = •••••••{PathologicFunctions, ...}
        Activities = •••••{MedicalProcedures, ...}
          PathologicFunctions = •••••••{Diseases, ...}
    Structures = •••{BiologicStructures, NonBiologicStructures, [Perceivers], [Systems], [Components], [Parts]}
      BiologicStructures = ••••{AbnormalStructures, BodyParts, Organisms, [BiologicAgents]}
      NonBiologicStructures = ••••{ChemicalsByTaxa, [Substances], Artifacts, [NonBiologicAgents]}
      [Systems] = •••••{BodySystems, ...}
      [Roles] = •••{SocialRoles, Agents, Patients, ...}
        SocialRoles = ••••{Family, Community, Group, Population, HealthcareOperator, ...}
  ReferentialRole = ••{Dispositions, Layers, Norm, Completeness, Self, Intention}
    Dispositions = •••{Space, Time, Matter, Amount}
      Space = ••••{Connectedness, Part, Positions, Schemas, Path}
        Connectedness = •••••{Boundedness, Unboundedness, Surface, Link, Host, Hole, Continuity, ...}
        Part = •••••{Part, Component, Element, Whole, Manifold, [Host], [Hole], ...}
      Time = ••••{Intervals, Plans, Clocktick, Cycle, Atemporality, ...}
        Intervals = •••••{Points, ...}
      Matter = ••••{PhysicalStates, Morphologies, PhysicalCategories}
        PhysicalStates = •••••{Gaseous, Liquid, Plastic, Solid}
        Morphologies = •••••{Textures, Shapes, SpatialAbnormalities, ...}
        PhysicalCategories = •••••{Color, Consistency, Density, Size, Temperature, Thickness, Volume, Weight}
      Amount = ••••{None, Unit, Two, Some, Many, All, Ordinal, Scale, Measurement}
    Layers = •••{MaterialLayer, BiologicLayer, PsychoSocialLayer}
      BiologicLayer = ••••{CellularLayer, MolecularLayer, OrganismLayer}
    Norm = •••{Normality, Abnormality, ...}
  Text = ••{Terms, Statements, Texts}

References

[1]Coté RA (ed.) — SNOMED Systematised Nomenclature of Medicine — College of American Pathologists, Skokie, 1979
[2]Falasconi S Stefanelli M — A Library of Medical Ontologies — in *Workshop on Comparison of Implemented Ontologies*, ECAI 94
[3]Gabrieli E — A New Electronic Medical Nomenclature — *Journal of Medical Systems*, 3, 6, 1989
[4]GALEN Project — documentation available from the main contractor Rector AL, Medical Informatics Group, Dept. Computer Science, Univ. Manchester, Manchester M13 9 PL, UK
[5]Gangemi A, Steve G, Rossi Mori A — Cognitive Design for Sharing Medical Knowledge Models — in *MEDINFO-95* (to appear)
[6]Guarino N, Poli R — Formal Ontology in Conceptual Analysis and KR — Dordrecht, Kluwer, 1995
[7]Humphreys BL, Lindberg DA —The Unified Medical Language System Project— in Lun KC et al. (eds.) *MEDINFO 92*. Elsevier Science Publishers, Amsterdam 1992
[8]Lakoff G — Women, Fire, and Dangerous Things — University of Chicago Press, 1987
[9] Langacker R — Subjectification — *Cognitive Linguistics*, 1, 1, 1990
[10]Rector A — Compositional Models of Medical Concepts: Towards Re-usable Application-Independent Medical Terminologies — in Barahona P & Christensen JP (eds.): *Knowledge and Decisions in Health Telematics*, IOS Press, 1994
[11]RossiMori A, Gangemi A, Galanti, M — The Coding Cage — in *Proceedings of Medical Informatics in Europe*, Freund Publishing House, 466-471, 1993
[12]Talmy L — How Grammar Structures Concepts — *Proceedings of 16th LW Symposium: Philosophy and the Cognitive Sciences, Wien, 1993*
[13]WHO — International Classification of Diseases 9th revision — WHO, Geneva, 1977

A Graph-based Approach to the Structural Analysis of Proliferative Breast Lesions.

Vincenzo Della Mea, Nicoletta Finato, Carlo Alberto Beltrami
Dept. of Pathology, University of Udine, Italy

1 Introduction

A particularly difficult problem in the field of histopathology is the classification of proliferative breast lesions, i.e. the diagnosis among epitheliosis, atypical hyperplasia and carcinoma *in situ*.

Page identified a set of rules for the discrimination of these pathologies [Pag86]. These rules are divided in three different subsets, involving respectively cytologic features, architectural characteristics and anatomic extent of the lesion.

In this paper the focus of attention is the second subset of rules, i.e. those that capture the way in which the cells occupy the space differently depending on the grade of malignancy of the lesion. The use of these rules implies a form of spatial reasoning on the lesion, with the aim of recognizing and quantifying spatial properties of the whole lesion or regions of it.

A modelization of the problem involving the use of perceptual graphs is proposed.

2 The Model

Our technique starts from a histologic image representing the transverse section of a single duct, because the aforementioned rules are referred to that objects (fig. 1).

A normal duct is a "pipe" with walls made by a double layer of cells; the internal layer may be involved in the lesions. The proliferative pathology causes a gradual filling of the internal cavity of the duct. The completeness of the filling is not relevant for the diagnosis; the fundamental character is the way in which the proliferating cells occupy the internal space, generating some structures (the so-called lumina, bridges, solid area) instead of others, or the same structures with different characteristics.

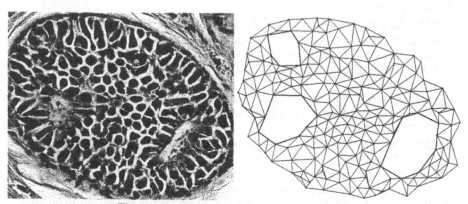

Fig.1. A breast duct with its graph representation

In our approach the lesion is represented by means of the neighbourhood relationship between nuclei as a primary element that allows the recognition of the spatial properties at a more abstract level. A hierarchical representation based on three main graphs (neighbourhood, planar and dual graphs) comes from an initial set of nodes which represents the cell nuclei of the lesion. Other graph-teorethical approaches have been studied for similar histopathological problems [Kay89, Kay92, Ray93, Van92].

Each node is associated to the coordinates of the center of gravity of a cell nucleus in the image plane, and eventually to some cytologic features. Initially, a neighbourhood graph is constructed, and then is used to generate a planar graph. Another step is made to obtain the dual graph. The last two graphs provide a core for the structural analysis of the lesion. Each graph may be partitioned into subgraphs representing some structures of the lesion, or using criteria related to cytologic uniformity to detect clusters of cells [Kay92, Vin89, Zah71].

The algorithms for the generation of the planar and dual graphs have been studied and tested.

This structural representation may be included in a decision support system able to generate a diagnostic suggestion starting directly from an histological image. In such a system, four modules are to be studied: a perception module, a module for the generation and evaluation of the structural representation, a module for the management of the diagnostic knowledge (learning and use), and a man/machine interface. A prototype of the system for the structural analysis of lesions has been developed using Prolog and Pascal (SANE, Structural ANalysis Environment). At this stage of development, the prototype comprehends only some of the required capabilities.

3 Results and conclusions

For the preliminary analysis of the system, we studied images representing 70 breast ducts. Only a subset of graph features were taken into account (38), and some of these revealed significative differences among the three diagnostic classes (5) and also between benign and malignant classes (15) [Bel95]. The significative features were used to study the performance of some automatic generators of decision trees.

Up to now the results obtained by the prototype system are encouraging. The possibilities of this technique suggest a deeper exploration. Moreover, the knowledge based module that realizes the decision support to the diagnosis should be improved. This approach seems also to be suitable for other similar classification problems.

4 References

[Bel95]   Beltrami CA, Della Mea V, Finato N: Structure Analysis of Breast Lesions using Neighbourhood Graphs. Anal. Quantit. Cytol. Histol., in press.

[Kay89]   Kayser K, Stute H: Minimum spanning tree, Voronoi's tesselation and Johnson-Mehl diagrams in human lung carcinoma. Path Res Pract 185:729-734,1989.

[Kay92]   Kayser K et al.: An approach based on two-dimensional graph theory for structural cluster detection and its histopathological application. J Microsc. 165(2):281-288, 1992.

[Pag86]   Page DL: Cancer risk assessment in benign breast biopsies. Hum Pathol 17:871-874,1986.

[Ray93]   Raymond E et al.: Germinal center analysis with the tools of mathematical morphology on graphs. Cytometry 14:848-861,1993

[Van92]   Van Diest PJ, Fleege JC, Baak JA: Syntactic structure analysis in invasive breast cancer: analysis of reproducibility, biologic background, and prognostic value. Hum Pathol 23:876-883,1992.

[Vin89]   Vincent L: Graphs and Mathematical Morphology. Signal Processing 16:365-388, 1989.

[Zah71]   Zahn CT: Graph-Theoretical Methods for Detecting and Describing Gestalt Clusters. IEEE Trans. on Comp. c-20/1:68-86, 1971.

A Workstation for Clinical Decision Support in a Local Area Network for Cardiology

Alessandro Taddei, Manila Niccolai, Mauro Raciti, Claudio Michelassi,
Michele Emdin, Paolo Marzullo, Carlo Marchesi°

CNR Institute of Clinical Physiology, via Trieste 41, 56126 Pisa, Italy

°Dept. Information and Systems, University of Florence, Florence, Italy

A project for the realization of an integrated and networked clinical management-information system (LAN-C) for the Department of Cardiology (DOC) is in progress at our institute. Heterogeneous (alpha-numerical, signal and imaging) data, obtained from the DOC equipments (ECG analysers, Echographs, Rx-imaging, Nuclear Medicine, CCU and Cath-lab systems), are processed and stored into the database. The LAN-C was designed using '486-based computers, interconnected through Token Ring network, with OS/2 operating system, DB2/2 SQL database and Presentation Manager as user interface.

To be of practical use and welcomed by busy health professionals, the system for clinical data acquisition and archiving realised by LAN-C calls for means by which tasks such as record review and interpretation, workflow management, drug prescription, guidelines driven decisions, information retrieval and clinical audit are made easier. These functions have been conceived for a special workstation, namely the *Integration Unit,* which is provided with the access to the whole set of clinical information related to the patient. It basically includes a set of applications aimed at:

- *Protocol driven care.* Clinical protocols, developed through a consensus at least among the clinicians of the same DOC, are used as non-prescriptive guidelines allowing to standardise the delivery of patient care.
- *Information retrieval.* In addition to the LAN-C database, access is also provided to external data banks through local and wide area networks. Particularly important is the consultation of DOC archives and the access to bibliographic and pharmaceutical information.
- *Data processing.* A number of applications are made available for in-depth analysis of clinical data, retrieved from either the database or other sources.
- *Clinical decision support.* Knowledge-based systems are made available for supporting the diagnosis and the treatment of specific pathologies.

The tool Smart Elements (Neuron Data), which integrates object- and rule-based knowledge representation with cross-platform graphical user interface design, was applied for developing the *Integration Unit.* Initially a protocol for the diagnosis of the coronary artery disease (CAD) and a statistical analyser for follow-up data were developed. The CAD protocol was defined by the staff of the DOC of our institute (Fig.1) and realised by the tool Smart Elements by a set of rules, applied to patient data. Relevant information for CAD diagnosis is related to: patient history, risk factors, physical examination, biohumoral data, chest RX-film, diagnostic and exercise ECG, ambulatory ECG, echo and radionuclide imaging. A particular concern has been reserved to the design of a 2-D graphic representation, able to synthesize the heart function. A planar projection of a 3-D silhouette of the left ventricle of the heart is plotted (bull's-eye diagram). Eleven myocardial regions are differentiated: apical region, distal and proximal portions of anterior, septal, inferior and lateral regions. Each region is evaluated in terms of contractility and of level of perfusion using a color map. Both contractility and perfusion are usually tested in different patient conditions (e.g. at rest and during stress) and by more techniques (e.g. echo, radionuclide). Thus, a number of bull's-eye diagrams are obtained (Fig.2). Data related to cardiac patients, followed up after discharge from hospital, are usually analysed to identify potentially important prognostic variables. The IMAGE database, created at our institute and containing almost 8,000 cases, has been applied for the estimation of patient survival (linking Smart Elements with the BMDP package and MS-Excel).

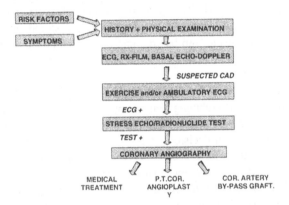

Fig. 1. Simplified version of CAD diagnostic protocol

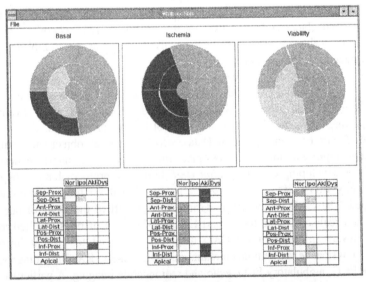

Fig. 2. Wall motion representation by bull's eye diagrams during echo-test in three conditions (basal, ischemia and viability).

Knowledge-based Education Tool to Improve Quality in Diabetes Care

Salzsieder, E., Fischer, U., Hierle, A.[1], Oppel, U.[1], Rutscher, A., Sell, C.

Institut für Diabetes der Ernst-Moritz-Arndt-Universität Greifswald, Mathematisches Institut der Ludwig-Maximilians-Universität München[1]

1 Introduction
Assurance and improvement of quality in Diabetes care necessitate the application of the state-of- the-art knowledge both in diagnostic and therapy. Therefore education and training in self-management of blood glucose control has become a permanent task for all people involved in diabetes care. Especially in primary health care there is actually a remarkable deficit in available knowledge concerning modern aspects of therapy in insulin-dependent (IDDM) diabetes mellitus.

2 Aim
It was therefore the aim of this study to develop and to realize a knowledge-based education tool that comprises the present knowledge about therapy of IDDM and which enables the user, mostly general practitioners (GPs), to accumulate this knowledge by an interactive learning by doing procedure.

3 Method
Considering a validated model of the glucoregulatory system a PC program has been designed which comprises 6 main components:
- representation of dose-dependent action profiles of different insulin formulations,
- visualisation of resorption profiles of food staffs,
- visualisation of action profiles of muscular activities in terms of equivalents of insulin action,
- presentation of the effects of different therapeutic measures in terms of daily blood glucose profiles,
- predicting of the metabolic effects of self-selected therapeutic regiments on daily blood glucose by performing model-based simulation procedure, and
- adaptation of the simulation programme to the individual metabolic situation and evaluation of the most suitable individual therapeutic regime by causal probabilistic network approach.

4 Results
The first version of the PC-program KADIS (Karlsburger Diabetes Management System) has been written in Turbo Pascal and runs under WINDOWS on any IBM compatible computer. It can easily be handled by any GP without any special experience in medical informatics.

After initialisation the program provides the user information about dose-dependent patterns of insulin action on glycaemia (figure 1) of different insulin formulations (short-, intermediate-, long-acting), individualised insulin action equivalents of physical exercise and the absorption rates of various food staffs.

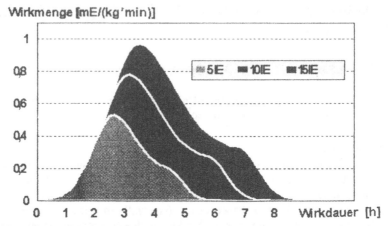

Fig. 1. Dose-dependent insulin action profiles after subcutaneous application of 5, 10 or 15 IU of short-acting regular insulin.

The user can also select different concepts of daily metabolic management (conventional or intensified conventional insulin therapy with or without considering exercise) which are visualised by presenting 24 hours profiles of glycaemia (figure 2). Optionally, profiles of insulin action, of absorption rates of food, and of insulin action equivalents of exercise during 24 hours can also be viewed. The educational approach is completed by means of a simulation tool in that the apparent effects of self-selected therapeutically strategies on 24 hours blood glucose profiles can be model-aided predicted and may be compared each other.

In a preliminary study, especially the interactive work with the program KADIS was much appreciated by those users who are members of diabetes care teams or responsible for training and education of patients or medical students.

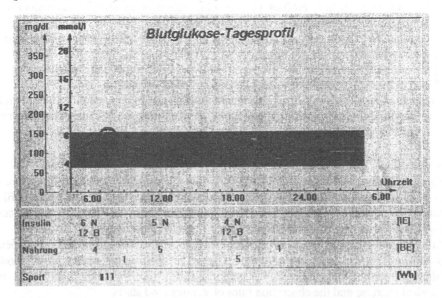

Fig. 2. Model-aided simulation of 24 hours profiles of glycaemia using KADIS.

NEPHARM: A Pharmacokinetic Database for Adjusting Drug Dosage to Impaired Renal Function

Keller F, Arnold R, Frankewitsch T, Zellner D, Giehl M*.

University Hospital, Medical Department, Division of Nephrology, Ulm.
*University Hospital Banjamin Franklin, Body Counter, Berlin

Background

The available electronic information systems on drug related knowledge provide the information only in a free-text manner, not in an informatically structured form. In addition, there is no consensus about the parameters or the concept how drug kinetics should be recorded (Benet 1990).

Preparatory Work

We have documented 10,000 data sets for 1,500 drugs out of 1,200 primary references. These parameters are recorded in a raw database. Drug dosage adjustment is derived from a unifying pharmacokinetic concept. An estimate is made of renal function, of the linear dependence of drug clearance on renal function, and of accumulation kinetics. The meta-analytic parameter with variance are obtained out of the pharmacokinetic database. The Bayesian objective function is used for an individual parameter estimate.

General Concept: We have based the system on a unifying concept of pharmacokinetics described by three pharmacokinetic terms namely half-life (T½), volume (Vd), and clearance (Cl). For these corresponding parameters the model-dependent and model-independent approaches are equivalent (Swanson 1983). Only the elimination parameters, not the distribution parameters depend on renal function (Wolter 1994).

Meta-analysis: A robust meta-analysis of pharmacokinetic parameters can be obtained by the median with the 0.95 confidence interval of the published values. More diversified methods will be developed to be applied to the variable data matrices.

Bayesian Function: From the meta-analyzed parameters (P) with estimates of variances (σ^2) and at least one measured drug concentration (y) an individual parameter estimate can be obtained by minimizing the Bayesian objective function (Proost 1992).

Prospective Steps and Aims

Presently, we have only dosage proposals in tabulated form for antiinfective theray in patients with impaired renal function, hemodialysis and hemofiltration. The next stage will be an idividualized parameter estimate derived from the Bayesian function (Figure). The proposed dosage adjustments will be evaluated by prospective clinical studies. For evaluating the performance and impact of our system, a model based statistical analysis will be prefered as compared to a randomized intention-to-treat analysis (Sheiner 1995).

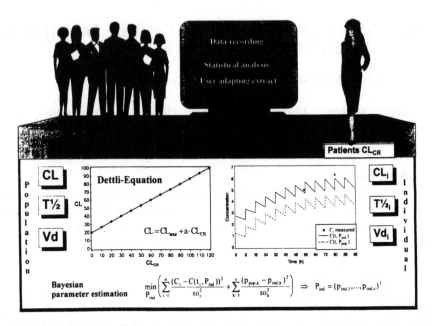

Figure. Dosage adjustment to individual parameters. The population-derived pharmacokinetic knowledge is used for an individual parameter estimate based on the Bayesian function.

References

Benet LZ, Williams RL. Design and optimization of dosage regimens: pharmacokinetic data. In (Ed) Gilman AG, Rall RW, Nies AS, Taylor P. Goodman and Gilman's: The Pharmacological Basis of Therapeutics. Pergamon, New York 1990: 1650 - 1735

Proost JH, Meijer DKF. MW/Pharm, an integrated software package for drug dosage regimen calculation and therapeutic drug monitoring. Comp Biol Med 1992, 22: 155 - 163

Sheiner LB, Rubin DB. Intention-to-treat analysis and the goals of clinical trials. Clin Pharmacol Therap 1995, 57: 6 - 15

Swanson DJ, Reitberg DP, Smith IL, Wels PB, Schentag JP. Steady-state moxalactam pharmacokinetics in patients: noncompartmental versus two-compartmental analysis. J Pharmacokinet Biopharm 1983, 11: 337 - 353

Wolter K, Claus M, Wagner K, Fritschka E. Teicoplanin pharmacokinetics and dosage recommendations in chronic hemodialysis patients and in patients undergoing continuous veno-venous hemodialysis. Clin Nephrol 1994, 42: 389 - 397

A Hybrid Architecture for Knowledge-Based Systems

Eleni Christodoulou

University of Cyprus, Department of Computer Science,
P.O.Box 537, Nicosia, Cyprus
e-mail:cseleni@turing.cs.ucy.ac.cy

Abstract

Research in knowledge-based systems has shown that using multiple knowledge representation models of a specific domain can result in more efficient and effective problem solving, thereby making complex problems tractable [1]–[3]. In a hybrid architecture, domain knowledge can be organized into separate modules, identified as *problem solvers*. A metalevel control unit, a *metareasoner*, coordinates the use of, and the commmunication between the different problem solvers. A problem solver is described as a unit implementing a specific knowledge representation formalism combined with reasoning methods. Each representation formalism embodies a different view of the same (or part of the same) body of knowledge. Some of these could be orthogonal views and others could simply be successively more abstract representations of essentially the same view. The definition of a problem solver as an integral unit that includes knowledge and processing methods considerably enhances the modularity perspective of knowledge-based systems. New modules, i.e. new problem solvers, could be added without influencing the existence of the other modules. Thus, a hybrid architecture can be dynamically extended and modified in accordance with domain requirements. A metareasoner contains knowledge about the individual problem solvers, such as preconditions for activating a given problem solver and a maximum time period for which a specific problem solver should be activated. A number of functions are used for the invocation of problem solvers, for the acquisition and manipulation of problem-case findings and for controlling the dependencies among different problem solvers. Thus, the role of a metareasoner is to activate useful problem solvers, potentially able to provide solution elements depending on the actual problem data currently available. Due to the large amount of knowledge usually included in knowledge-based systems, the presence of a metareasoner, that intelligently selects, at each time, a small subset of the knowledge to be activated (through the selected problem solver) turns out to be very helpful in reducing the search space for the solution, and enhances the system performance. We have demonstrated the use of a hybrid architecture in the development of a breast cancer diagnostic system [4]. Specifically, a knowledge-based system has been developed to assist histopathologists in the histological classification of a new breast cancer case. Such a classification is based on a number of diagnostic factors extracted from a manual microscopic analysis of breast biopsies. Our histopathologic advisory system makes use of three different problem solvers and recognizes the following breast cancer types: *ductal carcinoma, lobular carcinoma, medullary carcinoma, colloidal carcinoma, tubular carcinoma, invasive cribriform carcinoma and mucinous carcinoma.* The first solver represents the different histological breast cancer types by using their *most important* nuclear and cytoplasmic characteristics. It consists of different sets of rules. Every set of rules carries out inquiries on the necessary features of a specific histological type for breast cancer. The second solver is based on the definition of different classes. Each class represents a different histological breast cancer type and consists of *all the known* characteristics of the specific type. Each characteristic is associated with a factor of importance. A factor of importance gives the relation of a feature to the associated class (for example a feature is of high, middle or low importance for a specific class). The third solver consists of a number of already *experienced* breast cancer *cases* [5]. These cases are given as instances of the histological classes defined in the second solver. The purpose of this solver is to match a new case with an already existing case thus avoiding a complicated reasoning process for similar cases. In addition this supports the incorporation of new cases and the generalization of classes of cases which is an aspect of learning in the system. The entire problem solving process is directed by a metareasoner. This metareasoner communicates between the different solvers and decides about the best model to be used for a specific problem case. This decision is taken dynamically depending on the actual problem case characteristics and on the computational expenses of the different models. Normally, the least computationally expensive model (in our case the rule-based model) is chosen first. If at any execution step the chosen model can not produce an acceptable result, the next least expensive model is tried (in our case the deep-model), and so on. This problem solving strategy gets for every diagnostic case the best possible answer in the most efficient timewise way. The metareasoner is also responsible for transmitting new learned and existing information effectively between the multiple model representations and maintaining the consistency of the knowledge.

References

1. R. Simmons and R. Davis *The Roles of Knowledge and representation in Problem Solving,* in J.M. David and J.P. Krivine and R. Simmons, editors, *Second Generation Expert Systems,* Springer Verlag, 1993
2. M. DeJongh and J. Smith *Integrating models of a domain for problem solving,* Expert Systems and their applications, General Conference, Second Generation expert systems, Avignon 91, volume 2, May 27-31, 1991
3. G. Lanzola, M. Stefanelli *Computational Model 3.0,* AIM Project report: A General Architecture for Medical Knowledge-Based Systems, December 30, 1993
4. Christodoulou E. *An Integrated Decision Support System for the Domain of Breast Cancer,* Proceedings of the International Conference Neural Networks and Expert Systems in Medicine and Healthcare, 23-26 August 1994, Plymouth, UK
5. J.L. Kolondner *Case-Based Reasoning,* Morgan Kaufmann Publishers, 1993

Representing Medical Context Using Rule-Based Object-Oriented Programming Techniques

Michel DOJAT[1], François PACHET[2]

[1]INSERM Unité 296, Faculté de médecine, 94010 Créteil, France.

[2]LAFORIA-IBP, Boîte 169, Université Paris 6, 75252 Paris Cedex 05, France.

e-mails: {dojat, pachet} @laforia.ibp.fr

Abstract: We have studied the representation of context in an environment integrating object-oriented programming and rule-based programming. We propose three mechanisms that effectively contribute to represent context in medical applications.

1 Introduction

In medical reasoning the notion of context is central for interpreting physiological parameters and is relevant at all levels of expertise. More precisely according to [Br93] we consider two main dimensions for context: - 1) Context as a *"situation"*: This aspect encompasses context for the interpretation of data, context to account for various temporal phases in the course of the patient's disease, or context to account for the specificity of the clinical team - 2) Context as a *"set of beliefs"*: It is often important to explicitly represent the underlying assumptions of the medical doctor regarding the patient's status, such as the diagnosis established, the theoretically predicted evolution or the expected evolution in function of the current treatment. We aim at constructing effectively medical systems that take context into account and claim that *embedded rule-based object-oriented programming* is particularly well adapted to our objectives. Recently, we built the NéOpus system [Pa95] that extends the class/instance paradigm of an object-oriented language (Smalltalk-80) with production rules. Using NéOpus [Do92], we have identified three main mechanisms of objects/rules combination: *class inheritance*, *natural typing* and *rule base inheritance*, which contribute to represent context in medical knowledge.

2 Class Inheritance to represent Context

The organising principle of *class inheritance* allows to describe medical information and create information structures comprising concepts, which are statically related through common property characteristics. Although class inheritance can hardly be compared with true classification mechanisms, it nevertheless allows us 1) to represent taxonomies of physical or conceptual entities and generalisation/specialisation relation (e.g. class `Intensivist` inherits from class `Clinician` which in turn inherits from `MedicalDoctor`) and 2) to regroup information. For example, in medical applications we need to specify symbolic interpretations for each parameter in the context of a particular therapeutic state. All the symbols have not to be systematically specified in a new context. Inheritance factors out the information that do not change when a state transition is detected.

Well-structured problems of diagnosis are solved by the method of *heuristic classification* [Cl85]. In NéoGanesh [Do92], a knowledge-based system for ventilation management, class inheritance is used to represent the taxonomy of possible solutions. Each class representing possible ventilation states holds a set of constraints to which patient's data is matched for classification. The definition attached to each element of the taxonomy of solutions, as well as the thresholds used for interpreting physiological data, may vary according to the context (patient's characteristics, therapeutic state or clinician). Moreover, the refinement of the potential solutions may depend on the availability or the lack of certain data. New solution classes with constraints that explicitly refer to these various contexts are easily added to the initial taxonomy of potential solutions.

3 Natural Typing to Represent Context

Class inheritance and encapsulation may be combined to have a dynamic interpretation of rules. *Natural typing* [Pa95] allows the pattern-matcher to consider direct instances as well as instances of subclasses to be matched by rule variables for a given rule. This mechanism makes rules *context-dependent* where context is represented by the set of objects that match the rule. In mechanical ventilation for instance, various strategies [Do92], [Mi93] can be considered to adapt the rate of assistance reduction (*weaning*) as the patient's state improves or deteriorates. In NéoGanesh, these strategies are represented by different subclasses of class Intensivist. The subclasses redefine the methods understood by internist objects which appear in the condition parts of the weaning rules. The clinician has also to perceive significant changes between successive observed or expected patient's states. The perception of change is context-dependent: for instancce, the notions of *similarity* or *dissimilarity* between states depend on the class of the states compared. This is represented by methods attached to the corresponding subclasses of TemporalObject matched by rules that handle perception of change [Do94].

4 Rule Base Inheritance (RBI) to Represent Context

In NéOpus, each rule base may be defined as a sub-base of an existing rule base, thereby inheriting all its rules. In case of conflict, rules defined in the lowest sub-base will be selected. RBI provides a high level scheme for organising rules and allows to factor out common rules as well as to simplify the specification of control strategies. In NéoGanesh, we use RBI to gradually introduce context-dependency in temporal management. The root rule base TemporalManagement contains rules that define the perception of time in a use-neutral manner. They match general temporal objects (State). Sub-bases are introduced to refine and adapt this general-purpose temporal reasoning to the different steps of ventilation management. They gather rules that match specific objects (such as RespiratoryState). With RBI we introduce also a hierarchy of meta-bases for a declarative representation of control strategies.

We identified three mechanisms specific to object-oriented rule-based programming that proved particularly adapted to represent various dimensions of context-dependency in medical knowledge. These techniques were applied successfully to the construction of a real-world fully operating system.

[Br93] Brézillon P. *Proc IJCAI 93 Workshop on Using Knowledge in its Context.* Technical Report 93/13, LAFORIA-IBP, University Paris 6, 1993.

[Cl85] Clancey W.J. *Heuristic classification.* Artificial Intelligence 1985 ; 27 : 289-350.

[Do92] Dojat M, Pachet F. *NéoGanesh: an extendable knowledge-based system for the control of mechanical ventilation.* Proc IEEE-EMBS Paris 1992 ; 920-921.

[Do95] Dojat M, Sayettat C. *A realistic model for temporal reasoning in real-time patient monitoring.* Applied Artificial Intelligence (to appear) 1995.

[Mi93] Miksch S, Horn W, Popow C. and Paky F. *VIE-VENT: Knowledge-based monitoring and therapy planning of the artificial ventilation of new-born infants.* Proc AIME conference Amsterdam 1993 ; 218-229.

[Pa95] Pachet F. *On the embeddability of production systems in object-oriented languages*, Journal of Object-Oriented Programming (to appear) 1995.

Integration of Neural Networks and Knowledge-Based Systems in Medicine

Ultsch, A. [1], Korus, D.[1], Kleine, T.O. [2]

[1]Department of Mathematics / Informatics
[2]Department of Neurochemistry, Center of Nervous Diseases
University of Marburg
Hans-Meerwein-Straße / Lahnberge
D-35032 Marburg

Abstract Knowledge-Based Systems are used in medical diagnoses. They have the advantage to give an explanation of a diagnosis. But a main problem when dealing with Knowledge-Based Systems is the acquisition of knowledge. Artificial Neural Networks deal with knowledge in a subsymbolic form. Incomplete and imprecise data can be processed by approximating not linear relations in data. In a laboratory or medical system the integration of the neural network system into the decision making process may be required. We realised this by building a hybrid system consisting, first, of graphical visualisation methods and second, a machine learning module generating rules out of the neural network. The rules are presented in a form, which can be understood by humans and used in Knowledge-Based Systems.

Keywords: Knowledge-Based System, Neural Network, decision making, visualisation, machine learning.

Integration of Neural Networks and Knowledge-Based Systems

Knowledge-Based Systems are used in medical diagnoses. They have the advantage to give an explanation of a diagnosis. This is very important especially in the domain of medicine where the user wants to have the diagnosis proved. But a main difficulty when dealing with Knowledge-Based Systems is the acquisition of the domain knowledge. There are several problems with it. It is difficult to transform the explicit and implicit knowledge of the expert's domain, which also partly consists of own experience, in a form which is suitable for a knowledge base. The knowledge can also be inconsistent or incomplete. A second problem is that Knowledge-Based Systems are not able to learn from experience or to operate with cases not represented in the knowledge base.

Artificial Neural Networks deal with knowledge in a subsymbolic form. They can solve non-linear problems often better than conventional methods and are capable to approximate non linear relations in data. In addition, incomplete and imprecise data can be processed. Neural networks learn in a massively parallel and self-organising way. Unsupervised learning neural networks, like Kohonen's self organising feature maps [Kohonen89], learn the structure of high-dimensional data by mapping it on low dimensional topologies, preserving the distribution and topology of the data. But large neural networks can only be interpreted with analysing tools. We developed a visualisation method, the so called U-Matrix methods, to detect the structure of large two-dimensional Kohonen maps. It generates a three-dimensional landscape on the map, whereby valleys indicate data which belongs together and walls separate subcategories [Ultsch/Siemon90].

In a laboratory or medical system the integration of the neural network system into the decision making process may be useful. The knowledge of neural networks, however, is in this form not communicable; i.e. it is necessary to transform the knowledge into a form, which, first, can be understood by humans and second, can be processed by knowledge based systems. Knowledge based systems have the advantage that they can give an explanation of a diagnosis. By integrating both paradigms, knowledge based systems and neural networks, the disadvantages of both approaches can be redressed.

We are developing a hybrid system REGINA (rule extraction and generation in neural architecture) which consists of several parts. An unsupervised learning neural network maps the (preprocessed) data space onto a two-dimensional grid of neurons, whereby it preserves the distribution and topology of the input space. But only together with a visualisation module, called U-Matrix methods, we are able to detect structure in the data and classify it. A three-dimensional coloured landscape will be generated in which walls separate distinct subclasses and subcategories are represented by valleys. A machine learning algorithm sig* extracts rules out of the learned neural network [Ultsch/Li93]. In distinction to other machine learning algorithms like ID3 our algorithm considers the attributes by selecting those which are relevant for the classification. This corresponds to the proceeding of a medical expert. The rules can be used as a knowledge base for an expert system. Also fuzzy rules can be extracted out of the neural network.

426

Application in Medicine

In order to test our hybrid system we applied it to two medical applications. First, we used it to diagnose acidosis diseases. The data set consists of 11 attributes originating from the blood analysis. Several classification methods according to [Deichsel/Trampisch85] were used to explain these data. The Neural Network together with the U-Matrix method was able to classify the data into the subcategories healthy, lacacidemia, metabolical acidosis, respiratory acidosis and one patient with cerebral deficiency (Fig.1a). With our rule generation module

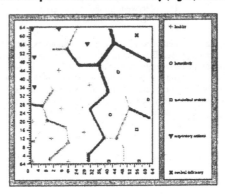

Fig. 1 U-Matrix and classification of acidosis data

sig* we extracted rules out of the Neural Network, which were described by 4 or 5 attributes resembling more closely the decisions made by medical experts [Ultsch/Li93]. Second, we used a data set with patients suffering from different types of the blood disease anaemia. Here no classifications were known a-priori. Deviations of blood values were indicators for a diagnosis of anaemia diseases (Fig.1b). The extracted rules are quite similiar to the diagnosis rules in a medical text book [Müller/Seifert89]. But additional rules were also found and could be verified by medical experts [Ultsch92].

In near future results from CSF analysis [Kleine89] to diagnose different forms of meningitis will be presented. The problems of the course of disease (time series) and multi-diseases (multi-clustering, pattern segmentation) will be also considered in this application. Further applications of our system lie in the area of environment and in the area of industrial processes.

References

[Deichsel/Trampisch85] Deichsel, G.; Trampisch, H.J.: Clusteranalyse und Diskriminanzanalyse, Fischer 1985.

[Kleine89] Kleine, T.O.: New diagnostic methods for inflammation of the human central nervous system by cerebrospinal fluid analysis, J. Clin. Chem. Clin. Biochem. 27, 1989, 895-932.

[Kohonen89] Kohonen, T.: Self-Organization and Associative Memory, Springer 1989.

[Müller/ Seifert89] Müller, F.; Seifert, O. Taschenbuch zur medizinisch-klinischen Diagnostik, Springer 1989.

[Ultsch92] Ultsch, A.: Self-Organizing Neural Networks for Knowledge Acquisition, Proc. Europ. Conf. Artificial Intelligence, Wien 1992, 208-210.

[Ultsch/Li93] Ultsch, A., Li, H.: Automatic Acquisition of Symbolic Knowledge from Subsymbolic Neural Networks, Intl. Conf. Signal Proc., Beijing 1993, Vol.2, 1201-1204.

[Ultsch/Siemon90] Ultsch, A., Siemon, H.P.: Kohonen's Self Organizing Feature Maps for Exploratory Data Analysis, Proc. Intern. Neural Networks, Kluwer Academic Press, Paris 1990, 305 - 308.

Generated Critic in the Knowledge Based Neurology Trainer

Frank Puppe [1], Bettina Reinhardt[1], Klaus Poeck[2]

[1]Würzburg University, Informatics Institute, Am Hubland, D-97074 Würburg, Germany
[2]Aachen University, Neurologic Clinic, Pauwelsstraße 30, D-52057 Aachen, Germany

1. Introduction

Tutoring systems for medical education have become quite popular (see e.g. [2]). While many of them are based on the hypertext / hypermedia technique consisting of links between predefined windows, the idea of intelligent tutor systems (e.g. [12, 10]) is to generate the contents of the windows from underlying domain and didactic knowledge. For example, case oriented tutoring systems can be built in both ways: The patient case can be presented with a hypermedia system, where the sequence and the contents of the windows are prepared specifically for this case. Another approach is building a knowledge base capable of solving cases and using the correctly solved cases for tutorial purposes. Didactic knowledge is required for generating the incremental presentation of the case and for providing the user with feedback on his or her actions. While the costs for building hypermedia based training systems are directly proportional to the number of cases included, knowledge based training systems require a large initial effort to build and test the knowledge base and then only minimal costs for adding any number of new cases. The first system exploring this approach was GUIDON [1], a tutoring system on top of the expert system MYCIN. Insights gained from this work are, that a general problem solving method like backward chaining of rules in MYCIN severely restricts the explainability for tutorial purposes and that an unstructured rule format makes it difficult for the students to differentiate the key clause in the rule precondition from context and activation clauses.

While GUIDON needed additional knowledge for tutorial purposes, the general lesson is, that tutorial systems can be built on top of expert systems, but the requirements concerning the structure of the problem solving method and the contents of the knowledge base are considerably higher. Commercially available knowledge based training systems are the tutor versions of ILIAD [5] and QMR [6]. They avoid the problems of GUIDON/MYCIN by a much simpler knowledge representation.

The general architecture of knowledge based tutorial systems is quite obvious: In addition to the basic components of expert systems - including knowledge base and problem solving, knowledge acquisition, explanation and interviewer component - specific tutorial features are case presentation and critic components; (see e.g. [3]). The main issues involved in designing such systems are practical evaluations regarding the influence of various problem solving methods, the sophistication of the knowledge representation, the case presentation technique depending on the amount of information to be presented, and the critic component depending on the users' time and motivation. In the following, the basic ideas for building and using a knowledge based neurology training system are described.

2. Building and Using the Training System

Developing a new training system with the diagnostic expert and tutor system shell box D3 [4, 11] only requires building a knowledge base and adding cases with the interviewer component. The knowledge base contains sufficient knowledge about the hierarchical or heterarchical structure of the findings to ensure automatic generation of (textual) case presentations in several modes, ranging from detailed presentations of findings to concentration on the key diagnostic elements for faster use. Early experiments [8] showed, that the basic practical problem is providing mechanisms for the user to maintain an overview on the vast amount of case data. Our solution is visualizing the heterarchical structure of findings in D3 [9] with graphical hierarchies, where each node can be expanded or closed one level at a time by a mouse click similar to the user file selection in the Macintosh Finder. When starting a new case, the training system presents initial data about the patient. The user then selects tests to investigate his or her hypotheses. The system provides comments on demand for the users' actions and can also criticize his or her justifications. Critic of hypotheses is generated by comparing them with hypotheses inferred by the system from the same data the user has interpreted so far. Criticizing justifications is more difficult, because the system bases its conclusions on intermediate (pathophysiological) concepts derived from the raw data. Because the user is unaware of these intermediate concepts, they have to be compiled out for rating how much individual raw findings support hypotheses. The critic of the test choices is easily generated from the explicit representation of that knowledge in the knowledge base, where for each diagnosis a sequence of tests useful for its exploration is specified. When the user selects a test, the system compares it with its own choices (also considering second rate choices) based on the suspected diagnoses in the present stage of the tutorial session. The user can also justify test selection by reference to his or her suspected hypotheses, which the system criticizes with respect to the correctness of the suspected hypotheses and how well they can be investigated by the selected tests.

A large knowledge base has been built in the domain of neurology covering the diagnostic part of a standard textbook [Poeck 94] and is available with a voucher in that textbook. Building and testing the knowledge base took the author of the textbook two years with an average of about 1 to 2 hours per day. The knowledge base contains the main 120 diagnoses in neurology, about 1500 highly structured findings, about 300 intermediate concepts (finding abstractions) and more than 2500 rules. A typical session for an experienced user takes about 5 to 10 minutes per case. Currently (March 1995), more than 300 persons have requested the neurology trainer based on the voucher in the 9. edition of the textbook after four months of its publication and the delivery of the software has just started. Our plan for the next version of the neurology trainer includes a more sophisticated but still generated multimedial case presentation with graphics, fotos, videos, sounds and historical recordings. This gives the user the opportunity, not only to interprete findings but also to recognize them.

Literature

[1] Clancey, W.: Knowledge-Based Tutoring: The GUIDON-Program, MIT Press, 1987.
[2] Eysenbach, G.: Computer-Manual, Urban&Schwarzenberg, 1994.
[3] Fontaine, D., Beux, P., Riou, C. and Jacquelinet, C.: An Intelligent Computer-Assisted Instruction System for Clinical Case Teaching, Meth. Inform. Med. 33, 433-445, 1994.
[4] Gappa, U., Puppe, F., and Schewe, S.: Graphical Knowledge Acquisition for Medical Diagnostic Expert Systems, Artificial Intelligence in Medicine 5, 185-211, 1993.
[5] Lincoln, M. et al.: ILIAD Training enhances Students Diagnostic Skills. J. med. Systems 15, 93-110, 1991.
[6] Miller, R. and Masarie, F.: Use of the Quick Medical Reference (QMR) program as a tool for medical education, Meth. Inform. Med. 28, 340-345, 1989.
[7] Poeck, K.: Neurology, (in German), Springer, 9. Auflage, 1994.
[8] Poeck, K. und Tins, M.: An Intelligent Tutoring System for Classification Problem Solving, in Ohlbach, H.J.: GWAI-92: Advances in Artificial Intelligence, Springer, LNAI 671, 210-220. 1993.
[9] Puppe, B. and Puppe, F.: A Knowledge Representation Concept Facilitating Construction and Maintenance of Large Knowledge Bases, Methods of Information in Medicine 27, 10-16, 1988.
[10] Puppe, F.: Intelligent Tutoring Systems (in German), Informatik-Spektrum 15, 195-207, 1992.
[11] Puppe, F., Poeck, K., Gappa, U., Bamberger, S., & Goos, K.: Reusable Components for a configurable Diagnostics Shell, (in German), Künstliche Intelligenz 2/1994, 13≅18, 1994.
[12] Wenger, E.: Artificial Intelligence and Tutoring Systems, Morgan Kaufman, 1987.

An Approach to Analysis of Qualitative Data with Insufficient Number of Quantization Levels

Natalya Polikarpova

Scientific Council "Cybernetics" of the Russian Academy of Sciences

Vavilov str.40, Moscow, 117967, Russian Federation

Extended Abstract

Learning algorithms, used at preliminary stage in decision making tasks, often exploit the idea of distances and/or proximities between objects and features of learning sample (see, for example, algorithms of estimates calculation (AEC) [3]). This allows to estimate the quality of information at hand, to introduce reasonable weights, probabilities, dependencies etc. For qualitative data, such as medical ones, metrics, based on weighted Hamming metric are usually used. The weights of features play an important role, because, being introduced in appropriate way, they allow to make a conclusion about informative power of features.

One of the problems in calculating distances for qualitative data is that often we aren't sure that the number of quantization levels of a feature is sufficient for the task at hand. For example, it is known, that a patient had hallucinations (the research was initiated by psychiatrists, and hallucination ambiguity was specially emphasized). But it is not fixed how often did they take place, in what form and etc. There is no opportunity to recover such information and also it is not known if it is essential. If two patients have this syndrome, in any metric it gives zero contribution to the distance between them. And it may happen that the difference in the type of hallucination is essential so it can be useful to introduce some method, which allows to take into account such ambiguity.

A possible approach of dealing with the problem using Multiple Correspondence Analysis (MCA) model for data representation [1] is proposed. The data are represented by an $n \times m$ (0,1) matrix Y, where n is the number of objects and m is the total number of gradations of all features, $m = \sum_{i=1}^{p} m_i$, p - the number of features. Each object is assigned a row in the matrix Y with exactly p units and $m-p$ zeros. The distance between objects is $\Delta_{ij}^2 = \frac{1}{p} \sum_{h=1}^{p} \sum_{s=1}^{m_h} \frac{n}{n_s^h} (y_{is}^h - y_{js}^h)^2$,where n_s^h is the number of objects, for which the feature h has exhibited the gradation s.

Having MCA model as a starting point, let us fix the "suspicious" gradation of ith feature. In the matrix Y it is represented by a column in submatrix Y_i with n_1 units (for simplicity let "1s" occupy the rows with numbers $1 - n_1$). Suppose that we really have k gradations instead of one. As we don't know, which objects have the same gradation, and which objects - different ones, we assume that objects may be combined into groups in arbitrary manner. The number of all possible ways of combining n_1 objects into k groups is equal to $\sigma(n_1, k)$, - the Stirling number of the second kind. For each possible partition into k quantization levels one can form distance matrices with elements Δ_{ij}^{2*}. In the table (second column) these elements are shown. They are calculated using the elements of original MCA-based distance matrix. Now the matrix $P= E(\Delta_{ij}^{2*})$ can be constructed, where new gradations are taken into account as the average of distance matrices over all possible partitions. The elements of the matrix $P= E(\Delta_{ij}^{2*})$ are shown in the third column.

$i>n_1, j>n_1$	$\Delta_{ij}^{2*}=\Delta_{ij}^2$	$p_{ij}^2=\Delta_{ij}$
$i\leq n_1, j>n_1$	$\Delta_{ij}^{2*} =\Delta_{ij}^2 -\dfrac{n}{pn_1} +\dfrac{n}{pn_l}$, n_l - the cardinality of the block containing ith object	$p_{ij}^2=\Delta_{ij}^2 +\dfrac{n(k-1)}{pn_1}$
$i\leq n_1, j\leq n_1$	a) if the ith and jth objects are in the same block, then $\Delta_{ij}^{2*} = \Delta_{ij}^2$; b) if ith and jth objects are in different blocks, then $\Delta_{ij}^{2*} =\Delta_{ij}^2 +\dfrac{n}{pn_s} +\dfrac{n}{pn_l}$, where n_s and n_l are the cardinalities of the blocks containing ith and jth objects, respectively.	$p_{ij}^2=\Delta_{ij}^2 +\dfrac{2n(k-1)}{p(n_1-1)}$

The inference of formulas for the elements of P can be found in [2]. The idea is the following. When $i\leq n_1, j>n_1$, then the random variable $\xi=1/s$, $(s = 1,2,..., n_1-k+1)$ is considered, $\xi=1/s$ when the block containing ith object has cardinality s.

$$P(\xi=1/s)=\frac{\binom{n_1-1}{s-1}\sigma(n_1-s,k-1)}{\sigma(n_1,k)} ; \qquad E(\xi)= \sum_{s=1}^{n_1-k+1}\frac{\binom{n_1-1}{s-1}\sigma(n_1-s,k-1)}{s\sigma(n_1,k)} =$$

$$\frac{\sigma(n_1+1,k)-\sigma(n_1,k-1)}{\sigma(n_1,k)} =\frac{k}{n_1}. \text{ As a result, } E(\Delta_{ij}^{2*})= \Delta_{ij}^2 +\frac{n(k-1)}{pn_1}.$$

Thus, the metric is introduced, which allows for variable contribution to distance between the objects with ambiguous gradation of a feature, including original zero contribution for $k=1$.

References

1. Lebart L., Morineane A., Warwick K.M., Multivariate Descriptive Statistical Analysis., Correspondence Analysis and Related Techniques for Large Matrices, N.Y. Chichester, etc., J.Wiley&Sons 1985 p.231
2. Polikarpova N., Scatterplots for Qualitative Data: An Approach to Feature Extraction for Visual Display, Pattern Recognition and Image Analysis, Vol.4, N1, 1994, pp.39-47.
3. Yu.I.Zhuravlev, I.B.Gurevitch, Pattern Recognition and Image Recognition, Pattern Recognition and Image Analysis, Vol.1, N2, 1991, pp.149-181.

Inductively Learned Rule for Breast Cancer Domain with Improved Interobserver Reproducibility

Dragan Gamberger

Ruđer Bošković Institute, 41000 Zagreb, Croatia

Medical domains represent a difficult task for inductive learning systems because of inherent noise that can not be completely eliminated even with the very careful preparation of the learning cases. The methods and algorithms for handling noise in learning domains are already well defined but the problem is not completely solved yet. Additionally, the application of some computer-aided diagnosis systems has shown that the classification accuracy can be significantly reduced when the system moves to other places and/or when it is used by other users. The phenomena are known as an interobserver reproducibility problem [1]. Only partially the problem can be solved by rigid data entry procedures or by automating the process of data collection.

Recently, the inductive learning system ILLM (Inductive Learning by Logic Minimization) was developed and successfully applied on an environmental domain [2]. It has three main differences to the well known propositional inductive learning systems: selection algorithm for important literals, form of generated rules and elimination of noise by iterative exclusion of 'suspicious' examples (cases).

In the work the publicly available Wisconsin Breast Cancer Database (version of 15 July 1992) for classifying fine needle aspirates (FNA) generated at Wisconsin Hospitals, Madison, by Dr. W.H. Wolberg is used. The database contains 699 cases (458 benign, 241 malignant) characterized with following 9 attributes: clump thickness, uniformity of cell size, uniformity of cell shape, marginal adhesion, single epithelial cell size, bare nuclei, bland chromatin, normal nucleoli, mitoses. All attributes are quantitative, represented by integers 1-10, with 1 closest to normal. Although the attributes are coded with maximal possible care, all applied methods detected the presence of noise. The presence of noise in the learning domain can be demonstrated also by rather simple inspection. In the domain there are 40 pairs of cases in which malignant case has smaller or equal values for all attributes than the benign case. All these pairs disappear if only 8 cases are removed from the domain. Additional 'suspicious' examples could be found by the search for malignant/benign pairs in which malignant attribute values are not more than 1 greater than the corresponding benign values.

The application of ILLM in its basic form resulted in the rule built of following 5 literals: A(clump thickness > 5), B(uniformity of cell size > 2), C(uniformity of cell shape > 2), D(single epithelial cell size > 2), E(bare nuclei > 3).

$$\text{malignant IF } CD \lor AB \lor E[A \lor B[C \lor D]]$$

In the rule $\vee$ is the sign of OR operation while AND operation is not explicitly written.

The rule is not correct for 3 malignant and 18 benign cases (sensitivity 0.988 and specificity 0.961). The 8 previously mentioned 'suspicious' cases are in this group although the applied algorithm was completely different. The good property of the rule is that it is rather simple so that it can be used directly both by physicians and expert systems. The selected literals point out the importance of some attributes together with significant boundary values. The relative great number of false positive cases suggests that the most significant noise comes from other illnesses or anomalies that can produce similar symptoms.

The properties of the previous rule are similar to the ones of the rules generated by other systems [1]. But specifically for medical application it is interesting to generate rules of as great sensitivity as possible even if it results in unproportionally lower specificity. The reason is that false negative classifications in medicine are much more dangerous than false positive ones.

The idea of improved interobserver reproducibility is to generate two rules for the domain, one for positive (malignant) cases and the other for negative (benign) ones. Application of this rule pair for a new case can produce four different answers: only benign true, only malignant true, both malignant and benign true, and both false. The answer is decisive (only benign true or only malignant true) or nondecisive (both are true or both are false). The intention is to generate such rules that will give decisive answers of great prediction accuracy even when the attribute values are not completely correctly coded while the probability of nondecisive answers should be as low as possible.

The generated rule pair by the ILLM system for breast cancer domain gives no decisive answer on in total 49 learning cases plus 6 cases in which one of the answers is not defined because of unknown input values (7.87%). It is interesting to notice that all 21 cases that are incorrectly predicted (excluded as potentially incorrect by the ILLM first step) in the basic case are among these 49 cases. If the cases are completely correctly coded, the rule pair will give the decisive answer for about 93% of cases (sensitivity 0.892 and specificity 0.950). If the cases are coded with maximal error of +/-1 in all input values, the decisive answers will be obtained for only 26.5% of cases. But in both situations the decisive answer will never be incorrect.

Complete generated rule of increased sensitivity and the rule pair with improved interobserver reproducibility can be obtained from the author. The breast cancer domain can be obtained from machine learning repository at University of California, Irvine by anonymous ftp.

References

1. W.H.Wolberg, O.L.Mangasarian, *Computer-Designed Expert Systems for Breast Cytology Diagnosis*. Analytical and Qauntitative Cytology and Histology, Vol.15, February 1993, pp.67-74.
2. D.Gamberger,S.Sekušak,A.Sabljić *Modelling Biodegradation by an Example-Based Learning System*. Informatica, Vol.17, 1993, pp.157-166.

Development and Evaluation of a Knowledge-Based System to SupportVentilator Therapy Management

N. Shahsavar[a]* and O.Wigertz[a]

[a]Department of Medical Informatics, Linköping University, S-581 85, Linköping, Sweden.

* Corresponding author, e-mail:NosSh@ami.liu.se

Abstract

Development of knowledge-based systems requires a greatdeal of research to find the best solutions for knowledge representation,knowledge acquisition, knowledge-base maintenance, system integration andevaluation. This abstract outlines a summary from the design, development, implementation andevaluation of VentEx, a knowledge-based decision-support and monitoringsystem we have built and applied in ventilator therapy [1-5]. Our work covers the whole development process from the prototype to an integratedon-line system (see figure 1).

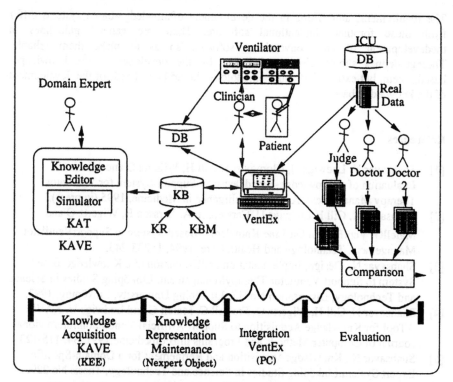

Figure 1: Overview of the main activities applied for thedevelopment.

Research on new technology in general and knowledge-based technology inparticular requires close co-operation between those who will to use it andthose who are developing it. A knowledge-based system is hard to get intoregular use without collaborationbetween the domain experts (those who are the source of the expert knowledge) and the knowledge engineers. The qualityof support for decision-makers should be at least of the same quality asthe expert can produce. Furthermore, using high level technologies andmethodologies in prototyping environments equipped with the latest and most expensive workstations is not thewhole solution for a development problem. A prototype environment canfacilitate testing the feasibility of different design concepts, knowledgestructures and user interfaces because the environment is more flexible than the production environment, but thefunctionality of new methods should be examined in the real and integratedenvironment. Thus a medical knowledge-based system should be integratedinto the clinical environment andshould also be evaluated in that environment. Evaluation of medicalknowledge-based systems is essential for measuring the potential of thesystem and its impact on patients and the organization.

We believe that the potential of knowledge-basedsystem technology depends on the degree to which it can produce high qualitydecision-support. Also, the knowledge-base needs to be supplied withcomprehensive expert knowledge, not only during the development of theknowledge-base but also in the clinical environment. Thus an active, continuous process of refinement should be organized bythe domain experts themselves for refinement and extension of theknowledge-base.

The requirements and efforts in the development ofknowledge-based systems differ from those forother conventional software. There are general guidelines in thedevelopment of any conventional software so as to make them reliable. Theseguidelines should also be followed by the developer of the knowledge-basedsystem, but extra effort and methodology should be utilized for the development of the knowledge-base.

References

[1]   Shahsavar N, Ludwigs U, Blomqvist H,Gill H, Wigertz O and Matell G. Evaluation of a Knowledge-BasedDecision-Support System for Ventilator Therapy Management. ArtificialIntelligence in Medicine, 1995 (in press).

[2]   Shahsavar N, Gill H,Ludwigs U, Carstensen A, Larsson H, Wigertz O and MatellG· VentEx: An On-Line Knowledge-Based System to SupportVentilator Management. Technology and Health Care, 1994, 1: 233-243.

[3]   Shahsavar N., Design, Implementation andEvaluation of a Knowledge-Based System to Support Ventilator TherapyManagement. Linköping Studies in Science and Technology, Dissertation No317, Linköping University, Linköping, 1993.

[4]   Shahsavar N, Gill H, Wigertz O, Frostell C, Matell G and Ludwigs U. KAVE: ATool for Knowledge Acquisition to Support Artificial Ventilation.International Journal of Computer Methods and Programs in Biomedicine,1991, 34: 115-123.

[5]   Shahsavar N., KnowledgeAcquisition and Refinement for a Domain-Specific Expert System.Linköping Studies in Science and Technology, Thesis No 229, LinköpingUniversity, Linköping, 1990.

A Neural Support to the Prognostic Evaluation
of Cardiac Surgery

R. Fiocchi°, A. Gamba°, R. Pizzi^, F. Sicurello*

° Department of Cardiac Surgery, Ospedali Riuniti di Bergamo (Italy)
* Istitute of Advanced Biomedical Technologies - National Research Council, Milano
(Italy)
^ Department of Electronic Engineering - University of Pavia (Italy)

Aim of our work was testing the worth of neural techniques as data analysis methods that could support and sometimes replace standard statistics (1).

To this purpose two neural models have been developed in order to set up a prognostic model for pediatric patients with congenital heart diseases . Processing the available data, coming from the Cardiac Surgery Department of the Osp. Riuniti di Bergamo, the networks are able to esteem the model's parameters and to forecast the new patients' prognoses.

The networks have been implemented on PC 486 using the S-PLUS statistical language, which makes available both tools for data analysis and useful math calculus structures.

According to the advice of the physicians of the Osp. Riuniti, we have chosen a model which links the patient's outcome (dead/alive, da) to its age at the first surgical intervention (age), its weight (weight) and the diameter of the applied shunt (diashunt):

$$da \sim age + weight + diashunt$$

After setting up in S-PLUS a Generalized Linear Model procedure to esteem the parameters' values using the standard statistical S-PLUS facilities, we experimented an alternative method for the parameters' esteem using a Boltzmann machine (2).

This network, due to G. Hinton (1984), performs the neural version of the simulated annealing technique (N. Metropolis 1953) (3): the idea is that, as in metallurgy, the global energy minimum can be reached by gently "shaking" the system with thermal agitation, in order to bypass the local minima.

The simulated annealing performed by the network can be summarized as follow:

1) Randomly select an initial vector x and a large value for the initial "temperature" T.
2) Select a vector x' which is the same as x, except that the value of a component x_i , chosen randomly, is changed. Then, let $\Delta = f(x') - f(x)$, where f is our cost function.
3) If $\Delta < 0$, then let x=x', and go to Step 5.
4) If $\Delta \geq 0$, then set x=x' with probability $exp(-\Delta/T)$,i.e. select a random number β between 0 and 1 , and if $\beta < exp(-\Delta/T)$, then set x = x', else leave x as before.
5) After M changes in x since the last change in temperature, set the value of T to αT, where α is a constant <1.
6) If the minimum value of f has not decreased more than ε (a small constant) in the last L iterations, then stop. Otherwise, go to Step 2.

The Boltzmann machine, usually conceived and applied for combinatorial optimisation, has been modified to accept real values and to better suit our specific problem .

Two versions of the Boltzmann machine have been implemented, a "vanilla" one and a second one that directly introduces the specific cost function in the processing of the system energy. This last version leads to convergence in definetely shorter time.

The problem of the evaluation of the Boltzmann machine's performances is a very complex one (4). In fact, in order to compare the goodness of fit of the model parameters obtained through different methods, it is necessary to show firstly that the cost function is in both cases a likelihood. In the case of a neural network, which performs simulated annealing in a nearly physical way, this is not an easy task and the proof of this issue is under way. Once settled this point, it will be easy to build the inverse hessian matrix for both the models and to decide on the best fit evaluating their minima according to the Rao's Theorem.

It has been also possible to implement by the way a prognostic network using the Hopfield-like structure underlying the Boltzmann machine: the Boltzmann machine is actually a Hopfield network endowed with a probabilistic rule. The Hopfield model (5) is a fully connected network with symmetrical weights and transfer function suitable for discrete values: we modified this model implementing a particular transfer function that could process real values.

Our network is able to forecast the values of any new patient's parameter starting from any set of other known parameter: for example, we can get his/her survival probability having his/her age, his/her weight and the diameter of his/her pulmonary artery.

The forecasting power of the underlying Hopfield-like network has been evaluated using a subset of the available data as a validation set, and shows the usual percentage of success of this kind of network (slightly less than 80 %). This performance can be considered fair if compared with that of a normally skilled physician, but it must be said that it is not exceptional by comparison with the performances that other more advanced neural network models (as for example recurrent backpropagation networks) could reach.

On the other hand, the performances of both networks are going to be further improved by adjusting their internal parameters and by running sets of new simulations.

REFERENCES

(1) Ripley B., Statistical Aspects of Neural Networks, in: O. Jensen, Networks and Chaos - Statistical and Probabilistic Aspects, Chapman & Hall London 1993

(2) Aarts E., Korst J., Simulated Annealing and Boltzmann Machines, Wiley New York 1989

(3) Press W., Teukolsky S., Vetterling W., Flannery B., Numerical Recipes in C, Cambridge University Press 1991

(4) Schwefel H.P., Numerical Optimization of Computer Models, Wiley Chichester 1981

(5) Hecht-Nielsen R., Neurocomputing, Addison Wesley 1990

DECISion-Support System for Radiological Diagnostic

G. Zeilinger* , J. De Mey+ , G. Gell* , G. Vrisk*

* Department of Medical Informatics, KF-University Graz, Austria
+ Department of Radiology, AZ-Vrije Universiteit Brussels, Belgium

Abstract

A software package to develop, maintain and evaluate decision-support systems for radiological diagnostic was devised. The programs, designed as WINDOWS-Applications for PC - versions for other hardware platforms are planned - empower physicians to build and refine consultation systems without the necessity of intervention by software engineers. Applications created with these software tools for particular radiological domains support radiologists at their diagnostic work by offering a question-answer dialog to promote a systematic check of appearance or absence of findings, and by informing about the agreement and/or disagreement of that finding selection with possible diagnoses. Example images of findings and diseases are displayed by request. Consultations can be stored for documentation and/or prospective system verification and improvements. At present, two applications for the radiological fields *High Resolution CT of the Lung* and *MR of the Traumatic Knee* were originated.

Introduction

DECIS (Decision Support) was a topic within the EU research program EurIPACS (European Integrated Picture Archiving and Communications Systems) with the purpose of building intelligent multimedia systems for computer aided diagnosis, in particular for radiological diagnostics. Starting point was the analysis of the diagnostic process in radiology to formulate demands on consulting programs. Then a developing strategy was conceived which gives the medical project partner relative autonomy in the development of consulting programs for particular radiological domains.

Support of the diagnostic process

The consulting program assists radiologists in the different steps of the diagnostic process: In the systematic *Looking for radiological signs*, the *Recognising of signs* (patterns) and the *Interpretation of radiological signs* in terms of diagnoses.

Systematic image analysis

To promote a systematic analysis of the image, a question-answer dialog for radiological signs (also demographic and clinical data, if they are accessible to the radiologist and important for a specific problem) is provided. The dialog is supported by an optical help function which retrieves on request example images for any sign.

Interpretation of signs

At any time during the dialog the user can induce a calculation (in the current version based on an adapted *Certainty Factor* model similar to the famous MYCIN-System) and a display of *evidence values* assigned to single diagnoses and diagnoses groups, which mirror the degree of congruence and/or incongruity by the current selected collection of present and absent symptoms (radiological signs). Changing his/her selection and restarting the calculation, the radiologist can estimate the influence and importance of a particular finding for the diagnosis.

Browsing

The radiologist can use the dialog to just find adequate example images for radiological signs or diseases in the *Image Archive*. The support of this less structured approach is important for the acceptance of the system in clinical practice.

Documentation

The entered finding aggregate can be stored into a *Case Archive*. Before that, the radiologist can add diagnosis(es) into the case record, together with data about its(their) reliability, and any free text note. Such *Case Archives* can be used to obtain statistical data about finding-diagnosis relations and to verify and improve the initial *Knowledge Base* of the system.

Tool-Shell

The basic idea was that comfortable and easy-to-use designer-tools enable the radiological experts to set-up and maintain the required *Data Bases* (e.g.: *Knowledge Base, Image Link Table, ...*) - characterising a particular radiological domain - without the necessity of intervention by computer specialists.

Fig. 1 shows the interplay of these tools with the final consulting program. Using the *Dialog Editor* Tool the radiologist defines the set of possible findings and different diagnoses simultaneously with the appearance of the question-answer dialog. The *Set-up Knowledge* Tool serves for the set up of the *Knowledge Base* which contains a numerical description of symptom-diagnosis relations. Explanatory texts about finding and diagnosis terms and example image stored in the *Image Archive* of the system can be linked to dialog items with *Linker* Tools. Another software tool is planned for the verification and updating of the initial *Knowledge Base* using case examples with confirmed diagnosis collected in the *Case Archive*.

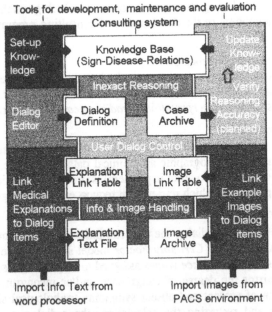

Fig. 1. System components

A Preliminary Investigation Into the Analysis of Electromyographic Activity Using a System of Multiple Neural Networks

P. Caleb[1], P.K. Sharpe[1] and R. Jones[2]

[1] Bristol Transputer Centre, UWE, Bristol, UK
[2] MS Research and Resources Unit, Biophysics Group, Bristol Royal Infirmary, UK

Abstract. Electromyography (EMG) is widely used by clinicians and therapists for diagnosis of certain neuromuscular disorders. This paper describes a preliminary examination of the application of multiple neural networks in the analysis of surface electromyographic activity. The multiple neural network system that is currently being developed enables the signal to be actively monitored by tracking changes in clinical assessment indicators such as force levels, dynamic changes in the force and fatigue. The system has been tested on a small number of subjects and shown promising results.

Electromyography (EMG) is used extensively by clinicians and therapists for diagnosis of neuromuscular disorders. Recordings using surface mounted or invasive (needle) electrodes may also be used to guide treatment after orthopaedic surgery, to correct posture and gait, and in sports medicine for training athletes. Due to the stochastic nature of the EMG signal and the high degree of intra- and inter-individual variability, traditional EMG analysis systems based on inferential statistics often lack robustness and adaptability, especially for surface recorded signal. The present study seeks to show that neural network based systems can successfully address these problems.

The multiple neural network system under development enables active monitoring of the signal by tracking changes in clinical assessment indicators such as force levels, dynamic changes in the force and fatigue. The use of a hybrid collection of networks has the advantage of allowing the integration of various connectionist data analysis techniques resulting in a more thorough and complete coverage of the available information.

The surface EMG signal used in this study is recorded from the Tibialis Anterior from subjects seated in a special assessment chair performing a voluntary ankle dorsiflexion. The raw EMG signal is rectified, filtered, windowed and feature extracted. The extracted feature data is passed on to the various neural network modules making up the system. The modules function concurrently, each analysing a different aspect of the information that can be derived from the signal. As the different extracted features are indicative of different characteristics of the signal, integrating the output of the different modules enables the analysis of the signal to give a much more extensive picture.

The first module is a multi-layered perceptron. It is trained using the error back propagation learning algorithm to indicate the level of force being exerted. The foot plate of the assessment chair is connected to strain gauges which give a force to frequency converted value. The outputs of the network are categorised as maximum force, 50 % of maximum force, negligible force and force belonging to an unmapped region. The limits for marking the ranges of these outputs are calculated statistically from the training set. As these values show inter-individual variations and long term time dependent intra-individual variations, the values used to construct a given network are stored in a knowledge base and used at the next stage by an inference engine for a higher level diagnosis and generation of a confidence measure.

The second module is a recurrent network with a single step time delay. This network is used to indicate dynamic changes in the signal as indicated by the changes in the level of force. The output from this network together with the output from the first network module are integrated with timing information to enable subject performance evaluations.

An additional module of the system, comprised of two stages, is concerned with muscle fatigue recognition. Muscle fatigue is indicated by changes in the Power Density frequency Spectrum of the signal, typically by a shift to lower frequencies. In the first stage, a neural network uses information from the periodogram of the windowed frequency transformed signal. The outputs from this network are input to the network in the second stage, a MLP. This MLP uses the output of the first stage together with information from the first and second modules of the system and timing information to distinguish between fatigued and unfatigued signal.

The system has been tested on a small number of subjects and shown promising results. However, in order to build a more robust and useful system we are currently in the process of building a larger subject data base and establishing a gold standard for various groups of subjects. This will enable the performance of the system to be clinically assessed.

References

1. Miller, A., et al.: Review of neural network applications in medical imaging and signal processing. Med. & Biol. Eng. Comp. Vol 14 (6), 1992
2. Binder-Macleod, S., et al.: Muscle Fatigue: Clinical implications for fatigue assessment and neuromuscular electrical stimulation. Physical Therapy, Vol. 73 (6), 1993
3. D'Allesio, T., et al.: On-line estimation of myoelectrical signal spectral parameters and nonstationarities detection. IEEE Trans. on Biomed.Eng. Vol. 40 (9), 1993
4. Mills, K. R.: Power spectral analysis of electromyogram and compound muscle action potential during muscle fatigue and recovery. J. Physiol. Vol. 326, 1982
5. Linssen, H.P., et al.: Variability and the Interrelationships of surface EMG parameters during local muscle fatigue. Muscle and Nerve, Vol 16 (8), 1993

Knowledge-Based System to Predict the Effect of Pregnancy on Progression of Diabetic Retinopathy

Sell, C.[1], Herfurth, S.[1], Rutscher, A.[1], Salzsieder, E.[1], Hierle, A.[2], Oppel, U.[2], Förster, M.[3], Müller, G.[3], DIADOQ-Group[4]

Institut für Diabetes Karlsburg und Augenklinik der Universität Greifswald[1], Mathematisches Institut der Ludwig-Maximilians-Universität Münch Institut für Medizinische en[2], Informatik und Biometrie der Universität Dresden[3], BMBF-Förderschwerpunkt MEDWIS[4]

1 Introduction

Despite the fact that pregnancy of a diabetic woman has lost today many of its previous frightening aspects, diabetic pregnancy remains a high-risk state for both the woman and her fetus. Without appropriate care before and during pregnancy, perinatal mortality may be more than 10% associated with an enhanced risk of development and progression of diabetic retinopathy. Therefore, it is absolutely essential to perform an individualised management including close-to-normal adjustment of metabolism, clear-cut commitment to early education and to have the possibility to predict the effect of pregnancy on the probability of progression of diabetic retinopathy in relation to different risk factors.

2 Aim

It was the aim of this study to elaborate and to validate a knowledge-based system which might be able to support the decision making process before conception by predicting the probability of progression of retinopathy during pregnancy according to the individual metabolic state and its related risk-profile.

3 Method

To create a knowledge-based system to support management before and during diabetic pregnancy a 5 step procedure was performed:
- selection of those risk factors in the pre-conceptional state which are related to progression of retinopathy during pregnancy,
- forming a decision model,
- selection and analysis of typical case records representing probabilities of different states of development of retinopathy during pregnancy,
- neuronal and causal-probabilistic network approach to establish a knowledge-based system to predict progression of retinopathy,
- validation of the knowledge-based system by implementing it into an appropriate computer program.

4 Results

Seven risk factors have been identified to be most important in predicting progression of retinopathy during diabetic pregnancy: current age and duration of diabetes, number of past pregnancies, glycosylated haemoglobin (HbA1$_c$), current status of diabetes related complications, smoking, and hypertension. The decision model for the period before conception was worked out concerning established recommendations in the management of diabetic pregnancies as applied in designated diabetes care centres (figure 1).

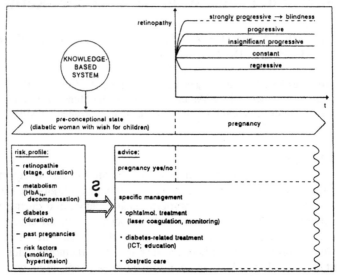

Fig. 1. Model of diabetes management before and during pregnancy considering the implementation of a knowledge-based system to predict the probability of pregnancy-related progression of retinopathy.

The case records were randomly selected from 3922 pregnant women that were treated and delivered in the Karlsburg Diabetes Clinic. Finally 232 case records formed the data base for the development and evaluation of the neuronal and causal-probabilistic network (figure 2) construction.

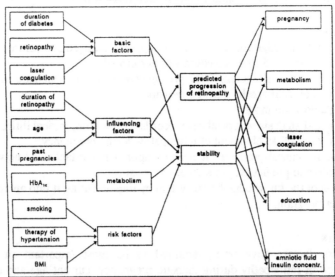

Fig. 2. Causal-probabilistic network to predict progression of retinopathy in diabetic pregnancy.

A Software to Evaluate
Multislices Radiotherapic Treatment Planning

R.Anselmi,G.Paoli,G.Ghiso,F.Foppiano,R.Martinelli,L.Andreucci.
Servizio di Biofisica - I.S.T. Istituto Nazionale Ricerca sul Cancro Genova-Italia.

Introduction

The goal of a radiotherapic treatment planning is to irradiate a target volume with a prescribed dose, by avoiding the irradiation of neighbouring healthy organs. Physicists and physicians duty is to choose a technique, to plan a treatment for every patient and to evaluate its goodness before the irradiation.

Aim of the work

The aim of this application is to give a reading key, both quantitative as qualitative, to evaluate how much the planned treatment is different from the "optimal" one. At the moment, every radiotherapic center has, at least, a bi-dimensional software system based on a significant number of slices (patient contour images), to compute dose distributions. Besides there are three-dimensional softwares that have introduced some new evaluations criteria. These ones are very expensive and need "high level" hardware configurations, but over all, they are based on integrated images (CT, NMR...) and the requested time to end the whole patient's treatment is too long to use them as routine. So we focused our efforts in the direction to realize a tool, user friendly and fast, that is able to use some of these criteria to upgrade the simpler bi-dimensional software.

Discussion

The application needs two input ASCII files for every slice: the first one includes information about number of contours and coordinates of each contour belonging the slice; the second one is a matrix dose distribution. In this way this application is indipendent by the specific dose calculation algorithm of used treatment planning system. A voxel is built on every matrix point using the pixel division of each slice and the slice-to-slice distance. The outputs are Dose Volume Histograms (DVH), one for each contour, that show, in a quantitative way, how and how much the whole organ is irradiated. This means to take out information intrinsic to the treatment planning system, but hidden if the slices are examined separately and in a qualitative way with the isodose curves visualization.

This software is written in C language, it doesn't need a special hardware configuration: a PC 386Dx/33MHz with 4Mbyte RAM is enough.

Outlooks

The daily use of this kind of application has interesting prospects, such as to train physicists and physician towards the use of three-dimensional treatment planning system. Besides it is an open system: this allow on the one hand to match it to several requirements of each center, on the other hand it is possible to implement it within a system for decision making support using artificial intelligence tecniques. So, it will be possible to support human euristic and qualitative planning evaluation with quantitative parameters whose importance is proportion to the increasing complexity of decisions.

TKR-tool: An Expert System for Total Knee Replacement Management

J. Heras[1] and R. P. Otero[2]

[1] Orthopaedic Surgery and Trauma Dpt.
Hospital General de Galicia
E-15701 Santiago de Compostela, A Coruña, Spain
[2] Dept. Computación, Fac. Informática
Universidade de A Coruña
E-15071 A Coruña, Spain

Abstract. TKR-tool is an expert system for Total Knee Replacement that gives advice and assists patient management at every stage of the TKR process from preoperatory and postoperatory evaluation to patient follow-up. The system uses temporal reasoning to manage the patient's history. Expert system advise includes patient category and scoring according to different studies, accurate parameters for TKR surgery success, postoperatory evaluation and prostheses risk failure.

Keywords: Applications of AI in Medicine, Temporal Reasoning, Integration of KBS and patient medical record

1 Introduction

Total Knee Replacement (TKR) is an orthopaedic surgery technique for the treatment of severe gonarthrosis. Different types of knee prostheses are available for TKR performing in daily practice. Constrained and semi-constrained, with the posterior stabilized option, are the basic profiles. Unicompartmental knee prosthesis are used as well as TKR in monocompartmental osteoarthritis of the knee [1]. Since 1988 we have been using the Miller-Galante I, Miller-Galante II, Insall-Burstein II and the Unicompartmental Miller-Galante as implants in knee replacement at the Department of Orthopaedic Surgery and Trauma of the Hospital General de Galicia in Santiago de Compostela (Spain). Each possible candidate for joint replacement is included in a clinical and radiological protocol that evaluates functional aspects of the patient, articular status of the knee and X-Ray assessment, to evaluate a general score for comparison with different studies and other authors [2] and also for an accurate preoperative plan

ning. A successful TKR outcome depends on several factors; one of them and mainly, is a correct alignement of the lower limb to restore to the central axis of the knee an accurate weight bearing central axis for a correct distribution of the loading stresses in both medial and lateral tibiofemoral compartments [3] [4]. Restoring patellar height and adequate tracking, getting a balanced stability of the colateral ligaments and optimize the thickness of femoral and tibial bone cuts, are factors that directly affect a successful result.

The correct adjustment of all these factors is not always easy, even though it is known how to do it, because many variables and limitations are implicated. This is what first motivated us to build an expert system to assist in carrying out the patient's preoperatory study, intraoperatory assessment and postoperatory evaluation. The success of this very specific ES has motivated us to augment the ES knowledge to assist in the whole process followed by a TKR patient from his initial evaluation as a possible TKR candidate, preop study, iop adjust, post-op evaluation and follow-up.

The availability of an own environment for ES development [5] with demonstrated adecuacy to the medical domain in the context of previous projects (ESPRIT P1 1592) [6] together with the possibility to adapt the environment if necessary, has influenced on the decision to use it for the development of the ES. The main characteristics of this environment include easy-to-use knowledge representation formalism, medical oriented user's interface, temporal reasoning, included a temporal data base and standard access to external data bases. There are also compatible versions for MS-Windows and Unix/X windows systems.

2 The Expert System

The expert system we present here, TKR-TOOL (Total Knee Replacement Tool) has over one thousand production rules -equivalent expressions- with two thousand different parameters for each patient and almost all of these parameters with temporal extension. The system, executed in 10 sec, gives advice and assists patient management at every stage of the TKR process. We can consider several parts in the ES, each one dedicated to a different temporal stage in the management of a TKR patient: preoperatory clinical and X-Ray evaluation, intraoperatory evaluation, postoperatory clinical and X-Ray evaluation and patient follow-up. Some of the advices and working areas of the ES are: 1) Patient's category of the American Knee Society [2];2) Clinical scoring system of the Hospital for Special Surgery, American Knee Society and Rush Presbiteriam Medical Center-Chicago [7];3) Infection risk [8];4) Optimization of the bone cuts at the femoral and tibial aspects to allow and restore kinematics of the patella with the appropiate polyethylene thickness into a "clasical" or an anatomic alignement of the knee; 5) Quality of components fixation and its quantitative relationship at the bone-implant interface [9];6) Polyethylene risk failure at the tibial and patellar component [10] [11]

The system has been evaluated at the hospital and it is being used on a daily practice basis since 1992.

3 Conclusion

We have obtained many profits during the development process of the ES and in the use of the ES at the hospital:1) Increase in the quality of the results; 2)

Increase in the quality and in the ease of the handling of the TKR patient;3) Very complete patients' history data base with high consistency and quick access to data;4) Opening-up of new fields to augment the knowledge and increase the control of the TKR problem and their solutions by the orthopaedics surgeons of our department.

Other additional use of the system is for continuous training of specialists in TKR management. Currently we are expanding the scope of the ES in two main areas: automatic measurements on a digitalized image of the X-Ray film (preoperatory and postoperatory) with direct input to the ES and inclusion of machine learning techniques in the ES for assist in TKR research.

Acknowledgements This research was supported in part by the Government of Galicia (Spain), grant XUGA10503B/94.

References

1. Chesnut W.J. Preoperative diagnostic protocol to predict candidates for unicompartamental arthroplasty. *Clinical Orthopaedics*, (273):146–150, 1991.
2. Insall J.N., Dorr D.D., Scott R.D., and Scott W.N. Rationale of the knee society clinical rating system. *Clinical Orthopaedics*, (248):13–14, 1989.
3. Johnson F., Leitf S., and Waugh W. The distribution of load across the knee: A comparison of static and dinamic measurements. *Journal of Bone and Joint Surgery (Br)*, (62-B):346–349, 1980.
4. Cooke T.D.V., Scudamore R.A., Bryant J.T., Sorbie C., Siu D., and Fisher B. A quantitative approach to radiography of the lower limb: Principles and applications. *Journal of Bone and Joint Surgery (Br)*, (73-B):715–720, 1991.
5. Otero R.P. *MEDTOOL, una herramienta para el desarrollo de sistemas expertos.* PhD thesis, Universidad de Santiago, 1991.
6. Barreiro A., Otero R.P., Marín R., and Mira J. A modular knowledge base for the follow-up of clinical protocols. *Methods of information in Medicine*, (32):373–381, 1993.
7. Edwall F.C. The knee society total knee arthroplasty roentgenographic evaluation and scoring system. *Clinical Orthopaedics*, (248):9–12, 1989.
8. Rand J.A., Bryan R.S., Morrey B.F., and Westholm F. Management of infected total knee arthroplasty. (205):75, 1986.
9. Rosemberg A.G., Bardem R.M., and Galante J.O. Cemented and ingrowth fixation of the miller-gallante prosthesis: Clinical and roentgenographic comparison after three-to-six-year follow-up studies. *Clinical Orthopaedics*, (260):71–79, 1990.
10. Engh G.A., Dwyer K.A., and Hanes C.K. Polyetylene wear of metal-backed tibial components in total and unicompartmental knee prostheses. *Journal of Bone and Joint Surgery (Br)*, (74-B):9–17, 1982.
11. Bartel D.L., Burstein A.H., Toda M.D., and Edwards D.L. The effects of conformity and plastic thickness on contact stresses in metal-backed implants. *J. Biomech Eng*, (107):193–199, 1985.

Author Index

Adamson, K.	409	Cherubino, P.	251
Alexiou, A.	165	Chittaro, L.	79
Alpay, L.	307	Christodoulou, E.	421
Amano, H.	393	Combi, C.	397
Andersen, S.K.	151	Copin, P.	165
Andreassen, S.	151	Cross, S.S.	239
Andreucci, L.	443	DIADOQ-Group	441
Anselmi, R.	443	Darmoni, S.J.	231
Arnold, R.	419	Das, A.K.	3
Auckenthaler, R.	165	De Mey, J.	437
Bak, A.	221	Degoulet, P.	42
Banks, G.	129	Del Rosso, M.	79
Barahona, P.	103	Della Mea, V.	413
Baud, R.H.	42, 307, 331	Densow, D.	265
Bellazzi, R.	185	Diamantini, C.	367
Beltrami, C.A.	413	Diamond, L.W.	221, 251
Berney, J.-P.	165	Dojat, M.	79, 423
Bernuzzi, G.	91	Downs, J.	239, 355
Berthoud, M.	165	Droy, J.-M.	231
Bianchi, N.	367	Emdin, M.	415
Bichindaritz, I.	395	Engelbrecht, R.	343
Binsted, K.	29	Entenmann, G.	343
Black, N.	409	Falasconi, S.	173
Blanc, T.	231	Favard, R.	319
Blinowska, A.	405	Finato, N.	413
Blokhina, S.I.	407	Fiocchi, F.	435
Borst, F.	165	Fischer, B.	265
Boscher, L.	209	Fischer, U.	417
Brauer, W.	343	Fliedner, T.M.	265
Brigl, B.	401	Förster, M.	343, 441
Caleb, P.	439	Foppiano, F.	443
Carenini, G.	129	Frankewitsch, T.	419
Cawsey, A.	29	Gamba, A.	435
Chassery, J.M.	379	Gamberger, D.	431

Gangemi, A.	411	Lang, K.	276
Garbay, C.	379	Lanzola, G.	173
Gell, G.	437	Larizza, C.	91
Ghiso, G.	443	Lemaitre, D.	42
Giehl, M.	419	Leroy, J.	231
Gierl, L.	209	Locatelli, F.	295
Grau, A.	401	Lovis, C.	307, 331
Hájek, P.	403	Marchesi, C.	415
Hacke, W.	401	Martinelli, R.	443
Harmancová, D.	403	Marzullo, P.	415
Harrison, R.F.	239, 355	Massari, P.	231
Haux, R.	401	McClelland, I.	409
Heindl, B.	209	Michel, P.-A.	42, 331
Hejlesen, O.K.	151	Michelassi, C.	415
Heras, J.	444	Miksch, S.	197
Herfurth, S.	441	Monti, S.	129
Hierle, A.	417, 441	Mori, A.R.	411
Horn, W.	197	Moustakis, V.	276
Ironi, L.	115	Müller, G.	441
Jean, F.-C.	42	Muncer, D.	221
Jones, R.	29, 439	Musen, M.A.	3
Joubert, M.	319	Nguyen, D.T.	221, 251
Juge, C.	42	Niccolai, M.	415
Kanoui, H.	319	Nowlan, A.	307
Keller, F.	419	Ohmann, C.	276
Kennedy, R.L.	355	Ohyama, K.	393
Keravnou, E.T.	67	Olkhovsky, I.F.	407
Kessler, C.	221	Oppel, U.	417, 441
Kindler, H.	265	Otero, R.P.	444
Kirchner, K.	343	Pachet, F.	423
Kleine, T.O.	425	Paky, F.	197
Korus, D.	425	Paoli, G.	443
Koschinsky, T.	343	Park, I.A.	251
Kuroda, T.	393	Patel, V.	139
Kurzynski, M.	405	Pinciroli, F.	397
Lagana, M.	165	Pittet, D.	165

Pizzi, R. 435
Poeck, K. 427
Polikarpova, N. 429
Pollwein, B. 209
Popow, C. 197
Pozzi, G. 397
Puppe, B. 282
Puppe, F. 427
Quaglini, S. 295
Raciti, M. 415
Ralph, P. 221
Ramoni, M. 139
Rassinoux, A.-M. 42
Rebouillat, L. 165
Rector, A.L. 17
Reinhardt, B. 427
Ringleb, P. 401
Riva, A. 139, 185
Rohner, P. 165
Roux, M. 53
Rush, T. 307
Rutscher, A. 417, 441
Safran, E. 165
Salzsieder, E. 417, 441
Sauvan, V. 165
Scherrer, J.-R. 42, 165, 307, 331
Schmid, G. 209
Schmidt, R. 209
Schulthess, P. 165
Sell, C. 417, 441
Shahar, Y. 3
Shahsavar, N. 433
Sharpe, P.K. 439
Sheridan, B. 221
Shütz, T. 343
Sicurello, F. 435

Smart, J.F. 53
Solomon, D. 307
Spinu, C. 379
Stefanelli, M. 91, 115, 139, 173, 295
Steiner, Th. 401
Steve, G. 411
Taddei, A. 415
Tamino, P.B. 251
Tanaka, H. 393, 399
Thurler, G. 165
Tsumoto, S. 393, 399
Tu, S.W. 3
Ultsch, A. 425
van Elk, P.J. 276
Vrisk, G. 437
Waschulzik, T. 343
Wigertz, O. 433
Wozniak, M. 405
Yang, Q. 276
Zeilinger, G. 437
Zellner, D. 419

Lecture Notes in Artificial Intelligence (LNAI)

Vol. 764: G. Wagner, Vivid Logic. XII, 148 pages. 1994.

Vol. 766: P. R. Van Loocke, The Dynamics of Concepts. XI, 340 pages. 1994.

Vol. 770: P. Haddawy, Representing Plans Under Uncertainty. X, 129 pages. 1994.

Vol. 784: F. Bergadano, L. De Raedt (Eds.), Machine Learning: ECML-94. Proceedings, 1994. XI, 439 pages. 1994.

Vol. 795: W. A. Hunt, Jr., FM8501: A Verified Microprocessor. XIII, 333 pages. 1994.

Vol. 798: R. Dyckhoff (Ed.), Extensions of Logic Programming. Proceedings, 1993. VIII, 360 pages. 1994.

Vol. 799: M. P. Singh, Multiagent Systems: Intentions, Know-How, and Communications. XXIII, 168 pages. 1994.

Vol. 804: D. Hernández, Qualitative Representation of Spatial Knowledge. IX, 202 pages. 1994.

Vol. 808: M. Masuch, L. Pólos (Eds.), Knowledge Representation and Reasoning Under Uncertainty. VII, 237 pages. 1994.

Vol. 810: G. Lakemeyer, B. Nebel (Eds.), Foundations of Knowledge Representation and Reasoning. VIII, 355 pages. 1994.

Vol. 814: A. Bundy (Ed.), Automated Deduction — CADE-12. Proceedings, 1994. XVI, 848 pages. 1994.

Vol. 822: F. Pfenning (Ed.), Logic Programming and Automated Reasoning. Proceedings, 1994. X, 345 pages. 1994.

Vol. 827: D. M. Gabbay, H. J. Ohlbach (Eds.), Temporal Logic. Proceedings, 1994. XI, 546 pages. 1994.

Vol. 830: C. Castelfranchi, E. Werner (Eds.), Artificial Social Systems. Proceedings, 1992. XVIII, 337 pages. 1994.

Vol. 833: D. Driankov, P. W. Eklund, A. Ralescu (Eds.), Fuzzy Logic and Fuzzy Control. Proceedings, 1991. XII, 157 pages. 1994.

Vol. 835: W. M. Tepfenhart, J. P. Dick, J. F. Sowa (Eds.), Conceptual Structures: Current Practices. Proceedings, 1994. VIII, 331 pages. 1994.

Vol. 837: S. Wess, K.-D. Althoff, M. M. Richter (Eds.), Topics in Case-Based Reasoning. Proceedings, 1993. IX, 471 pages. 1994.

Vol. 838: C. MacNish, D. Pearce, L. M. Pereira (Eds.), Logics in Artificial Intelligence. Proceedings, 1994. IX, 413 pages. 1994.

Vol. 847: A. Ralescu (Ed.) Fuzzy Logic in Artificial Intelligence. Proceedings, 1993. VII, 128 pages. 1994.

Vol: 861: B. Nebel, L. Dreschler-Fischer (Eds.), KI-94: Advances in Artificial Intelligence. Proceedings, 1994. IX, 401 pages. 1994.

Vol. 862: R. C. Carrasco, J. Oncina (Eds.), Grammatical Inference and Applications. Proceedings, 1994. VIII, 290 pages. 1994.

Vol 867: L. Steels, G. Schreiber, W. Van de Velde (Eds.), A Future for Knowledge Acquisition. Proceedings, 1994. XII, 414 pages. 1994.

Vol. 869: Z. W. Raś, M. Zemankova (Eds.), Methodologies for Intelligent Systems. Proceedings, 1994. X, 613 pages. 1994.

Vol. 872: S Arikawa, K. P. Jantke (Eds.), Algorithmic Learning Theory. Proceedings, 1994. XIV, 575 pages. 1994.

Vol. 878: T. Ishida, Parallel, Distributed and Multiagent Production Systems. XVII, 166 pages. 1994.

Vol. 886: M. M. Veloso, Planning and Learning by Analogical Reasoning. XIII, 181 pages. 1994.

Vol. 890: M. J. Wooldridge, N. R. Jennings (Eds.), Intelligent Agents. Proceedings, 1994. VIII, 407 pages. 1995.

Vol. 897: M. Fisher, R. Owens (Eds.), Executable Modal and Temporal Logics. Proceedings, 1993. VII, 180 pages. 1995.

Vol. 898: P. Steffens (Ed.), Machine Translation and the Lexicon. Proceedings, 1993. X, 251 pages. 1995.

Vol. 904: P. Vitányi (Ed.), Computational Learning Theory. EuroCOLT'95. Proceedings, 1995. XVII, 415 pages. 1995.

Vol. 912: N. Lavrač, S. Wrobel (Eds.), Machine Learning: ECML – 95. Proceedings, 1995. XI, 370 pages. 1995.

Vol. 918: P. Baumgartner, R. Hähnle, J. Posegga (Eds.), Theorem Proving with Analytic Tableaux and Related Methods. Proceedings, 1995. X, 352 pages. 1995.

Vol. 927: J. Dix, L. Moniz Pereira, T.C. Przymusinski (Eds.), Non-Monotonic Extensions of Logic Programming. Proceedings, 1994. IX, 229 pages. 1995.

Vol. 928: V.W. Marek, A. Nerode, M. Truszczyński (Eds.), Logic Programming and Nonmonotonic Reasoning. Proceedings, 1995. VIII, 417 pages. 1995.

Vol. 929: F. Morán, A. Moreno, J.J. Merelo, P.Chacón (Eds.), Advances in Artificial Life. Proceedings, 1995. XIII, 960 pages. 1995.

Vol. 934: P. Barahona, M. Stefanelli, J. Wyatt (Eds.), Artificial Intelligence in Medicine. Proceedings, 1995. XI, 449 pages. 1995.

Vol. 941: M. Cadoli, Tractable Reasoning in Artificial Intelligence. XVII, 247 pages. 1995.

Lecture Notes in Computer Science

Vol. 905: N. Ayache (Ed.), Computer Vision, Virtual Reality and Robotics in Medicine. Proceedings, 1995. XIV, 567 pages. 1995.

Vol. 906: E. Astesiano, G. Reggio, A. Tarlecki (Eds.), Recent Trends in Data Type Specification. Proceedings, 1995. VIII, 523 pages. 1995.

Vol. 907: T. Ito, A. Yonezawa (Eds.), Theory and Practice of Parallel Programming. Proceedings, 1995. VIII, 485 pages. 1995.

Vol. 908: J. R. Rao Extensions of the UNITY Methodology: Compositionality, Fairness and Probability in Parallelism. XI, 178 pages. 1995.

Vol. 909: H. Comon, J.-P. Jouannaud (Eds.), Term Rewriting. Proceedings, 1993. VIII, 221 pages. 1995.

Vol. 910: A. Podelski (Ed.), Constraint Programming: Basics and Trends. Proceedings, 1995. XI, 315 pages. 1995.

Vol. 911: R. Baeza-Yates, E. Goles, P. V. Poblete (Eds.), LATIN '95: Theoretical Informatics. Proceedings, 1995. IX, 525 pages. 1995.

Vol. 912: N. Lavrač, S. Wrobel (Eds.), Machine Learning: ECML – 95. Proceedings, 1995. XI, 370 pages. 1995. (Subseries LNAI).

Vol. 913: W. Schäfer (Ed.), Software Process Technology. Proceedings, 1995. IX, 261 pages. 1995.

Vol. 914: J. Hsiang (Ed.), Rewriting Techniques and Applications. Proceedings, 1995. XII, 473 pages. 1995.

Vol. 915: P. D. Mosses, M. Nielsen, M. I. Schwartzbach (Eds.), TAPSOFT '95: Theory and Practice of Software Development. Proceedings, 1995. XV, 810 pages. 1995.

Vol. 916: N. R. Adam, B. K. Bhargava, Y. Yesha (Eds.), Digital Libraries. Proceedings, 1994. XIII, 321 pages. 1995.

Vol. 917: J. Pieprzyk, R. Safavi-Naini (Eds.), Advances in Cryptology - ASIACRYPT '94. Proceedings, 1994. XII,

Vol. 918: P. Baumgartner, R. Hähnle, J. Posegga (Eds.), Theorem Proving with Analytic Tableaux and Related Methods. Proceedings, 1995. X, 352 pages. 1995. (Subseries LNAI).

Vol. 919: B. Hertzberger, G. Serazzi (Eds.), High-Performance Computing and Networking. Proceedings, 1995. XXIV, 957 pages. 1995.

Vol. 920: E. Balas, J. Clausen (Eds.), Integer Programming and Combinatorial Optimization. Proceedings, 1995. IX, 436 pages. 1995.

Vol. 921: L. C. Guillou, J.-J. Quisquater (Eds.), Advances in Cryptology – EUROCRYPT '95. Proceedings, 1995. XIV, 417 pages. 1995.

Vol. 922: H. Dörr, Efficient Graph Rewriting and Its Implementation. IX, 266 pages. 1995.

Vol. 923: M. Meyer (Ed.), Constraint Processing. IV, 289 pages. 1995.

Vol. 924: P. Ciancarini, O. Nierstrasz, A. Yonezawa (Eds.), Object-Based Models and Languages for Concurrent Systems. Proceedings, 1994. VII, 193 pages. 1995.

Vol. 925: J. Jeuring, E. Meijer (Eds.), Advanced Functional Programming. Proceedings, 1995. VII, 331 pages. 1995.

Vol. 926: P. Nesi (Ed.), Objective Software Quality. Proceedings, 1995. VIII, 249 pages. 1995.

Vol. 927: J. Dix, L. Moniz Pereira, T. C. Przymusinski (Eds.), Non-Monotonic Extensions of Logic Programming. Proceedings, 1994. IX, 229 pages. 1995. (Subseries LNAI).

Vol. 928: V.W. Marek, A. Nerode, M. Truszczyński (Eds.), Logic Programming and Nonmonotonic Reasoning. Proceedings, 1995. VIII, 417 pages. (Subseries LNAI).

Vol. 929: F. Morán, A. Moreno, J.J. Merelo, P. Chacón (Eds.), Advances in Artificial Life. Proceedings, 1995. XIII, 960 pages. 1995. (Subseries LNAI).

Vol. 930: J. Mira, F. Sandoval (Eds.), From Natural to Artificial Neural Computation. Proceedings, 1995. XVIII, 1150 pages. 1995.

Vol. 931: P.J. Braspenning, F. Thuijsman, A.J.M.M. Weijters (Eds.), Artificial Neural Networks. IX, 295 pages. 1995.

Vol. 932: J. Iivari, K. Lyytinen, M. Rossi (Eds.), Advanced Information Systems Engineering. Proceedings, 1995. XI, 388 pages. 1995.

Vol. 933: L. Pacholski, J. Tiuryn (Eds.), Computer Science Logic. Proceedings, 1994. IX, 543 pages. 1995.

Vol. 934: P. Barahona, M. Stefanelli, J. Wyatt (Eds.), Artificial Intelligence in Medicine. Proceedings, 1995. XI, 449 pages. 1995. (Subseries LNAI).

Vol. 935: G. De Michelis, M. Diaz (Eds.), Application and Theory of Petri Nets 1995. Proceedings, 1995. VIII, 511 pages. 1995.

Vol. 936: V.S. Alagar, M. Nivat (Eds.), Algebraic Methodology and Softwasre Technology. Proceedings, 1995. XIV, 591 pages. 1995.

Vol. 937: Z. Galil, E. Ukkonen (Eds.), Combinatorial Pattern Matching. Proceedings, 1995. VIII, 409 pages. 1995.

Vol. 938: K.P. Birman, F. Mattern, A. Schiper (Eds.), Theory and Practice in Distributed Systems. Proceedings, 1994. X, 263 pages. 1995.

Vol. 941: M. Cadoli, Tractable Reasoning in Artificial Intelligence. XVII, 247 pages. 1995. (Subseries LNAI).